S0-BQI-235

TOPICS IN STEREOCHEMISTRY

VOLUME 7

A WILEY-INTERSCIENCE SERIES

ADVISORY BOARD

TOPICS IN STEREOCHEMISTRY

EDITORS

NORMAN L. ALLINGER

Professor of Chemistry
University of Georgia
Athens, Georgia

ERNEST L. ELIEL

Professor of Chemistry
University of North Carolina
Chapel Hill, North Carolina

VOLUME 7

INTERSCIENCE PUBLISHERS
a Division of JOHN WILEY & SONS
New York • London • Sydney • Toronto

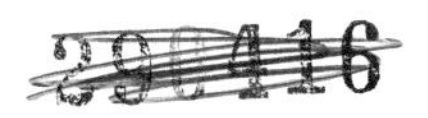

Library of Congress Catalog Card Number: 67-13943

ISBN 0-471-02471-6

Printed in the United States of America

10 9 8 7 6 5 4 3 2 1

INTRODUCTION TO THE SERIES

During the last decade several texts in the areas of stereochemistry and conformational analysis have been published, including *Stereochemistry of Carbon Compounds* (Eliel, McGraw-Hill, 1962) and *Conformational Analysis* (Eliel, Allinger, Angyal, and Morrison, Interscience, 1965). While the writing of these books was stimulated by the high level of research activity in the area of stereochemistry, it has, in turn, spurred further activity. As a result, many of the details found in these texts are already inadequate or out of date, although the student of stereochemistry and conformational analysis may still learn the basic concepts of the subject from them.

For both human and economic reasons, standard textbooks can be revised only at infrequent intervals. Yet the spate of periodical publications in the field of stereochemistry is such that it is an almost hopeless task for anyone to update himself by reading all the original literature. The present series is designed to bridge the resulting gap.

If that were its only purpose, this series would have been called "Advance (or "Recent Advances") in Stereochemistry." It must be remembered, however, that the above-mentioned texts were themselves not treatises and did not aim at an exhaustive treatment of the field. Thus the present series has a second purpose, namely to deal in greater detail with some of the topics summarized in the standard texts. It is for this reason that we have selected the title *Topics in Stereochemistry*.

The series is intended for the advanced student, the teacher, and the active researcher. A background of the basic knowledge in the field of stereochemistry is assumed. Each chapter is written by an expert in the field and, hopefully, covers its subject in depth. We have tried to choose topics of fundamental import, aimed primarily at an audience of organic chemists but involved frequently with fundamental principles of physical chemistry and molecular physics, and dealing also with certain stereochemical aspects of inorganic chemistry and biochemistry.

It is our intention to bring out future volumes at approximately annual intervals. The Editors will welcome suggestions as to suitable topics.

We are fortunate in having been able to secure the help of an international board of Editorial Advisors who have been of great assistance by suggesting topics and authors for several articles and by helping us avoid duplication of topics appearing in other, related monograph series. We are grateful to the Editorial Advisors for this assistance, but the Editors and Authors alone must assume the responsibility for any shortcomings of *Topics in Stereochemistry*.

N. L. Allinger
E. L. Eliel

January 1967

PREFACE

Volume 7 of TOPICS IN STEREOCHEMISTRY begins with a chapter entitled "Some Chemical Applications of the Nuclear Overhauser Effect" by R. A. Bell and J. K. Saunders. The nuclear Overhauser effect, first applied in nmr spectroscopy by F. A. L. Anet and A. Bourn in 1965, has proved to be a highly useful tool for determining the distance between two nearby protons in a molecule. The principles involved in such a determination are outlined in the first chapter which also provides many examples of the use of the NOE technique. A knowledge of the distance between specific protons is often the key to assigning configuration or conformation to a molecule.

There are many ways to synthesize 1,2-epoxides. Oxidation of alkenes with any of a large number of reagents is perhaps the most commonly used method; 1,3-elimination reactions are also widely useful. In addition, there are other synthetic methods that depend upon various reactions of aldehydes and ketones (such as the Darzens reaction, the reaction of a carbonyl compound with a diazoalkane, etc.). In the second chapter, G. Berti discusses the stereochemical aspects of epoxide synthesis by all of the methods.

Carbonium ions, because of their common occurrence as intermediate reactions, have long been intriguing species to the chemist. Because of the relatively low stability of such ions, it has been difficult or impossible to study their structures by the methods applicable to stable substances, such as X-ray crystallography, electron diffraction, and the like. With the recent refinement of the so-called *ab initio* calculations (i.e., quantum mechanical calculations on electronic structures, including all the multicentered electronic integrals, and carried close to the so-called Hartree-Fock limit), it has become computationally feasible to deal with relatively simple carbonium ions to obtain information on geometry and energy in which one may place reasonable confidence. In the third chapter, the structural conclusions derived from such calculations for a number of small carbonium ions are discussed by V. Buss, P. von R. Schleyer, and L. C. Allen.

In the final chapter, isomerizations about double bonds are discussed by H. Kessler and H. Kalinowski. Simple ethylenes have very high rotational barriers as long as the molecules remain in the singlet ground state. Conjugated compounds may have formal double bonds whose pi bond orders are much less than one, and whose torsional barriers are correspondingly smaller. These and other types of double bonds are considered in the chapter.

The carbon-nitrogen double bond is examined in detail, particularly with respect to rotation versus inversion mechanisms for imines and related compounds.

This volume is being produced in "cold type," that is, by typewriter composition. We hope that the product will be pleasing and that the resulting economy in price may be extended to future volumes.

N. L. Allinger
E. L. Eliel

May 1972

CONTENTS

TOPICS IN
STEREOCHEMISTRY

VOLUME 7

A WILEY-INTERSCIENCE SERIES

SOME CHEMICAL APPLICATIONS OF THE NUCLEAR OVERHAUSER EFFECT

R. A. BELL

McMaster University
Hamilton, Ontario, Canada

J. K. SAUNDERS

Université de Sherbrooke
Sherbrooke, Quebec, Canada

I. INTRODUCTION

Nuclear magnetic double resonance has been of particular value to chemists as an aid to spectral assignment and thence to structure elucidation (1). The second applied field may vary in strength from a high power level, where complete decoupling effects are observed, to relatively low power levels, where line perturbations or spin tickling effects are found. One application that utilizes a relatively low power level is the *nuclear Overhauser effect* (NOE). Here the second radio-frequency (rf) field is applied to saturate the spins of one nucleus, or set of nuclei, and the weak observing field is used to monitor the lines of the remaining nuclear spins in the molecule. These lines may, or may not, show changes in their intensities (generally increases but not necessarily so), and it is these changes that constitute the nuclear Overhauser effect phenomenon. There are three subcategories of the effect which should be distinguished. (*a*) A redistribution of intensity among the individual transitions of a line, with no change in the total intensity, is referred to as the *generalized Overhauser effect* (1). (*b*) A change in the total intensity of a line of a nuclear spin in the same molecule as the spin being saturated is termed an *intramolecular nuclear Overhauser effect*. (*c*) A change in the total intensity of a line of a nuclear spin in another molecule from the spin being saturated is referred to as an *intermolecular nuclear Overhauser effect*. We are concerned in this review largely with (*b*), the intramolecular NOE, which has been studied most extensively for protons, and which was first reported on in 1965 by Anet and Bourn (2). The intermolecular NOE was also noted in 1965 by Kaiser (3), but subsequent years have seen very little recorded work on this effect, and it will therefore occupy a minor role in this review. No comment at all is made on the generalized NOE.

As will become evident, the intramolecular NOE between two spins, one being saturated and the other observed, is greatly dependent upon their distance apart (for protons, 3.5 Å or less). The phenomenon therefore leads to information about intramolecular distances, and the success of the method in contributing to our knowledge of molecular geometry has recently been underscored by two publications (4,5) in which quantitative estimations of molecular geometry have been made from NOE data. It should be stressed here that the NOE is strictly related to the relaxation properties of the nuclear spins, and the spins are *not required* to be coupled by the normal J-type, or scalar spin-spin coupling. The only criterion for the observation of an NOE is that the nuclear spins should be spatially proximate, and the dominant relaxation mechanism be a mutual magnetic dipole-dipole interaction. The presence or absence of a normal J-type coupling between the spins is irrelevant to the NOE phenomenon.

In the following we present a brief introduction to the relevant theory from which quantitative deductions of intramolecular distances can be obtained, and comment in some detail on a selection of organic compounds that have been studied and that illustrate the value of NOE measurements in structural and stereochemical problems.

II. THEORY

A. Classical Mechanics Approach

The nuclear Overhauser effect can be treated in a classical and pictorial manner by consideration of the Bloch phenomenological equations as modified by Solomon (6) and Abragam (7). For two weakly coupled nuclear spins, A and B, a perturbation of the spin system results in a return to equilibrium of the magnetization in the direction of the applied field following the simple rate laws:

$$\frac{dM_z^A}{dt} = -R_A \, (M_z^A - M_0^A) - \sigma_{BA} \, (M_z^B - M_0^B) \qquad [1]$$

$$\frac{dM_z^B}{dt} = -R_B \, (M_z^B - M_0^A) - \sigma_{AB} \, (M_z^B - M_0^B) \qquad [2]$$

To prevent confusion we adopt the notation used by Noggle and Schirmer (5) with $R_A = 1/T_{AA}$, where T_{AA} is the longitudinal relaxation time of nucleus A containing both intramolecular and intermolecular components, and $\sigma_{AB} = 1/T_{AB}$, where T_{AB} is the time constant representing the interaction of nucleus A on nucleus B.* The terms M_z^A and M_0^A represent the instantaneous and equilibrium values of the magnetization in the z direction. When equilibrium is established after saturation of nucleus A, $dM_z^B/dt = M_z^A = 0$ and eq. [2] can be represented as

*For a more detailed description of these time constants in terms of transition probabilities see ref. 6.

$$\frac{M_z^B - M_0^B}{M_0^A} = \frac{\sigma_{BA}}{R_B} = f_B(A) \qquad [3]$$

The notation $f_B(A)$ will be taken to mean the fractional enhancement of the B nucleus when nucleus A is saturated. The NOE as a percentage is then given by NOE (%) = 100 X $f_B(A)$. In the limit of extreme narrowing for the *isolated* two-spin system, where A and B have spin numbers I and S, respectively, the fractional enhancement is given by

$$f_B(A) = \frac{\sigma_{BA}}{R_B} = \frac{I\ (I + 1)}{2S\ (S + 1)} \frac{\gamma_1}{\gamma_S} \qquad [4]$$

and thus if A and B are both protons the maximum NOE possible is 50%.

In order to assess the correlation between NOE and molecular geometry R_B and σ_{BA} must be expressed in terms of internuclear distances. This can be achieved by expressing the transition probabilities in terms of simple forces. Thus, provided the dominant relaxation process is the dipole-dipole interaction, σ_{AB} is represented in the limit of extreme narrowing and under isotropic conditions (8) by

$$\sigma_{AB} = \frac{h^2\ \gamma_A^2\ \gamma_B^2}{8\pi^2\ r_{AB}^6}\ \tau_r \qquad [5]$$

where r_{AB} is the internuclear distance between nuclei A and B and τ_r is the isotropic rotational correlation time. The term R_B is more complex as both intramolecular and intermolecular contributions must be assessed. The total can be computed using

$$R_B = (\rho_{BB})_{\text{intra}} + (\rho_{BB})_{\text{inter}} \qquad [6]$$

and under similar conditions as for σ_{AB} above, (ρ_{BB}) *intra* can be represented by

$$(\rho_{BB})_{intra} = \frac{h^2\ \gamma_A^2\ \gamma_B^2}{4\pi^2\ r_{AB}^6}\ \tau_r \qquad [7]$$

In the case where more than one nuclear spin is effectively relaxing B intramolecularly, the expression for $(\rho_{BB})_{intra}$ can be obtained by adding like terms to eq. [7] such that

$$(\rho_{BB})_{intra} = \frac{h^2}{4\pi^2}\gamma_B^2\tau_r\sum_i \frac{\gamma_i^2}{r_{Bi}^6} \qquad [8]$$

provided that cross-correlation terms can be neglected, a valid procedure for dilute solution (8).

The intermolecular contribution to R_B can be expressed as shown in eq. [9] assuming that intermolecular solute-solute interactions are negligible and that only solute-solvent interactions with solvent nuclei of the first solvation sphere need be considered. Thus

$$(\rho_{BB})_{inter} = \frac{h^2}{10\pi^2}\gamma_B^2\,\tau_t\sum_S \gamma_S^2\,S(S+1)\frac{1}{b^6} + \rho^* \qquad [9]$$

where τ_t is the isotropic translational correlation time, Σ_S is the summation over all solvent nuclei of spin S, b is the internuclear distance between spin B and spin S, and ρ^* represents all intermolecular interactions other than dipole-dipole which contribute to the relaxation of B.

Combining eqs. [5], [7], and [9] gives on rearrangement

$$\frac{R_B}{\sigma_{AB}} = 2 + \left[\frac{4}{5}\frac{1}{\gamma_A^2}\frac{\tau_t}{\tau_r}\sum_S \gamma_S^2\,S(S+1)\frac{1}{b^6} \quad \frac{8\pi^2\rho^*}{h^2\gamma_A^2\gamma_B^2\tau_r}\right]r_{AB}^6 \qquad [10]$$

which simplifies to

$$\frac{1}{f_B(A)} = \frac{R_B}{\sigma_{AB}} = 2 + Ar_{AB}^6 \qquad [11]$$

provided that the ratio τ_t/τ_r, the distance of closest approach b, and the interaction designated ρ^* are all independent of the solute (4).

Equation [11] strictly applies to a two-spin system isolated from all other spins in the molecule, a situation which is seldom realized in practice, and thus it should be modified to allow for the presence of other intramolecular spins. If we assume that cross-relaxation processes can be neglected, then

$$\frac{1}{f_B(A)} = 2 + Gr_{AB}^6 \qquad [12]$$

where

$$G = \frac{2}{\gamma_A^2}\sum_i \frac{\gamma_i^2}{r_{B_i}^6} + A$$

which must be constant for the existence of a correlation between $f_B(A)$ and internuclear distances. Qualitatively we can see that at a specific distance there will be the dipole-dipole interaction between the two spins in question combined with all other relaxation processes. The latter must be considered constant, and independent of the type of processes involved. If other spins in the molecule significantly relax spin B the term A will be very small, whereas if the interaction with other spins is small, A will be significant; i.e., there is a general compensation between the two terms and we may take their sum as being roughly constant. This explanation is simplified and a more quantitative treatment will be given elsewhere (9). Equation [11] is a combination of two separate equations which take into account the 0.5 maximum present in the dependence of an NOE on the internuclear distance. When r_{AB} assumes small values, less than a certain specific value, r_{AB}^0 the NOE approaches the limiting value and eq. [12] becomes simply

$$f_B(A) = \frac{1}{2} \qquad [12a]$$

whereas when $r_{AB} > r_{AB}^0$

$$f_B(A) = G^{-1}\,(r_{AB})^{-6} \qquad [12b]$$

In compounds where more than one spin are efficiently relaxing the observed spin, B, the total NOE can be computed, provided cross terms are negligible, using

$$f_B(\text{total}) = f_B(A) + f_B(C) + f_B(D) + \ldots$$

with a maximum $f_B(\text{total}) = 0.50$ when A, B, and C are of the same species. In such examples the internuclear distance can be calculated using

$$\frac{f_B(A)}{f_B(C)} = \left(\frac{r_{BC}}{r_{AB}}\right)^6 \quad [13]$$

which is of particular value when the two spins are so close to the observed spin that the distances cannot be uniquely deduced using eq. [12].

In order to test the validity of eq. [12b] and to assess its application to structural problems, a series of relatively rigid compounds of unequivocal structure were examined (4) where single proton-proton interactions and proton-methyl group interaction existed. In all the compounds used for the latter study, the methyl group was rotating rapidly and the protons were considered to be acting as a net dipole from a point defined by the intersection of the C_3 axis of rotation with the plane containing the three protons of the methyl group (10). In both studies a plot of NOE against internuclear distance using logarithmic axes gave straight lines of slope -6 as predicted. The excellent linearity of the points infers that the gross approximation of considering G to be constant for different protons was a reasonable one. In order to be able to use the observed results as a guide to internuclear distances, a least-squares plot of NOE versus $(r_{AB})^{-6}$ was obtained, the results of which are reproduced in Figures 1 and 2. The standard deviation for intercept and slope were both less than 5% and the line passed through the origin for both cases.

The application of Figures 1 and 2 to molecules of unknown structure should, under the appropriate circumstances, enable specific internuclear separations to be obtained. As will be seen in the following sections such information is particularly valuable in deducing molecular geometries and subtle conformational effects.

B. Quantum Mechanical Approach

The necessary background theory for the following derivations will be dealt with only briefly here because a complete exposition is to appear shortly (11). If we assume (*a*) that cross-correlation effects can be neglected, (*b*) that the component being studied is at sufficient dilution that intermolecular solute-solute interactions are negligible, (*c*) that all spins are weakly coupled, and (*d*) that none of the spins being studied are involved in chemical exchange, then the NOEs of a multispin system are related (5) by

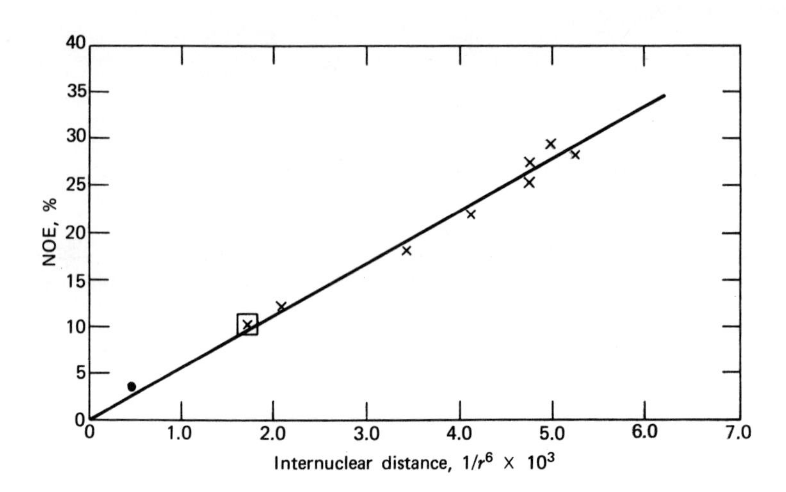

Fig. 1. NOE versus internuclear separation for H-H interactions. $A = 1.8 \times 10^{-4}$.

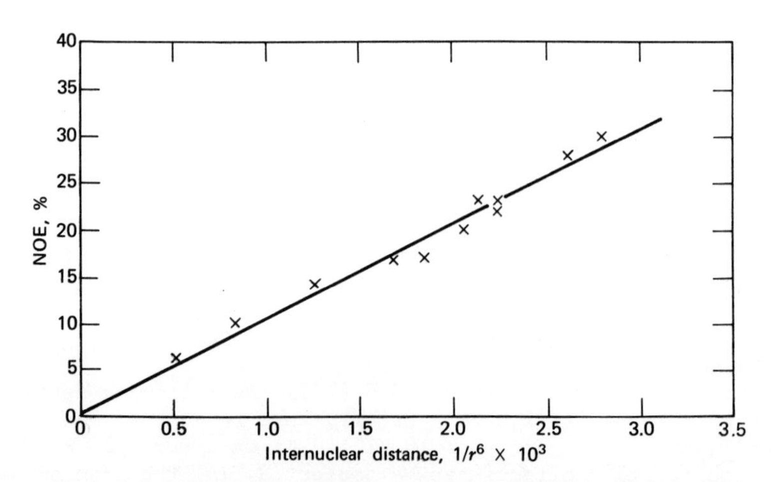

Fig. 2. NOE versus internuclear separation for CH_3-H interactions. $A = 0.98 \times 10^{-4}$.

$$f_A(B) = \sum_B \frac{\gamma_B \rho_{AB}}{2\gamma_A R_A} - \left(\frac{1}{2\gamma_A R_A}\right) \sum_C \gamma_C \rho_{AC} f_C(B) \qquad [14]$$

where A is the nucleus being observed, B are the nuclei being saturated, C represents all other spins in the system, and the other terms are as described in Sect. II-A.

As an example of eq. [14] we will consider the system consisting of three nonequivalent species, AMX. There are six forms of eq. [14] corresponding to the various A-M-X combinations, which when solved by pairs yield equations of the type

$$f_A(M) = \frac{\gamma_M}{\gamma_A} \left[\frac{\rho_{AX}\rho_{XM} - 2\rho_{AM} R_X}{\rho_{AX}^2 - 4R_A R_X} \right] \qquad [15]$$

From these equations, one can obtain equations of the form

$$\rho_{AX} = 2R_A \left(\frac{\gamma_M}{\gamma_A}\right) \frac{f_A(X) + f_A(M) f_M(X)}{1 - f_X(M) f_M(X)}$$

and if we assume that the correlation times τ_{AM} and τ_{AX} are equal, a valid assumption for rigid molecules, then

$$\left(\frac{r_{AM}}{r_{AX}}\right)^6 = \left(\frac{\gamma_M}{\gamma_X}\right)^3 \left[\frac{f_A(X) + f_A(M) f_M(X)}{f_A(M) + f_A(X) f_X(M)}\right] \qquad [16]$$

which when $\gamma_X = \gamma_M$ can be approximated (12) to

$$\left(\frac{r_{AX}}{r_{AM}}\right)^6 = \frac{f_A(M)}{f_A(X)}$$

which is the same as that derived using a classical treatment (eq. [13]). Conversely it is possible to predict the magnitude of the NOE from a knowledge of the internuclear distance using eq. [15]. This is achieved by assuming that one value defines the correlation time of all nuclei and that all the intermolecular interactions can be considered negligible. Thus for three nuclei of the same γ

$$f_A(M) = \frac{\left(\frac{1}{r_{AX}}\right)^6 \left(\frac{1}{r_{MX}}\right)^6 - 2\left(\frac{1}{r_{AM}}\right)^6 \left[\left(\frac{1}{r_{MX}}\right)^6 + \left(\frac{1}{r_{AX}}\right)^6\right]}{\left(\frac{1}{r_{AX}}\right)^{12} - 4\left[\left(\frac{1}{r_{AM}}\right)^6 + \left(\frac{1}{r_{AX}}\right)^6\right]\left[\left(\frac{1}{r_{MX}}\right)^6 + \left(\frac{1}{r_{AX}}\right)^6\right]} \qquad [17]$$

In a "linear" system the NOE between adjacent protons is positive, whereas that between next-nearest neighbors is negative (13). The alkaloid ochotensimine, *1*, has a four-spin "linear" system which gave rise to the results recorded in Table 1. As can be observed, some enhancements were not recorded. Noggle (12) attempted to calculate the missing data from the recorded NOEs using $f_A(X) = -f_A(B)f_B(X)$. Substitution of this expression into eq. [16] results in the numerator being zero, a limit approached only when $r_{AX} >> r_{AM}$, a condition that would give a zero NOE between A and X. A criterion for observing a negative enhancement is that both A and X be in close proximity to B; hence the above expression is only a rough approximation. The values so obtained are given in Table 1. Using 1.8 Å as

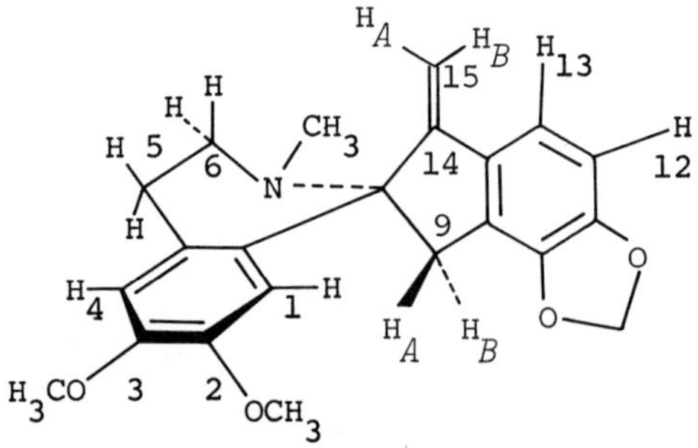

1

the known internuclear distance between H-15_A and H-15_B a value of 2.4 Å for the distance H-15_B,H-13 was calculated from eq. [13] or eq. [16]. Measurement of a molecular model gives the internuclear separation as 2.4 Å, illustrating that for this example both equations agree favorably with experiment (14).

Another example of interest is in the work of Fraser and Schuber (15), who used NOEs to advantage in the assignment of a number of bridged biaryls, including the benzthiepin *2*. In Table 2 are collected the NOE results obtained from experiment, those predicted using eq. [17], and those predicted using eqs.

Table 1 NOE Results for Ochotensimine, *1*

Observe	Irradiate		
	H-15_A	H-15_B	H-13
H-15_A	—	0.40	-0.03[a]
H-15_B	0.40	—	0.08
H-13	-0.08	0.24	—
H-12	0.03	-0.07	0.29[a]

[a]Values estimated from eqs. [13] or [16].

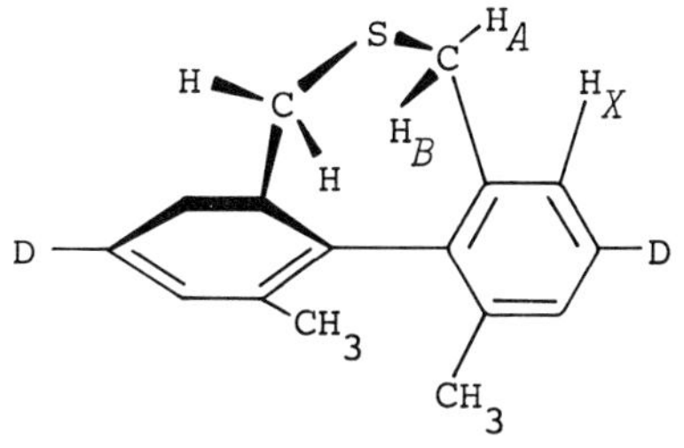

2

[12] or [13]; the internuclear distances were taken as

$$r_{AB} = 1.8\ \text{Å}, \qquad r_{AX} = 2.4\ \text{Å}, \qquad \text{and} \quad r_{BX} = 3.5\ \text{Å}$$

In almost all instances the correlation between observed and predicted values is well within experimental limits. One exception is $f_X(A)$ calculated using eq. [17], and this is a direct consequence of neglecting intermolecular interactions. Reversing the process, i.e., calculating the distance from the observed NOE data, gives reasonable agreement with the measured value. However, the experimental error involved using eq. [12] is much less than that using eq. [16], because in the latter case four NOE values are necessary, whereas with the former only one measurement is necessary.

Both theoretical treatments suffer from the same type of

Table 2 Predicted and Experimental NOEs for Compound *2*

	Predicted values		
	Eq. [12] or [13]	Eq. [17]	Experimental
$f_A(B)$	0.42	0.43	0.41
$f_B(A)$	0.50	0.50	0.48
$f_X(A)$	0.30	0.48	0.31
$f_A(X)$	0.08	0.07	0.08
$f_X(B)$		-0.14	-0.10
$f_B(X)$		-0.04	-0.05
r_{AX}	2.36±0.1	2.45±0.3	2.40

limitation; i.e., they consider the system to be isolated. Equation [12] takes into account intermolecular interactions with the spin lattice but does not consider the interdependence of the interaction between the two spins in question with other spins in the molecule. In addition, the approximation of considering G a constant and comparing it with the two-spin system is a gross one with little theoretical justification. However, in a large number of examples studied (14), correlation with Figures 1 and 2 has been independent of the number of spins involved in the relaxation of the observed spin. Thus, despite eq. [12] being suspect theoretically, it appears to be a good approximation experimentally. The quantum mechanical approach considers the effect of other spins on the observed NOE but eq. [16] does not allow for intermolecular interactions. This procedure is valid for determination of internuclear distances provided all three nuclei are on the rigid portion of the molecule and are equally affected by intermolecular interactions. One serious problem with eq. [16] is that four NOE measurements are required and this can be frequently difficult because of similar chemical shifts; furthermore, each NOE has its own inherent experimental error which leads to additional error in the distance assessment. Indeed the most serious drawback to the quantitative use of NOEs is the accuracy of the results; the best method to date has an error of at least ±1% and thus for $f_B(A) = 0.10$ the error is ±10%. Even with this error range, however, the quantitative use of NOEs can in many examples lead to useful information in structural and stereochemical problems.

C. Conformationally Labile Molecules

In systems where conformational lability exists additional problems arise. If the different conformations show distinctive chemical shifts the problem is fairly straightforward, unless the rate of conformation change is such that chemical exchange effects must be considered. This is illustrated by studying the NOE observed at the formyl proton of dimethylformamide (DMF) as a function of temperature (16), as shown in Table 3. The temperature at which the two methyl groups show essentially the same NOE is far below normal coalescence as the two methyl signals are separated by 0.15 ppm. The temperature dependence shown by the methyl group (*B*) is a result of increasing efficiency of the transfer of spin saturation by the methyl group at site *B* to site *A* as the rotation about the C-N bond increases with temperature. The consequence of these results is exceedingly important when NOE data are being applied to unknown substrates. The existence of an NOE in a conformationally labile system cannot be taken as sufficient evidence to infer a conformational preference.

Table 3 Variable-Temperature NOEs in Dimethylformamide

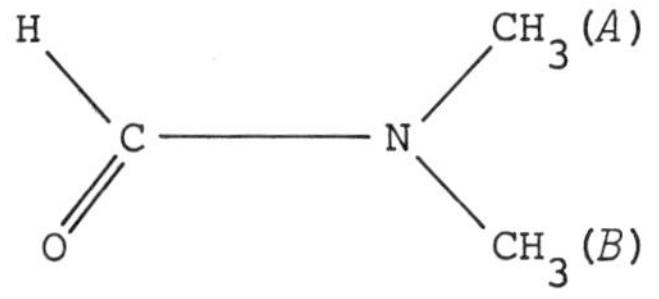

H–C(=O)–N(–CH_3 (*A*))(–CH_3 (*B*))

Temperature, °C	NOE: Saturate CH_3(*A*)	NOE: Saturate CH_3(*B*)
31	0.28	0.03
40		0.06
50	0.28	0.09
55	--	0.13
60	0.28	0.18
70	0.28	0.24
80	0.28	0.26
90	0.28	0.27-0.28

From the recorded data, a plot of log(NOE) versus $1/T$ gives a straight line. However, the energy value obtained includes the temperature dependence of the T_1's and also the energy of activation of the intermolecular motion. Nevertheless, in many systems where these data are obtainable, information on rotation rates can be gained from a study of the temperature dependence of the observed NOE. The relevant theory and equations for a two-site exchange system have recently been developed by Combrisson et al. (17).

In the case of very fast rotation, the group can be considered as a net dipole acting from a point designated as the mean of all possible conformations. This was used in Sec. II-A for the methyl group-proton interaction. To further the knowledge of NOEs in systems where rapid rotation is occurring, the compounds shown in Table 4 were studied. The distance corresponding to the NOEs $f_7(2)$ and $f_2(7)$ in *3* is 2.92 Å, which corresponds closely to that measured between H-2 and the position designated as X in the structures *3* to *8*. Thus H-7 is acting as a net dipole from this point, the mean position of all conformations, with regard to being saturated and also to being observed. The use of the same correlation time for H-7 as for H-2 is valid, as the net dipole will move as the rest of the molecule and will not be influenced by the rapid rotation about the C-1,C-7 bond. The compounds *5* and *7* show similar results except that in *7* the NOE from the CH_3 group is of the same order as that observed for the CH_2D group, which in turn is slightly less than twice that observed for the CHD_2 group. The reasons for this are quite complex and have been dealt with elsewhere (18). Table 4 shows in addition that replacement of D by Br at C-6 (compare *3* with *4*, and *5* with *6*) causes an increase in the observed NOE at H-2, implying a distortion with X remaining in the plane of the aromatic ring but moving closer to H-2; such a compression is fully expected on steric grounds. However, for the compounds *9* and *10* the reverse is the case; i.e., the NOE is decreased by the presence of a bromine at C-6. This is best explained by differences in population of the conformers in which H-2 and H-7 are cis or trans to one another. Obviously the position X will tend to remain in the plane of the ring but is now moved toward the bromine because the conformer with H-7,H-2 trans would be more populated than that with H-7,H-2 cis. Therefore, to designate position X, a weighted average must be taken. The reverse situation is of even greater interest since, with a knowledge of the NOEs and the possible conformations, information on conformer populations can be obtained in systems where rapid rotation is occurring.

Table 4 NOE Results for Some 6-Substituted Piperonyl Alcohol Derivatives

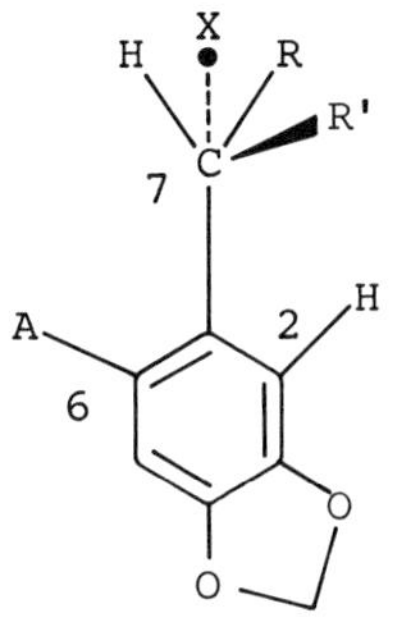

Compound	Irradiate	Observe	NOE, %
3: R = R' = A = D	H-7	H-2	9
	H-2	H-7	8
4: R = R' = D; A = Br	H-7	H-2	14
5: R = A = D; R' = H	H-7	H-2	17
6: R = D, R' = H; A = Br	H-7	H-2	24
7: R = R' = H; A = D	H-7	H-2	16
8: R = R' = D; A = Br	H-7	H-2	22
9: R + R' = O; A = H	H-7	H-2	13
10: R + R' = O; A = Br	H-7	H-2	10

D. Effect of Scalar Coupling

In both theoretical expressions in Sects. II-A and II-B (eq. [12] and eq. [14]) it was assumed that the spins involved were only weakly coupled; i.e., δ/J was large (δ = chemical shift difference). This "weak coupling" condition is required only between the observed and irradiated spins (5). In such weakly coupled systems the only effect of the J coupling is to distribute the intensity of the resonance among the lines of the resulting multiplet, because the populations of one spin are relatively independent of the other spin unless they are connected by some relaxation process. However, as the ratio δ/J approaches 1 as in a tightly coupled system, spin mixing in a multispin system becomes extensive and the energy levels are changed. In such tightly coupled systems, strong irradiation of one transition may affect all the energy level populations in the spin system, and thus it would no longer be valid to treat the system by simple theories as has been done here, but rather a full density matrix treatment should be employed (7).

The results (14,19) for a series of compounds, in which δ/J was varied from 150 to 2.5, are given in Table 5. The lower

Table 5 NOE Values for Compounds that Show Weak and Strong J-Coupling

		Geminal NOE		Total NOE	
Compound	δ/J	$f_A(B)$	$f_B(A)$	f_A(all)	f_B(all)
Ochotensimine	150	0.40	0.40	0.42	0.47
Methyl α-methacrylate (10% CS_2)	120	0.42	0.48	0.51	0.48
α-Methacrolein	45	0.42	0.39	0.49	0.48
Atisinone	37.5	0.34	0.47	0.47	0.47
5*H*-Dibenzo(*a*,*d*)-cycloheptyl-(*b*,*c*)epoxide	12.3	0.35	0.47	0.47	0.47
1,11-Dimethyl-5,7-dihydrodibenz-(*c*,*e*)thiepin (*2*)	3.4	0.42	0.47	0.49	0.47
9-Carbomethoxy-9,10-dihydroanthracene (*87f*)	2.5	0.35	0.40	0.46	0.47

limit of 2.5 was employed to ensure that the intensity of the observed line was not affected by direct irradiation at the power levels used. As can be noted the NOE observed between the geminal protons is independent of the value of δ/J even when this ratio is 2.5. Thus it is valid to use the simple theories expounded earlier in all systems where it is experimentally feasible to use the NOE method, including relatively tightly coupled systems.

An interesting sidelight of this experiment was the result obtained for α-methacrolein in both CS_2 and $CDCl_3$ solutions, as is shown in Table 6. As can be observed in the table, the re-

Table 6 Comparison of Results for Methyl α-Methacrylate and α-Methacrolein

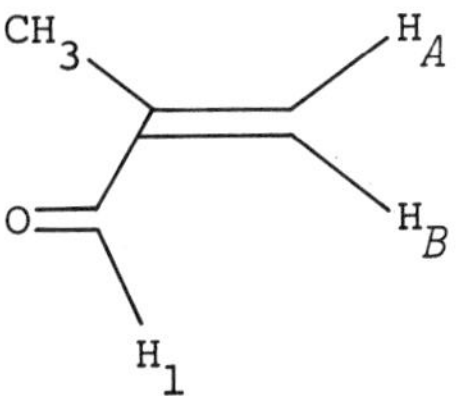

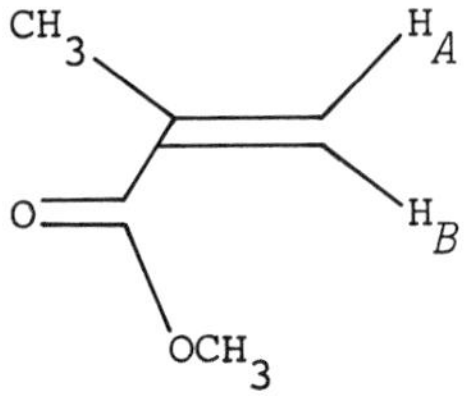

Proton j	Proton i	$f_i(j)$ α-Methacrolein		Methyl α-Methacrylate	
		5% DMSO-d_6	5% CS_2	5% CS_2	Pure liquid
CH_3	H_A	0.09	0.09	0.09	0.09
CH_3	H_B	0	0.06	0	0.06
H_B	H_A	0.39	0.30	0.42	0.28
H_A	H_B	0.42	0.28	0.48	0.29
H_1	H_A	0	0	--	--
H_1	H_B	0.07	0.06	--	--
OCH_3	H_A	--	--	0	0.05
OCH_3	H_B	--	--	0	0.05

sults for α-methacrolein in DMSO-d_6 and methyl methacrylate (20) in CS_2 can be compared, as can the results for α-methacrolein in $CDCl_3$ and methyl methacrylate as the pure liquid. In the pure liquid, intermolecular interactions should play a significant role in the relaxation processes, as is evidenced by the reduced NOEs between the geminal protons and also the NOE at H_B when the α-CH_3 group was saturated. By analogy, then, it is reasonable to assume that in $CDCl_3$ (and CS_2) α-methacrolein is in an associated form, although the exact nature of this species cannot be ascertained. These results emphasize why it is necessary in an intramolecular NOE study to minimize intermolecular interactions; but conversely they illustrate how NOEs can be employed in intermolecular interaction studies.

III. CHEMICAL APPLICATIONS

In the following sections a number of examples of the use of NOE measurements in the study of naturally occurring and synthetic compounds will be presented in order to illustrate the information available from the technique. It is not intended that the examples discussed should be an exhaustive literature survey, but rather that they should represent a wide array of structural types that cover the specific interests of as many chemists as possible. The detailed arrangement of the subsections is somewhat arbitrary, and is a reflection of the authors' principal interests rather than a particular logical design. A limited number of the examples are dealt with in some detail, since it is felt that the reader will be able to more fully appreciate the deductive logic used in applying NOE data from complex molecules. As an additional convenience, and to avoid the unnecessary use of decimal fractions, the NOE will be recorded in the sequel as a percentage enhancement in the form [H-1],H-2 = 9% which should be read as "saturation of the proton signal H-1 caused a 9% area increase of the signal for proton H-2.

A. Alkaloids

The spiroisoquinoline alkaloids have been subjected to extensive NOE study as the technique was found to be extremely useful in the structural and stereochemical elucidation of the member compounds (14,21-25). The first member of this class to be identified as a result of chemical and X-ray studies was ochotensimine, *1* (26). The NOE results obtained for *1* and the synthetic ketone *11* (27), a precursor synthesized during a total synthesis of *1*, are given in Table 7 and will be examined first. The conclusions obtained will be used to discuss the unknown alkaloids. Saturation of the two C-5 protons in compound *1* caused an increase in area of the singlet at 6.52δ as well as removing a small coupling, and saturation of the methoxyl signal at 3.84δ caused an area increase of the same singlet. The chemical shifts of the C-3 methoxyl, H-4, and the C-5 protons have thus been confirmed. The occurrence of 28% NOE between the C-2 methoxyl and H-1 defines their chemical shifts and exemplifies the NOE method as of special value in locating the position of methoxyl groups in aromatic rings, a task that otherwise can be extremely difficult. The benzylic protons attached to C-9 give

Table 7 NOEs in Ochotensimine *1* and Synthetic Spiroisoquinoline Alkaloid *11*

		NOE, %	
Irradiate	Observe	*1*	*11*
H-9_A	H-1	14	25
2-OCH_3	H-1	25	21
3-OCH_3	H-4	25	28
H-5(ax)	H-4	9	17
H-5(eq)	H-4	16	17
H-5[a]	H-4	25	26
N-CH_3	H-9_B	8	15
H-15_B	H-13	24	--

[a]Triple irradiation.

rise to an *AB* quartet and the NOEs [H-9_A],H-1 = 14% and [N-CH_3],H-9_B = 8% infer that the spatial proximity of H-1,H-9_A and N-CH_3,H-9_B must be as shown in Figure 3. The NOEs connected

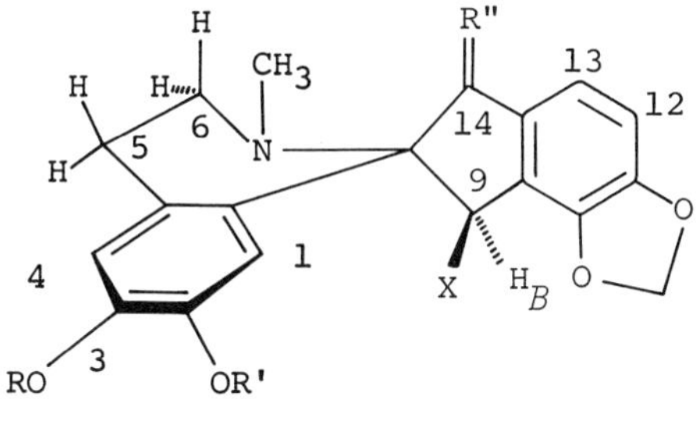

1: R" = CH_2 X = H_A R = R' = CH_3
11: R' = O X = H_A R = R' = CH_3

Fig. 3. Stereostructure of spiroisoquinoline alkaloids.

with the geminal protons on C-15 and the aromatic ring D have been discussed above in Sect. II. The synthetic ketone *11* shows similar results, and since the NOE data appear to correlate well with the structures of *1* and *11*, the data obtained for the remaining members of the class can now be interpreted in structural terms.

The compounds *12*, *13*, and *14*, whose gross structural de-

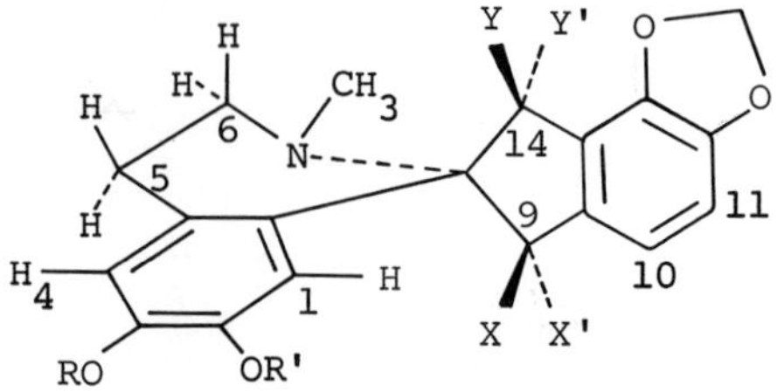

12:	X = X' = H	Y + Y' = O	R + R' = CH_2
13:	X = X' = H	Y = OH; Y' = H	R = R' = CH_3
14:	X = X' = H	Y = OH; Y' = H	R = CH_3; R' = H
15:	X = X' = H	Y = OAc; Y' = H	R = CH_3; R' = H
16:	X + X' = O	Y = OH; Y' = H	R + R' = CH_2
17:	X = H; X' = OH	Y + Y' = O	R = CH_3; R' = H
18:	X = H; X' = OH	Y = OH; Y' = H	R + R' = CH_2
19:	X = H; X' = OAc	Y = OAc; Y' = H	R + R' = CH_2

tails were obtained from other physical methods (21), including nmr and mass spectral data, yielded the NOE results collected in Table 8 (21). In compound *12* rings A and D both contain methylenedioxy groups, whereas in *13* and *14* only one methylenedioxy group is present, the position of which was in some doubt. Saturation of the C-5 protons in compound *12* caused an area increase in H-4, as did saturation of the low-field methoxyl group, and therefore the methoxyl must be attached to ring A. Compound *14*, which gave *13* when methylated with diazomethane, contains a methoxyl and phenolic function, the relative positions being defined by the observance of an NOE between the methoxyl and H-4. Assignment of the methylenedioxy at the 12,13-position rather than the 10,11-position, as in ochotensimine, was confirmed by the observance of an NOE between the C-9 protons and H-10 for both *12* and *13*. The alcohol *13* showed an OH stretching frequency in the infrared spectrum at $\nu_{max}(CS_2)$ 3560 cm^{-1} which did not change on dilution, implying hydrogen bonding. The coupling constant $J_{14,OH}$ = 11.0 Hz, obtained at -10°C, indicates a dihedral angle of 160-170° and this, together with the infrared, is explicable by assuming that the OH is hydrogen bonded to the nitrogen; i.e., Y' = OH. However, the observance of an NOE between the N-CH_3 group and H-14 and the lack of an NOE between H-14 and H-1 reverses the above assumption. Thus Y = OH with the hydrogen bonding occurring between the OH and the

Table 8 NOEs in Spiroisoquinoline Alkaloids *12*, *13*, and *14*

Irradiated	Observed	NOE, %		
		12	*13*	*14*
H-9_A	H-1	22	19	
H-9_A	H-10	10		
H-9_B	H-10	10		
H-9[a]	H-10	19	20	20
2-OCH_3	H-1		21	
3-OCH_3	H-4		23	24
H-5(ax)	H-4	9		
H-5(eq)	H-4	17		
H-5[a]	H-4	25	24	25
N-CH_3	H-9_B	7	7	
H-10[a]; N-CH_3[a]	H-9_B	14	14	
N-CH_3	H-14		14	15

[a]Triple irradiation.

aromatic ring (28). For *15*, the NOE results (Table 9) (22) de-

Table 9 NOEs in Spiroisoquinoline Alkaloids *15*, *16*, *17*, and *18*

Irradiate	Observe	*15*	*16*	*17*	*18*
H-9_A	H-1	20		4	13
H-9_A	H-10		20	9	9
H-9[a]	H-10	19			
3-OCH_3	H-4	24		25	
H-5(ax)	H-4	8			
N-CH_3	H-14	25	25		25
H-5	H-9			16	

[a]Triple irradiation.

fined the cis relationship between the N-CH_3 group and H-14, and also positioned the hydroxyl and methoxyl at C-2 and C-3, respectively. In both *16* (23) and *17* (24) the ring contains a ketonic and an alcoholic function, the relative positions of which, together with the configuration of the hydroxyl group, are defined by the NOE results (see Table 9). The two hydroxyls in ring C of the alkaloid ochrobirine, *18* (25), were shown to be trans to one another because of the NOEs [N-CH_3],H-14 = 25% and [H-9],H-1 = 13%, with the NOE [H-9],H-10 = 9% defining the position of the methylenedioxy group relative to the hydroxyl groups.

In the preceding discussion we have seen how the NOE can be of considerable use in gross structure elucidation. The results obtained, however, will yield further information and can be interpreted in terms of more subtle stereochemical effects as a consequence of changes in substituents on ring C of the spiroisoquinolines. Changes on rings A and D will be neglected as these will not affect significantly the conformations of rings B and C. The principal measure of the stereochemical changes will be the NOEs between H-1 and H-9 (or H-14).

The X-ray analysis of ochotensimine (26) performed on the N-methyl salt showed ring B in a twist-boat conformation. It is reasonable to assume that in solution the free base also exists with ring B in a twist conformation with the N-methyl group taking up two possible configurations, namely, α or β. Inspection of molecular models shows that when the N-CH_3 is β (cis to the exocyclic methylene) the distance H-1,H-9 is 1.8 Å, and that between H-15_A and either H-5(ax) or H-6(ax) is 1.9 Å, with the CH_3 group being approximately equidistant from C-9 and C-14. When the N-CH_3 is in the α configuration the distances are H-1,H-9 = 3.8 Å, H-5(ax),H-9_B = H-6(ax),H-9_A = 2.4 Å, and H-1,C-14 = 2.4 Å with the N-CH_3 again equidistant from C-9 and C-14. The distance H-1,H-9 obtained from the experimentally observed NOE is 2.8 Å, inferring that in this compound inversion about the nitrogen is rapid and that both conformations are approximately equally populated. When the substituent at C-14 is a carbonyl, as in *11* and *12*, the distance H-1,H-9 (2.5 Å) is less than that observed in *1*. Evidently this is a consequence of a lower steric requirement for the oxygen relative to the exocyclic methylene as the conformation with the N-CH_3 group β must be more populated in *11* and *12* than in *1*. In the alcohol *13*, it is possible to obtain hydrogen bonding to the aromatic ring with the N-CH_3 group either α or β, and hence it appears that in this compound both conformations are well populated. The acetylated analogue *15* shows a slightly higher H-1,H-9 distance (2.6 Å) which could be explained either by a change in conformer population or by small changes in ring B. In this compound interactions between the aromatic ring and the acetoxy substituent would be more severe than in *13*.

The two compounds *16* and *17* have the same functionality in ring C but because of differences in hydrogen bonding, resulting from opposite orientations of the hydroxyl group, the conformation adopted by ring B for the two compounds is completely different. In *16* the conformation is similar to that described above for *1* and *13* because the hydrogen bonding is between the OH and the aromatic ring. In *17*, however, for hydrogen bonding to exist between the OH and the nitrogen, the N-CH_3 group must be β and ring B must twist such that N-CH_3 moves toward C-9, and H-5 is moved, much closer to H-14. This causes ring C to be twisted such that C-9 moves closer to H-1, as is evidenced by a downfield shift of the signal assigned H-1. This twist is mirrored by only a very small NOE being observed between H-1 and H-9, as this distance has now increased dramatically over that found in *15*. Thus by using the distances H-5,H-14 and H-1,H-9, together with the magnitude of the dihedral angle between H-9 and the OH (70°), a reasonably accurate conformation can be deduced. Compound *18* contains two hydroxyls trans to one another, with one hydrogen bonded to nitrogen and the other to the aromatic ring. The NOE between H-1 and H-9 is less than that observed in compound *13* but greater than in *17*. As has been noted above, the bonding between OH and the nitrogen causes twisting such that the proton geminal to the hydroxyl moves further away from H-1. If this twist is as severe in *18* as in *17*, the C-14 OH is no longer in a position to bond to the π electrons in the aromatic ring. Ring B therefore undergoes only a slight deformation and twists such that the H-1,H-9 distance is greater than in *13* but less than that assumed in *17*, and in this conformation simultaneous hydrogen bonding can occur between the hydroxyls and the nitrogen and the aromatic ring. The general conclusions that can be drawn are that all the compounds have fairly similar stereochemistries, the differences being small and dependent generally on the steric requirements of the groups attached to ring C.

An interesting aspect of the nmr spectra of the acetoxy compounds *15* and *19* is that the acetoxy methyl group resonance is at higher field than normally encountered, and acetylation causes a marked downfield shift in the position of the H-1 resonance. This suggests that the acetoxy group lies across ring A such that the carbonyl group deshields H-1 and the aromatic ring A shields the methyl group. Low-temperature spectra of compound *15* show the methyl resonance to move downfield and the H-1 resonance upfield, suggesting that the conformation where the acetoxy group lies across ring A, although well populated at room temperature, is not the preferred conformation. Another interesting aspect of the nmr spectra of these compounds is that the C-3 methoxyl absorbs at lower field than the C-2 methoxyl, and as the respective NOEs are similar, this indicates that a relatively long-range shielding of the C-2 methoxyl by ring D is present.

The protoberberine alkaloids (Fig. 4) are related biogenet-

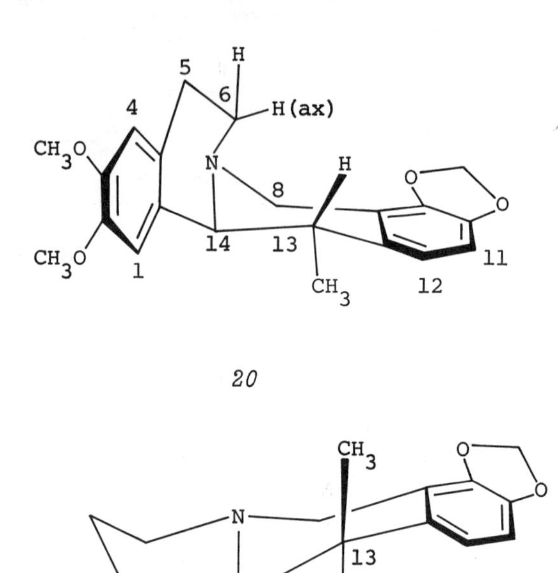

Fig. 4. Conformations of *cis*- and *trans*-protoberberines.

ically to the spiroisoquinoline alkaloids and have been the subject of extensive studies (29). Inversions about the nitrogen can cause the compounds to exist either as cis or trans quinolizidines with the trans being the energetically favored compound in the absence of steric interaction between substituents at C-1 and C-13. MacLean et al. (30) identified two alkaloids, one of each isomer, both of which were subjected to an NOE study (14) in order to gain insight into the stereochemistry of rings B and C and to confirm nmr spectral assignments. The data from this study, together with internuclear distances measured on molecular models and also the distances predicted from the NOE data using Figures 1 and 2, are given in Table 10.

Table 10 NOEs in Protoberberine Alkaloids *20* and *21*

		NOE %	
Irradiate	Observe	*20*[a]	*21*[a]
CH_3	H-12	18; 2.85 (2.85)	6; 3.55 (3.20)
CH_3	H-1	4; 3.60 (3.00)	--
H-13	H-12	7; 3.20 (2.80)	16; 2.65 (2.60)
H-13	H-1	--	27; 2.45 (2.05)
H-14	H-1	18; 2.55 (2.40)	8; 3.10 (2.90)
OCH_3	H-4	24	23
OCH_3	H-1	24	--
OH	H-1	--	12

[a]Quantities in parentheses are internuclear separations in Å measured from a Dreiding molecular model.

As can be seen, the distances predicted are incompatible with those measured, assuming that rings B and C adopt half-chair conformations. Compound *20* is a *cis*-quinolizadine and with rings B and C as half-chairs, a serious nonbonded interaction exists between H-6(ax) and H-13. To relieve this, ring B twists such that the distance H-6(ax),H-13 increases while H-5,H-13 decreases. Any attempt to alleviate the interaction by twisting ring C causes a serious 1,3-interaction between H-14 and H-8(ax) The twisting of ring B results in H-1 moving farther away from the methyl group but has little effect on the CH_3,H-12 and the H-13,H-12 distances. At a position where the protons H-6 and H-5 are equidistant from H-13, the predicted and measured distances become compatible (within ± 0.05 Å). Compound *21* is a *trans*-quinolizidine with the protons at C-14 and C-13 cis to one another. The most serious nonbonded interaction is that between H-13 and H-1. To alleviate this interaction, the molecule distorts as a whole (depicted in Fig. 4), increasing the H-13,H-1 distance and more markedly the distance between the CH_3 group and H-1. At a point where the distance between H-13 and H-1 is 2.40 Å, the other measured and predicted distances are also compatible. Thus *21* exists with the twist as shown in Figure 4.

In these compounds the stereochemistry about the N-C-14 bond can be ascertained by the magnitude of the chemical-shift difference between the two C-8 protons, as a study on a number of compounds has shown that the chemical shift difference is much greater for the trans isomer than the cis (30). MacLean et al. suggested that the low-field doublet of the *AB* is the equatorial proton, as this shows the least dependence on configuration at nitrogen. This was reinforced by the lack of an observable NOE at H-14 when the low-field doublet was saturated in both compounds. Unfortunately, the magnitude of the NOE between H-14 and H-8(ax) could not be determined although such an interaction does exist. This suggests that the assignment is in fact correct, albeit based on negative evidence.

During the identification of a number of benzophenanthridine alkaloids, an alkaloid was isolated which appeared to be a bis adduct of chelerytherine and acetone (31). As this compound was virtually insoluble in nmr solvents, the mono adduct of chelerytherine and acetone, *22*, was subjected to an NOE study, the results of which are given in Table 11. The proton H-10 was assigned on the basis of the area enhancement on saturation of the N-CH_3 signal; saturation of this group also showed that it was cis to the proton at C-11. Irradiation of H-7 resulted in an area increase of H-6, which is in turn coupled to H-5. Finally the spectral assignments of H-3 and H-4 were made by

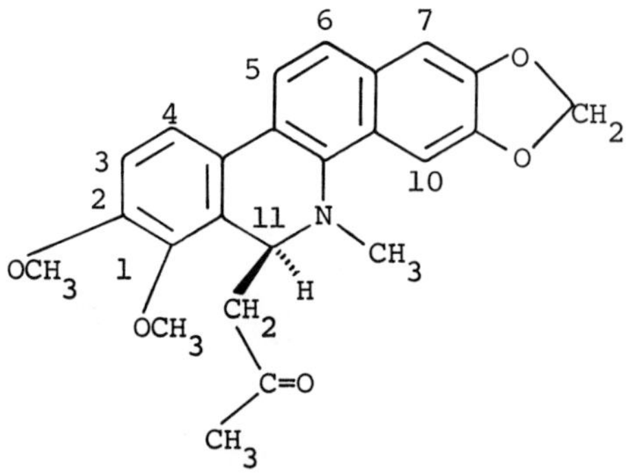

22

Table 11 NOEs in Chelerytherine-Acetone Adduct *22*

Irradiate	Observe	NOE, %
$N-CH_3$	H-10	30
$N-CH_3$	H-11	23
$2-OCH_3$	H-3	24
H-7	H-6	27

observing an NOE between the C-2 methoxyl and H-3. The assignments are in agreement with, and confirm, those of Slavik et al. (32), who made similar deductions from an examination of a series of benzophenanthridine alkaloids.

The structure of the *O*-methylated Hofmann degradation product of cancentrine, *24*, was obtained by an X-ray analysis (33)

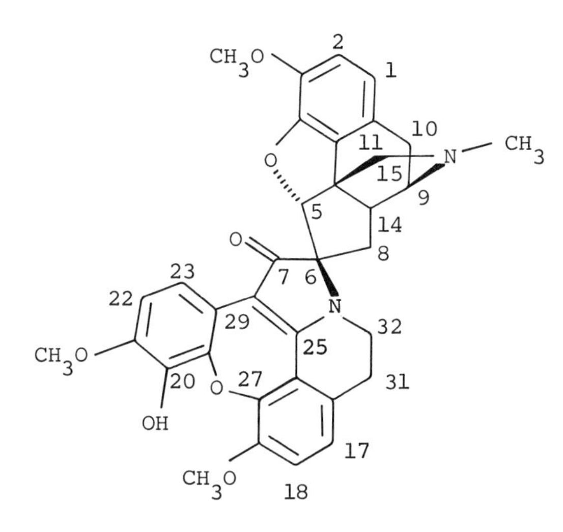

23

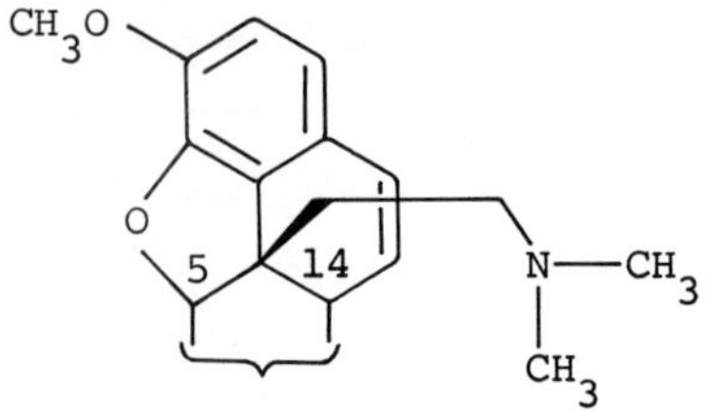

24

with the structure of the parent compound, *23*, being obtained by the use of nmr and NOE data. The point of attachement of the nitrogen bridge was obtained by comparison of the spectra of *24* and its acetylated analogue with other morphine-type compounds (34). The spectral position of the C-10 protons was thus defined and irradiation of these signals defined H-1. The NOE [C-3 OCH_3],H-2 = 25% and the coupling constant $J_{1,2}$ = 8.0 Hz defined the C-3 methoxyl and H-2. The protons H-17 and H-18 were assigned on the basis of [H-31],H-17 = 20%, [C-19 OCH_3],H-18 = 25%, and $J_{17,18}$ = 9.0 Hz. Acetylation of *24* caused the resonance assigned to H-23 to move downfield by 0.37 ppm, indicative of either an ortho or a para shift. The latter was confirmed by the observance of the NOE [C-21 OCH_3],H-22 = 24%. Therefore, all the aromatic signals have been assigned with the phenolic function existing at C-20.

The absolute configuration of the indole alkaloid *25* (35) at C-3a was confirmed, and that at C-8a inferred, by lengthy chemical techniques. NOEs gave a much more direct approach to the assignment of the ring fusion to the indole, and hence the stereochemistry at C-8a. Irradiation of either the C-3a methyl or the low-field N-CH_3 (N-8) signals caused an area increase of 15% at H-8a, whereas no NOE was observed between the N(1)-CH_3 and H-8a. This confirms the absolute configuration as R at C-8a and suggests that the two N-CH_3 groups are trans to one another with no inversion at N-8 occurring. The size of the NOE, however, between the N(8)-CH_3 and H-8a could also be observed if rapid inversion is occurring, with the isomer where the methyl group and the protons are trans being present in negligible amounts. Similarly a lack of NOE between the N-1 methyl group and H-8a does not rule out inversion, as the corresponding internuclear distance for equal population of two rapidly inverting isomers would be of the order of 3.5 Å, which is at the limit of NOE observance. Thus the ring fusion is confirmed as being cis, with the N-8 methyl group being

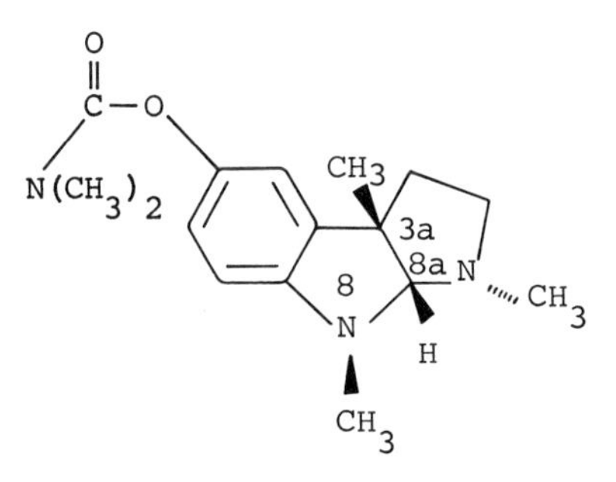

25

virtually exclusively in the configuration where it is cis to H-8a.

B. Terpenes and Steroids

One of the earliest studies on the use of NOE as an aid to structural elucidation in natural products was performed by Woods and co-workers (36) during identification of the diterpenoid ginkolides (Fig. 5). The results of this study are given

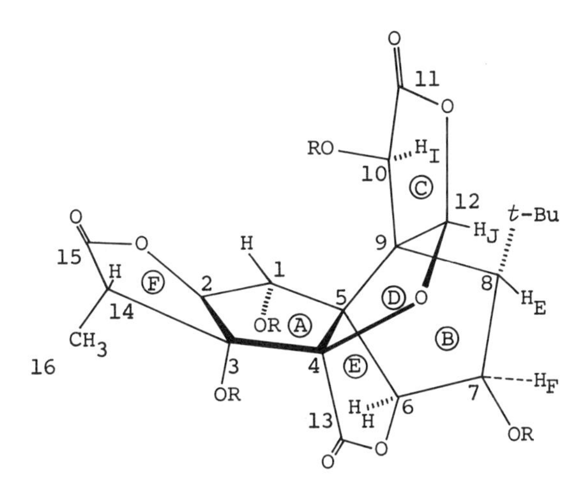

26: R = H
27: R = Ac

(*a*)

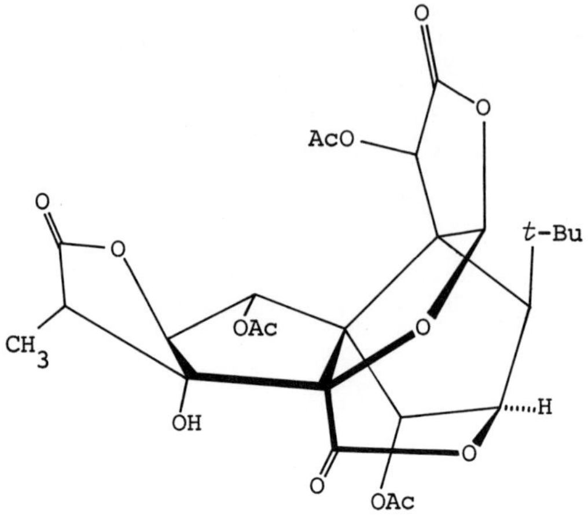

28

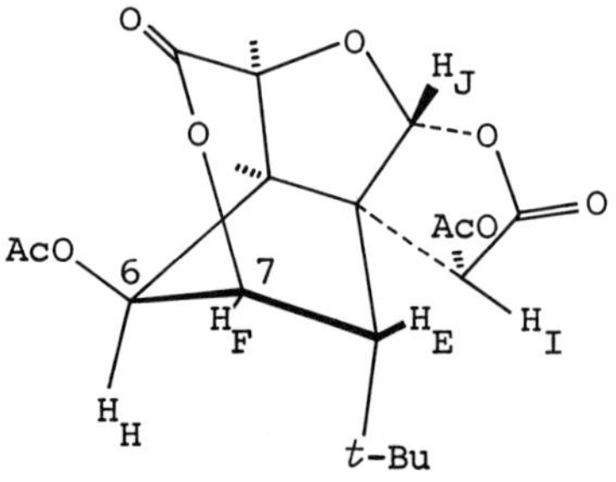

29

(*b*)

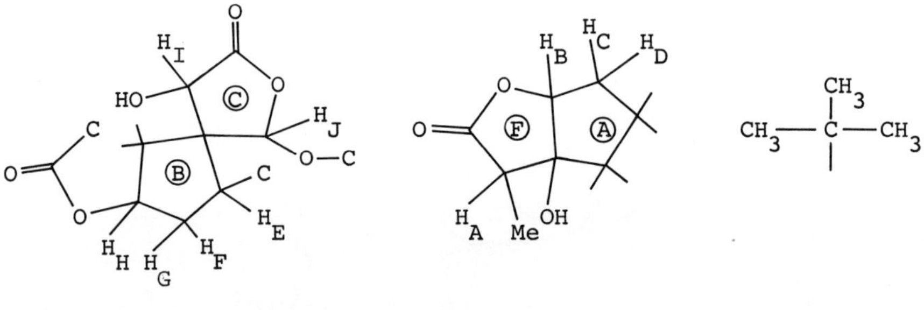

(*c*)

Fig. 5. (*a*) Stereostructure of ginkolide A and ginkolide 1,3,7-tri-*O*-acetate *27*. (*b*) Stereostructure of isomeric ginkolide 1,3,7-tri-*O*-acetate *28* and ginkolide tetraacetate *29*. (*c*) Partial structure of ginkolides derived from chemical and spectroscopic data.

in Table 12. Saturation of the *t*-butyl group caused significant

Table 12 NOEs in Ginkolide and Ginkolide Acetates

		NOE,[a] %			
Irradiated	Observed	*26*	*27*	*28*	*29*
t-Bu	I	20	33	30	27
t-Bu	J	20	22	0	0
t-Bu	F	4	16	24	19
t-Bu	E	--	6	14	13
t-Bu	H	0	0	12	6
E	J	0	0	7	10
J	E	0	0	20	23

[a]Data were recorded in trifluoroacetic acid solution.

increases in the areas of protons E, F, I, and J, inferring that these protons were close to the *t*-butyl group. This information was of great value in dealing with the structure and stereochemistry of the ginkolides. From some elegant chemical and nmr results, the partial structures shown in Figure 5*c* were obtained, and these rings could be fused to give eight possible structures. For the *t*-butyl group to interact with the four protons, this group cannot be attached at either C-5 or C-6. Thus four of the possibilities can be discarded, but the remaining four could still satisfy the NOE results. However, only one, namely that shown in Figure 5, explains both the NOE and nmr results on the ginkolide derivatives (36). The observance of an NOE between the *t*-butyl group and proton I indicates that the C-1 hydroxyl is cis to the C-9,C-5 bond. Mild acetylation of *26* gives the 1,3,7-tri-*O*-acetate which can be further acetylated to give the 1,3,7,10-tetra-*O*-acetate *27*, the NOE results of which are also recorded in Table 12. More vigorous acetylation of *26* yields an isomeric triacetate *28*, which can be further acetylated to the tetraacetate *29*. The isolactones are formed by *trans*-lactonization of lactone E from C-6 to C-7. The lack of an NOE between the *t*-butyl and proton J, and the observance of an NOE between the *t*-butyl and proton H are in complete agreement with this change in lactone terminus. The *trans*-lactonization in addition confirms that the C-6 and C-7 oxygen functions are cis.

The spatial arrangement of the molecule shows that the *t*-butyl and the four protons E, F, I, and J are in close proximity, provided the *t*-butyl group is in fact in a pseudoequatorial orientation. If the *t*-butyl group should adopt a pseudoaxial position, proton J is no longer close enough for an efficient interaction, whereas proton H is in close proximity. In addition, proton E would be moved closer to proton J. However, the natural products all show NOEs between the *t*-butyl and protons E, F, I, and J but not H, and also none between protons E and J, and must therefore exist as depicted with the *t*-butyl pseudoequatorial. On the other hand, the isoacetates *28* and *29* show NOEs [*t*-Bu], E, F, I, and H, but not J, with a significant interaction between protons E and J, and consequently exist with the *t*-butyl pseudoaxial. All these results were obtained by Woods in trifluoroacetic acid solution, a solvent which should be extremely efficient at relaxing the solute protons. It must be assumed here that the *t*-butyl group is insulating the I, J, E, F, and H protons from close approach of solvent molecules and therefore the major relaxation pathways remain the intramolecular dipolar interactions. Normally one would expect a serious decrease in the intramolecular NOE for solutes dissolved in solvents containing nuclei of high magnetic moment as ^{1}H or ^{19}F (3).

Recently a unique sesquiterpene, bilobalide, has been isolated (37) and shown with the aid of NOE measurements to be closely related to the ginkolides (38).

The structure of taxinine, *30* (Fig. 6), was proved by X-ray analysis (39). The results of the NOE study of taxinine and three related compounds, undertaken by Woods et al. (40) for comparison with the results of similar natural products, is given in Table 13. The positive* NOEs [CH_3-15α],H-2, [CH_3-15α],H-9, [C-12 Me],H-21, [C-12 Me],H-10, and [H-7],H-10 define four of the asymmetric centers of taxinine and suggest the configuration of the fifth at C-5. The stereostructure most consistent with these data is that shown in Figure 6. Three other taxinine derivatives were also examined and showed similar NOE values (see Table 13). The observance of the NOE [C-12 Me],H-21 again suggests the stereochemistry at C-5 which is further substantiated by coupling constant data. Compounds *32* and *33* have one constitutional asymmetric center, C-13, the stereochemistry at which can be assigned on the basis of the NOEs [C-15β Me],H-13 = 10-15%.

*Regrettably, in the original communication (40) only a range of 10-20% is indicated for the recorded NOEs.

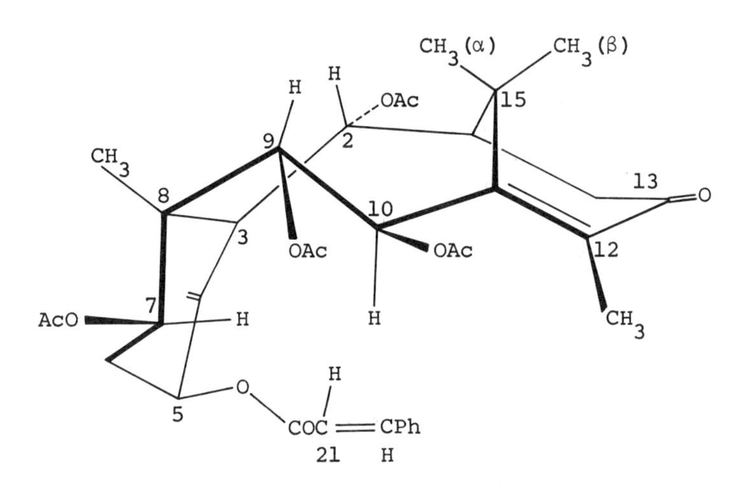

Fig. 6. Stereostructure of taxinine B, *30*.

Table 13 NOE Results for Some Taxinine Derivaties

		NOE[a]			
Irradiated	Observed	*30*	*31*	*32*	*33*
CH_3-15α	H-2	+	+	+	+
CH_3-15α	H-9	+	+	+	+
CH_3-12	H-10	+	+	+	+
CH_3-12	H-21	+	+	+	+
H-7	H-10	+	+	+	+
CH_3-15β	H-13			+	+

[a]The magnitudes of the NOEs were reported as between 10 and 20%.

The stereochemistry of the 6-bromo-7-oxoditerpenoids has recently been assigned on the basis of an X-ray structure determination (41) of *34* (Fig. 7) and has shown that in the solid-state ring B exists with the bromine in the 6α position. The NOE results of the related compound *35* (42) given in Table 14 illustrate that these compounds also exist in the same conformation in solution. The observance of the NOE [CH_3-18],H-6 = 18%

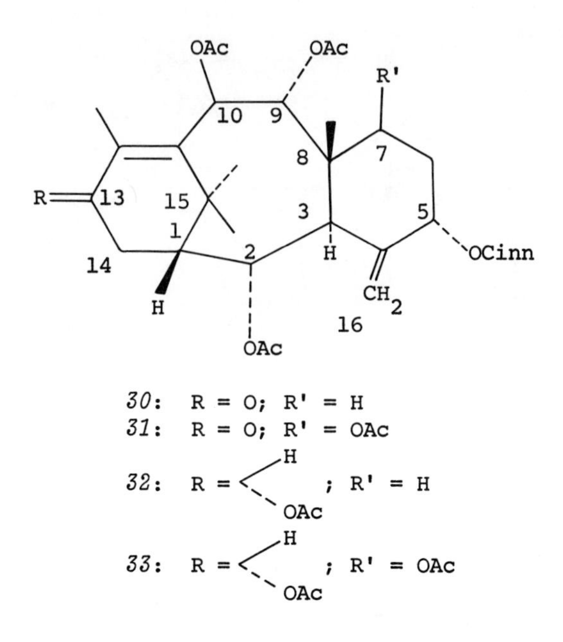

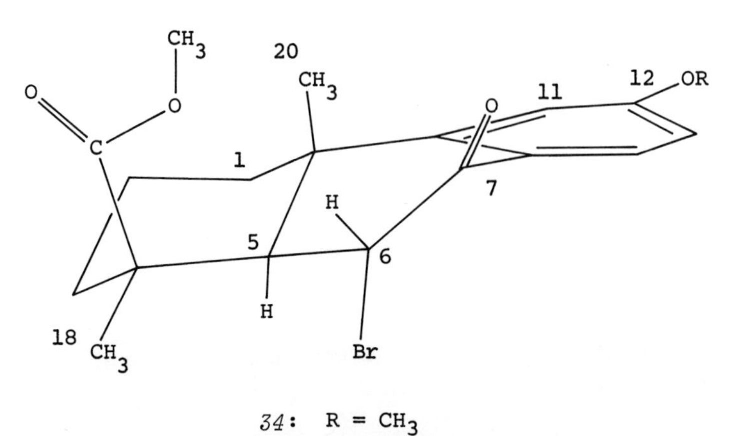

Fig. 7. Stereostructure of 6α-bromo-7-oxoditerpenoids, *34* and *35*.

can be explained only by placing the bromine axial with ring B adopting a classical boat conformation. The relative magnitudes

Table 14 NOEs in Methyl 12-Acetoxy-6-bromo-7-oxopodocarpate *35*

Irradiate	Observe	NOE, %
CH_3-20	H-6	7
CH_3-20	H-11	0
CH_3-18	H-6	18
H-1(ax)	H-11	16
H-1(eq)	H-11	25

of the area increases experienced by the aromatic proton H-11 when the two C-1 protons are saturated is also an indication of ring B existing in a boat conformation (14). The bromoketone can be readily converted into the lactone *36*, the NOE results for which are very similar to those for the bromoketone.

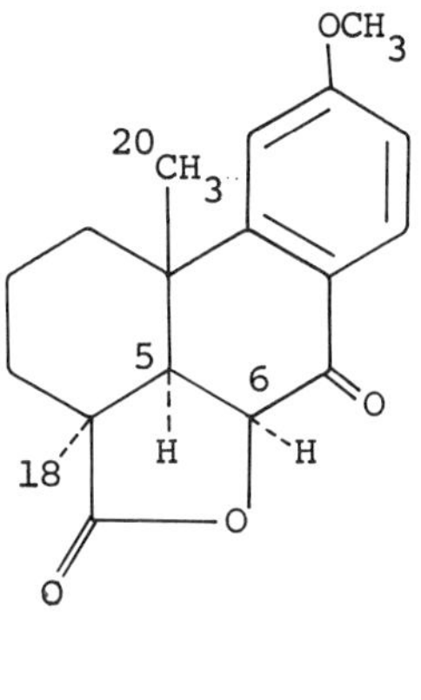

36

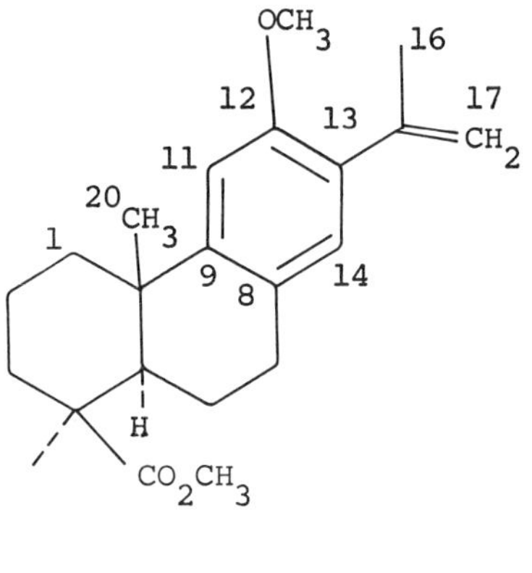

37

The NOE results (14) for the methyl *O*-methylpodocarpate derivative, *37*, (Fig. 8) are given in Table 15. One of the major points of interest was to see if an assessment could be made with regard to conformer population of the isopropylidene side chain. The most electronically stable conformation is the one in which the vinyl group is in the plane of the aromatic ring (*a* or *b*). However, these conformations suffer from various steric interactions, whereas conformations *c* and *d* are satis-

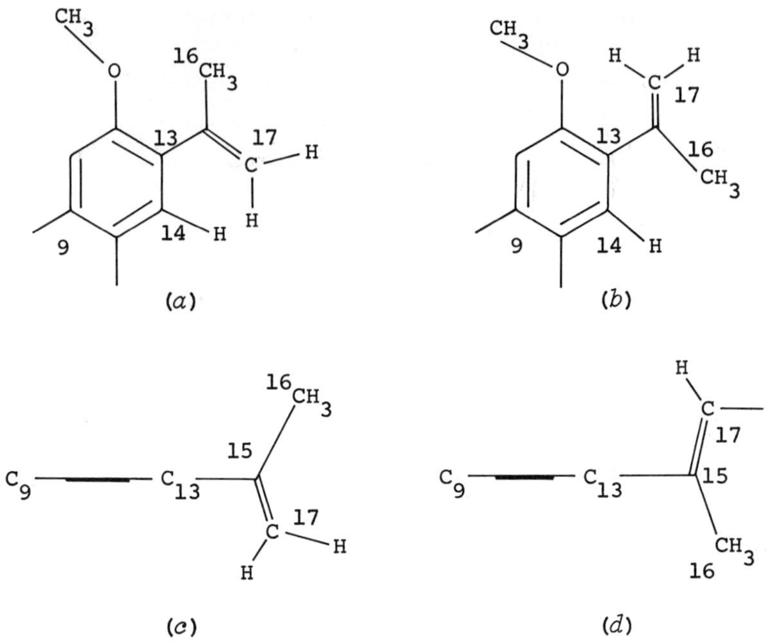

Fig. 8. (*a*), (*b*) Planar conformations of isopropylidene side chain in *37*. (*c*), (*d*) Conformations of isopropylidene side chain in *37* which are perpendicular to aromatic ring.

factory from a steric viewpoint but have no electronic stabilization. The NOE results are best viewed considering the isopropylidene as a rapidly rotating group with the protons acting as net dipoles from their respective distances along the C-13,C-15 bond. The remainder of the NOE results are quite straightforward. Ring B must be in the normal chair conformation. The observed NOE between the 20-methyl and H-11 is lower than normally observed, a consequence of the close proximity of the two C-1 protons. The ratio of the NOEs observed at H-11 on saturation of the two C-1 protons shows that ring A twists slightly from its chair conformation in order to relieve some of the nonbonded interaction between H-1(eq) and H-11.

The structure of the bromoketone *38* has been studied extensively (43) but no definite structural assignment could be made because of conflicting evidence. Inspection of a molecular model shows that the spectroscopic and chemical data could be explained by placing the bromine either at 11α with ring C in a chair conformation or at 13β with ring C in a twist

Table 15 NOEs in Methyl *O*-Methyl-13-Isopropylidenepodocarpate *37*

Irradiated	Observed	NOE, %
CH_3-16	H-17	9
CH_3-16	H-14	6
12-OCH_3	H-11	21
12-OCH_3	H-17	0
H-17	H-14	10
H-7(ax)	H-14	12
H-7(eq)	H-14	18
H-1(ax)	H-11	10
H-1(eq)	H-11	17
H-7(ax + eq)[a]	H-14	30
H-1(ax + eq)[a]	H-11	29

[a]Triple irradiation.

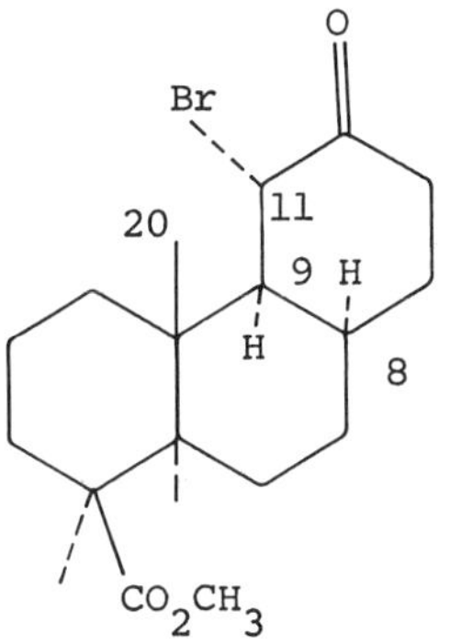

38

conformation, as shown in Figure 9 (*a* and *b*, respectively). Overhauser enhancement measurements represented a particularly simple solution to the problem (44). The NOE [CH_3-20],H-11 = 7% demonstrated that this proton must be sufficiently close to the 20-methyl for significant interaction, and measurement of a molecular model gives the 20-CH_3,H-11β distance as 2.8 Å and the 20-CH_3,H-13β distance as 3.6 Å. This latter distance is clearly too great for any observable interaction between the 20-methyl and H-13β. Additional virtues of the NOE data are shown in the

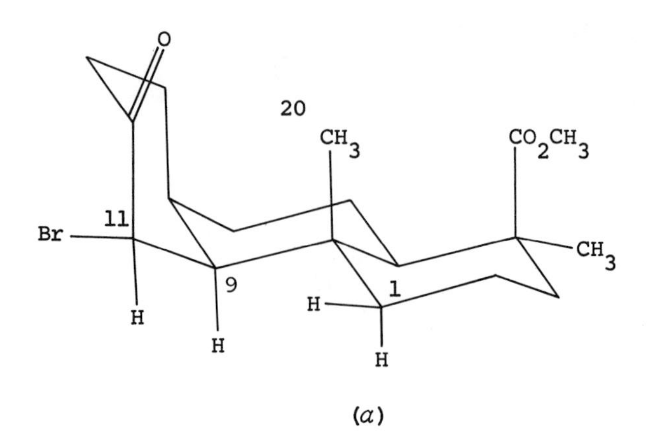

(*a*)

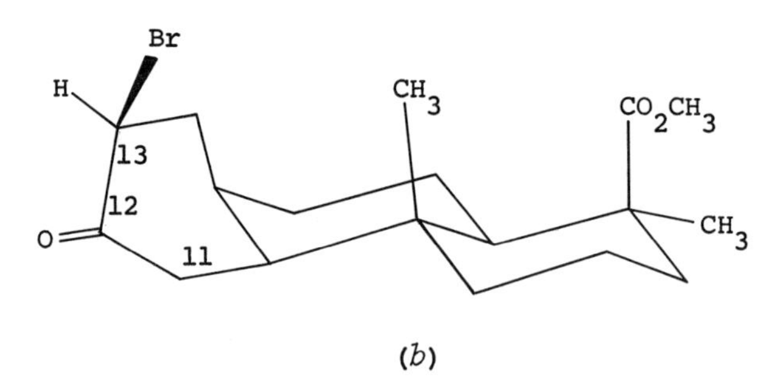

(*b*)

Fig. 9. (*a*) Stereostructure of methyl 11α-bromo-12-oxo-8α-podocarpan-19-oate. (*b*) Stereostructure of methyl 13β-bromo-12-oxo-8α-podocarpan-19-oate.

spectral assignments that could be made for the protons at C-9 and C-1 as irradiation of the H-9 signal removed a small coupling ($J_{9,11}$ = 2.0 Hz) and increased the area of H-11.

The more complex nagilactones C and D, *39a* and *39b*, respectively, examined by Itô et al. (45) have had light shed on their relative configurations from NOE measurements. In particular, the 3β orientation of the 3-OH follows from the NOE [CH_3-18],H-3α = 17% and the close proximity of H-11 and H-1 was demonstrated by the NOE [H-1],H-11 = 25%. The major evidence for the structure of these compounds and the related inumakilactones was obtained from coupling constant and solvent shift data on the free alcohols and their acetate derivatives.

The terpenoid bicyclic lactones *40* to *43* are an excellent illustration of the way in which information on steric

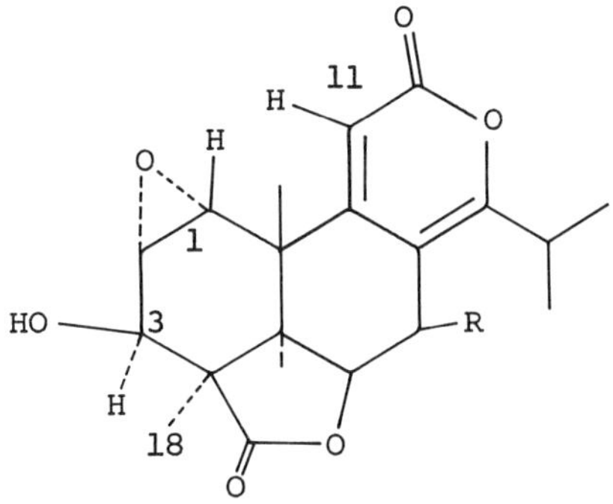

39a: R = OH
39b: R = H

compressions can be obtained. The deshielded proton at C-8 is conveniently located for examining NOEs (see Table 16). Unfortunately the number of adjacent protons at C-7 and C-6 makes any quantitative evaluations extremely difficult, but by measurement of the NOEs from the 20-methyl (46) it is possible to make qualitative assessments of the steric compressions. Thus the observed NOE of 18% at H-8 upon irradiation of the 20-methyl of *40* is far greater than would be expected from inspection of a molecular model and application of Figure 2 (Sect. II-A). The measured CH_3-20,H-8 internuclear distance is 3.32 Å, which corresponds to a maximum NOE of 8%, since it is reasonable to expect that H-7β and H-6β will both be contributing to the relaxation of H-8. We must conclude, therefore, that H-8 is much closer to the 20-methyl than the model would indicate; in fact a minimum of 2.9 Å would be necessary to give the observed NOE. The origin of this distortion is the intense steric compression between H-1 and H-11 (2.30 Å apart in the normal model) which can be relieved by a twisting of ring B, which in turn forces H-8 toward the 20-methyl (Fig. 10). A similar argument applies to the *cis*-lactone *41*, where an H-5,H-11α interaction further complicates the B ring distortions. The end result is again a twisting motion which leaves H-8 in closer proximity to CH_3-20. The two isomeric lactones *42* and *43* are readily distinguished by the presence of the NOE [CH_3-20],H-8 = 21% in *42* and the absence of an NOE in *43*. The interesting aspect of the NOE in *42* is that a molecular model shows it to be far too *small* (CH_3-20,H-8 distance, 2.55 Å), in contrast to the preceding two lactones. However, the equatorial-equatorial fusion of the lactone ring has actually relieved the H-1,H-11 interaction and

Table 16 NOE Results for Norditerpenoid, Bicyclic Lactones

Compound	Irradiated	Observed	NOE, %
40	CH_3-20	H-8β	18
	CH_3-20	H-11	0
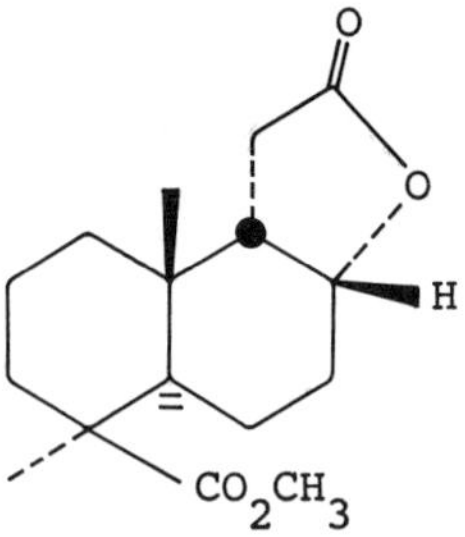 *41*	CH_3-20	H-8β	13
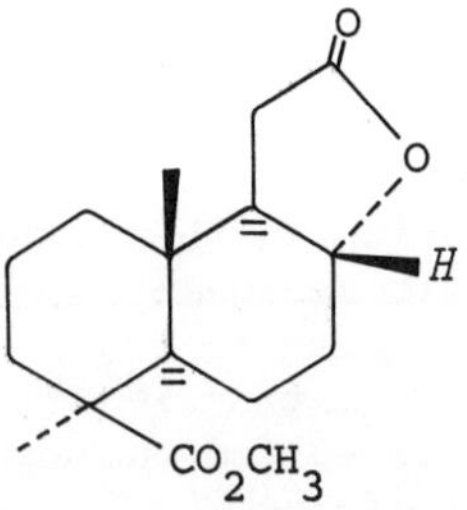 *42*	CH_3-20	H-8β	21

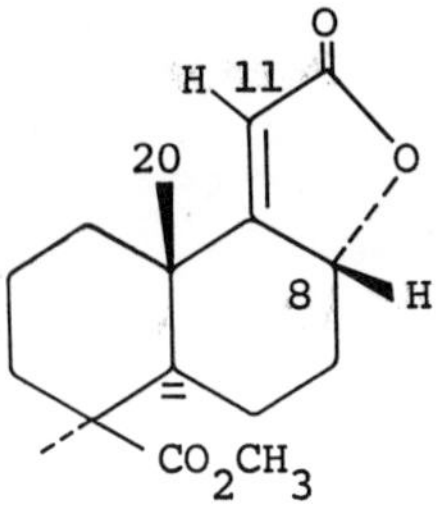

Table 16 Continued

Compound	Irradiated	Observed	NOE, %
43	CH_3-20	H-8β	0

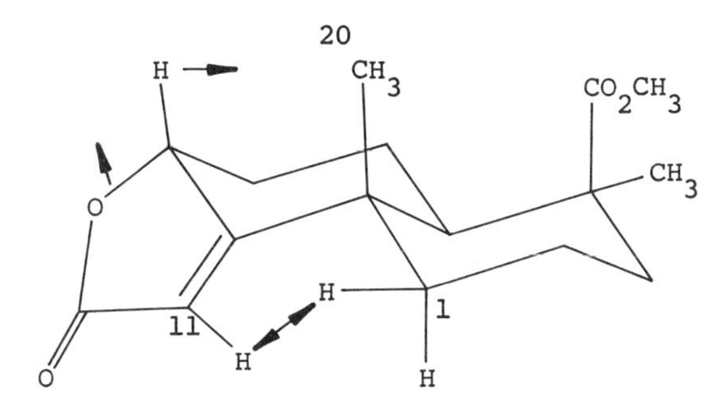

Fig. 10. Stereostructure of unsaturated lactone *40*. Arrows indicate nuclear motion caused by the H-11,H-1(eq) interaction.

left the CH_3-20,H-8 compression as the most intense in the molecule. Not unnaturally, therefore, the rings distort to move H-8 farther away from the 20-methyl and to lower the observed NOE. Lactone *42* appears to be one of the few examples of tricyclic diterpenoids where the H-1,H-11 compression is not the dominant one in the molecule, and it behoves one to beware of the assumption that all perhydrophenanthrenes suffer strong H-1,H-11 interactions.

As a group, the sesquiterpenoids offer a wide range of structural variety, and NOE measurements have found an increasing importance in recent structure elucidations. Tori et al. (47) on examination of molecular models suggested two possible conformations for the ten-membered ring sesquiterpenes zeylanine *44* and zeylanane *45* (Fig. 11). One conformation

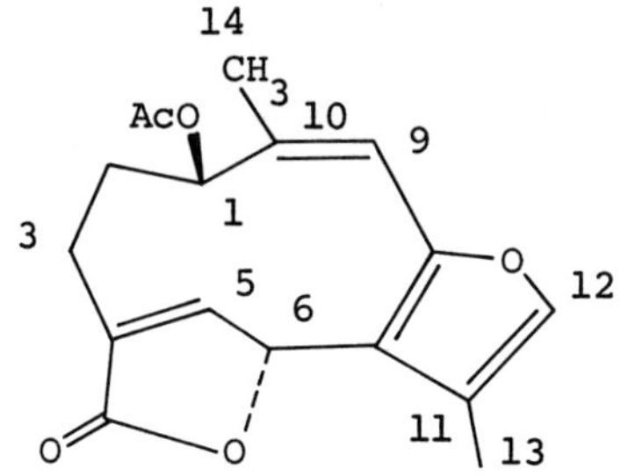

44 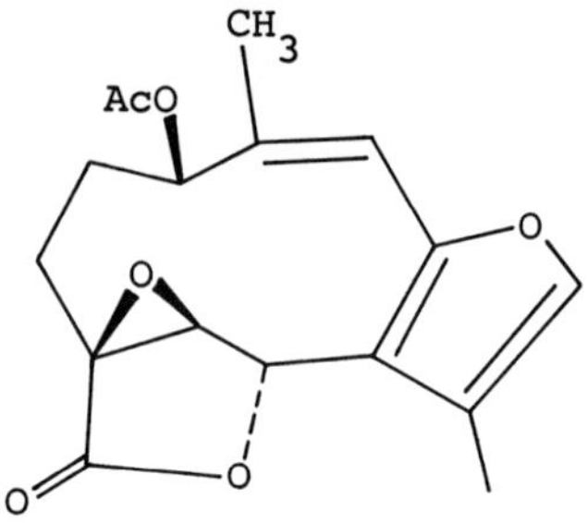 *45*

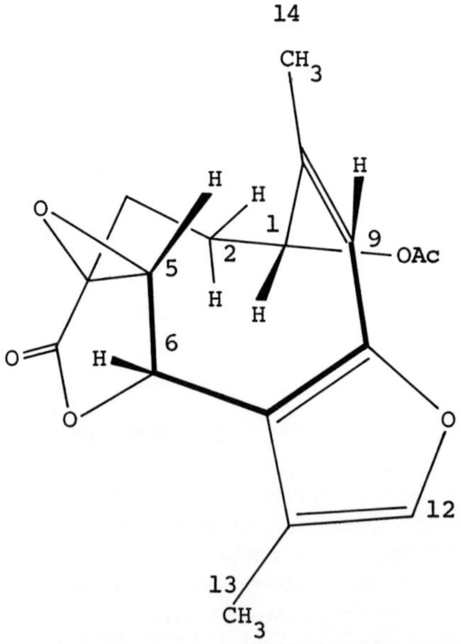

Fig. 11. Conformation of ten-membered ring in zeylanane *45*.

placed H-5 and CH_3-14 syn, whereas the alternative placed them anti. The observance of an NOE between H-5 and H-9 (Table 17)

Table 17 NOEs in Zeylanine *44* and Zeylanane *45*

Compound	Irradiated	Observed	NOE,[a] %
44	H-5	H-1α	2
	H-6β	H-5	11
	H-9	H-5	8
	CH_3-14	H-5	2
	H-5	H-6β	18
	H-9	H-6β	-4
	CH_3-13	H-6β	13
	H-5	H-9	6
	CH_3-14	H-9	20
	CH_3-13	H-12	19
45	H-5	H-1α	0
	H-6β	H-5	15
	H-9	H-5	10
	H-5	H-6β	14
	H-9	H-6β	-4
	CH_3-13	H-6β	17
	H-5	H-9	8
	CH_3-14	H-9	20
	CH_3-14	H-5	0
	CH_3-13	H-12	16

[a]Data recorded in $CDCl_3$ solution.

conclusively demonstrates that the ten-membered rings in *44* and *45* adopt a conformation such that H-5 and H-9 are in close proximity. Saturation of the 14-methyl group causes a small area increase of H-5 in *44* and has a negligible effect on H-5 in *45*. However, in *45* the NOE observed between H-5 and H-9 is larger than the corresponding value in *44*. These facts imply that the distance H-5,H-9 is greater and the distance CH_3-14,H-6 is less in *44* than in *45*. As the shape of the quartet of H-1α in *45* is virtually independent of magnetic field strength, the coupling constants $J_{1\alpha,2\alpha}$ and $J_{1\alpha,2\beta}$ can be obtained from a first-order analysis as 5.0 Hz and 12.0 Hz, respectively, and thus H-1α and H-2β are trans to one another. This fact, together with the magnitude of the NOE between the 13-methyl group and H-6β, defines the conformation of *45* as shown in Figure 11. The conformation of *44* is very similar, with small differences as noted above.

The tricyclic furanosesquiterpenoids *46a* and *46b* have both

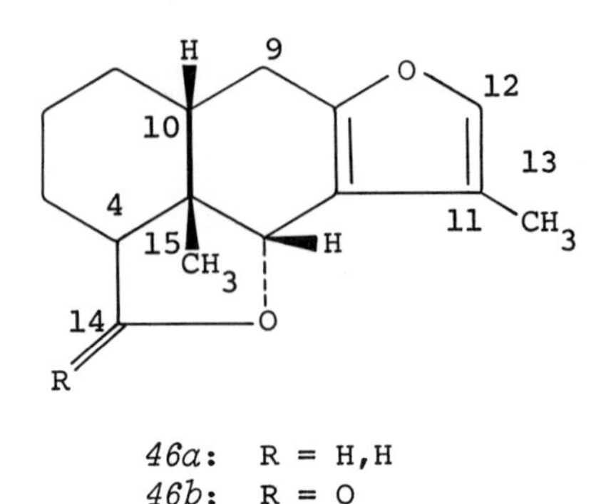

46a: R = H,H
46b: R = O

yielded interesting conformational information on examination of their NOE data. Furanoeremophilan-14β,6α-olide, *46a*, possesses the conformation shown in Figure 12 (48) by virtue of the 28%

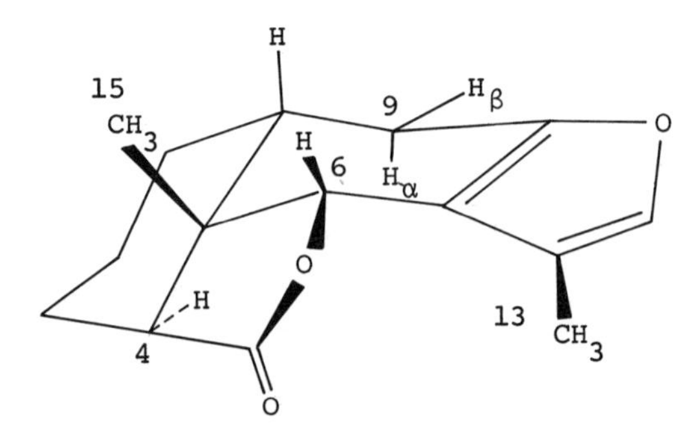

Fig. 12. Conformation of furanoeremophilan-14β,6α-olide *46a*.

NOE at H-6β on irradiation of CH_3-15 and the NOE [CH_3-13],H-6 = 6%. The stereochemistry at C-4 is β and is defined by the NOE of 8% at H-14β and -2% at H-14α when CH_3-15 is irradiated in *46b*. The overall cis geometry of the two six-membered rings follows from the spin-spin coupling $J_{9\alpha,10}$ = 1.3 Hz and $J_{9\beta,10}$ = 5.9 Hz, which indicates H-10 has a trans or an eclipsed conformation with respect to one of the C-9 methylene protons, and from the two large homoallylic couplings $J_{6,9\alpha}$ = 1.8 and $J_{6,9\alpha}$ = 1.2 Hz, which place H-6 in a quasiaxial conformation

(49). Since CH_3-15 is cis to H-6 it has a pseudoequatorial conformation, and the only possible conformation for H-10 is pseudoaxial and trans to H-9α. Confirmation of the physical data came from the transformation of *46a* into furanoeremophilane.

Isolinderalactone, *47*, is a ring-opened diene which can adopt four different conformations for the central six-membered ring (50). Spin-spin coupling data and the absence of an NOE

47

between CH_3-14 and H-9β demonstrated that the six-membered ring adopts a half-chair conformation with axially oriented CH_3-14, H-5, and H-9β (Fig. 13). The two protons, H-5 and H-6 are cis

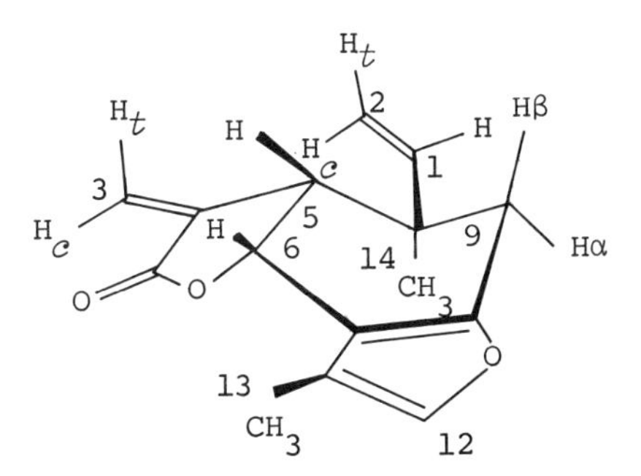

Fig. 13. Conformation adopted by isolinderalactone *47*.

as is shown by the NOE [H-5],H-6 = 9% and H-6 is adjacent to CH_3-13 from [CH_3-13],H-6 = 7%. Of special interest is the inference we may make about the conformation of the C-10 vinyl side chain. Thus there were NOEs for [H-5],H-1 = 6%,

[H-9α],H-1 = 12%, [H-9β],H-1 = 12%, and [CH_3-14],H-2*c* = 20%. These clearly imply that H-1 must reside in conformations where it is adjacent to H-5 and the two H-9 protons, whereas the cis vinylic proton H-2*c* will be juxtaposed to the axial CH_3-14. The NOE data is confirmed by the presence of a long-range coupling between CH_3-14. Isolinderalactone is, in addition, an excellent example of the care with which the absence of an NOE must be interpreted. The process [H-3*t*],H-5 gives a value of 8% but the reverse process [H-5],H-3*t* is 0%. Were it not possible to observe the former process one would be tempted to the conclusion that H-5 and H-3*t* are not proximate to each other. The rationale for the data is of course the overwhelming dominance of the geminal proton H-3*c* toward the relaxation of H-3*t*. The NOEs must then be in the ratio of the sixth power of the internuclear separations and in the present example, the NOE at H-3*t* on saturation of H-5 has dropped to within the limits of experimental error.

A most interesting correlation between the conformations adopted by the ten-membered ring germacrane sesquiterpenes and the stereochemistry of their ring-opened [3,3]-sigmatropic shift products has recently been proposed by Takeda and co-workers (51). It was noted that linderalactone, *48a*, listealactone, *48b*, and the related ether, *49*, all gave the antipodal elemane-type products (e.g., *50*) upon thermal rearrangement, but the

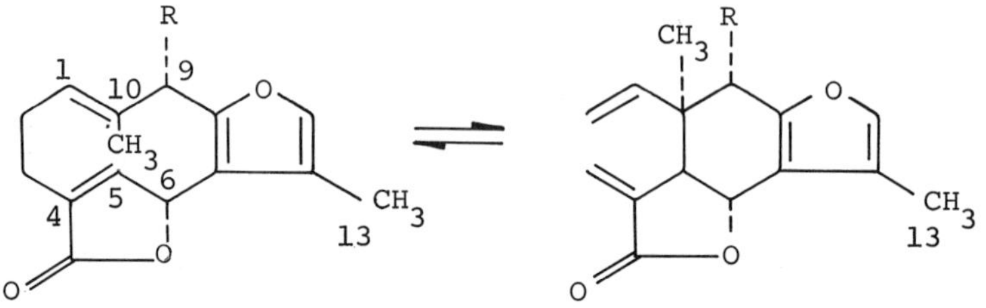

48a: R = H
48b: R = OAc

50

bicyclic ether *51* and dihydrotamaulipin-A, *52* (52) both gave the normal elemane-type products *53*. The NOEs present in linderalactone and listealactone, particularly [H-1],H-5 = 9%, [H-5],H-1 = 7%, and [H-9β],H-1 = 11%, implied the presence of a specific conformation of the ten-membered ring where the two double bonds are cross oriented (Fig. 14) and thus a chair-

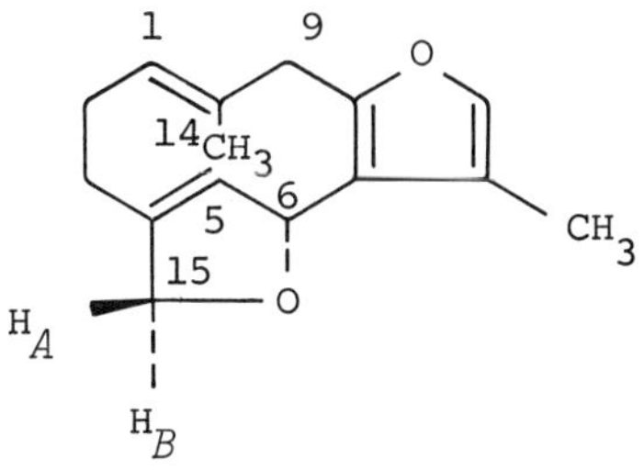

49

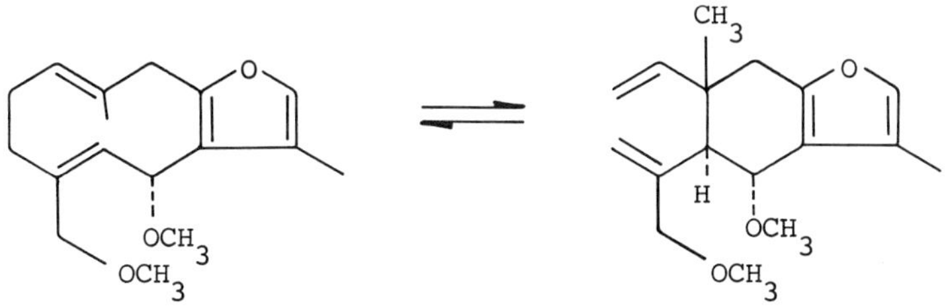

51 *53*

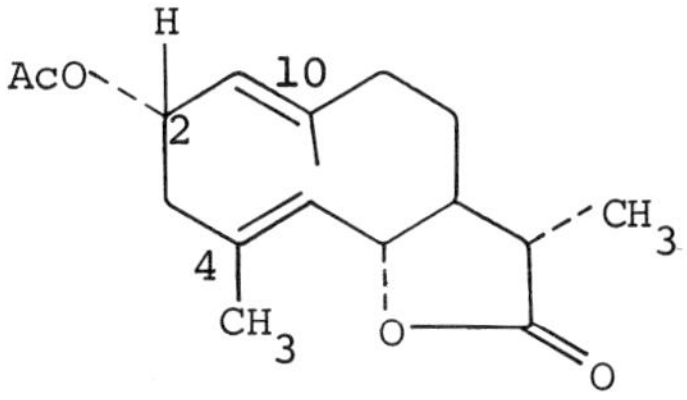

52

shaped transition state (53) for the [3,3]-sigmatropic shift will lead immediately to the antipodal elemane-type product. The NOEs present in the ether *49* were less definitive since the similarity in chemical shift of H-1 and H-5 made the observance of an NOE impossible. However, the NOEs [CH_3-14],H-15_A = 3%, and [H-9α],H-1 = 8%, plus a consideration of the long-range

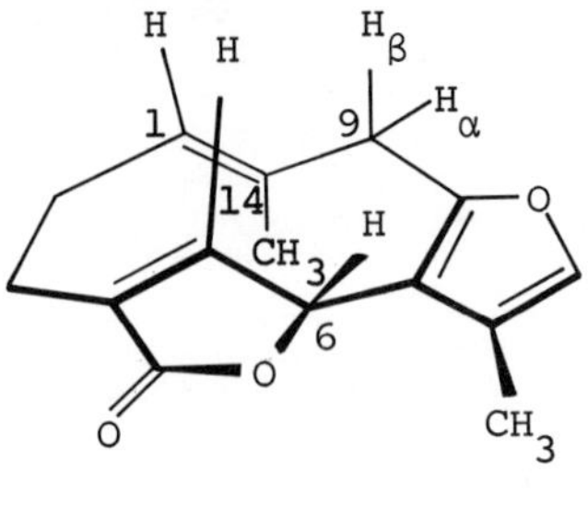

Fig. 14. Conformation adopted by linderalactone *48a*.

couplings present, led to the conclusion that the ten-membered ring must possess the conformation shown in Figure 15 where the

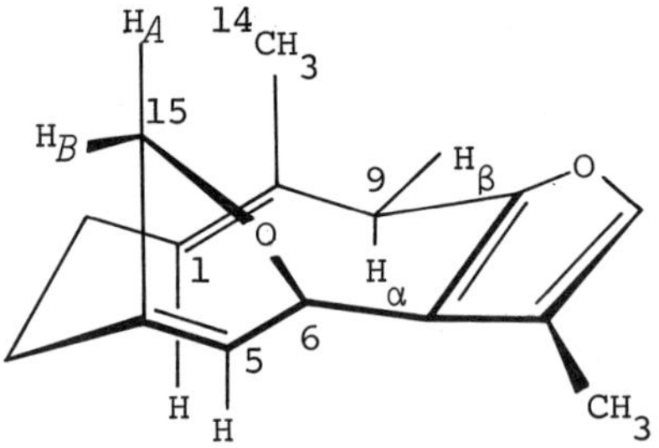

Fig. 15. Conformation adopted by ether *49*.

double bonds are cross oriented in the opposite sense to linderalactone. Sigmatropic rearrangement of this conformer via a chair transition state therefore leads to the normal elemane-type stereochemistry *53*. The methyl ether *51* was likewise shown to possess cross-oriented double bonds as for ether *49*.

The related natural product dihydrotamaulipin-A, *52*, examined by Bhacca and Fischer (52), showed the NOEs [CH_3-4],H-2 = 10% and [CH_3-10],H-2 = 15%, which again suggests a similar conformation as depicted for ether *49* in Figure 15. Sigmatropic rearrangement of *52* gives 2-acetoxysaussurealactone, the ring-opened compound with normal elemane stereochemistry (54). Other compounds in the sesquiterpene ten-membered ring series have been examined, and in the cases of furandienone and isofurandienone (55) it has again been possible to assign a specific conformation to the ten-membered ring. Neolinderalactone, *54*,

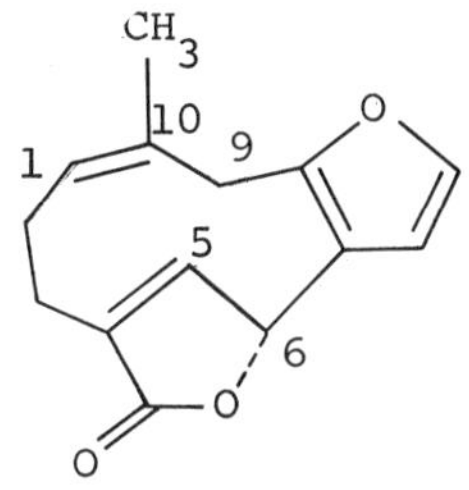

54

is a most interesting example since not only does it possess a *cis*-1,10 double bond (NOE [CH_3-10],H-1 = 15%), but at room temperature it clearly exists in two conformations. These were assigned on the basis of benzene-induced solvent shifts plus an NOE of 7% for [H-9α],H-5 for the major conformer. Line-shape analysis over the temperature range 55-85°C gave an activation barrier to inversion of about 10 kcal/mole, and both nmr and *cd* data gave the free-energy difference between the two isomers as 0.6 kcal/mole.

The steroids as a group have received surprisingly little attention in NOE studies. In the Vitamin D series tachysterol, *55*, (as the 3,5-dinitro-4-methylbenzoate) has been examined by Lugtenburg and Havinga (56) and the results are shown in Table 18. The particular NOEs [CH_3-19],H-6 = 35% and [H-9],H-6 = 10% implicate conformation *55A* as the principal one, but the NOE [CH_3-19],H-7 = 15% shows that conformer *55B* is also present. Clearly for more detailed information a variable-temperature study should be performed on this system.

The most impressive work in the steroid area has been that of Tori and co-workers (57). A selection of the compounds examined, together with the measured NOEs, is shown in Table 19. 6α-Methyl-11β-hydroxy-pregn-4-ene-3,20-dione, *56*, is an excellent example of how a 6α-methyl substituent can be quickly identified by the intense, 37% NOE for [CH_3-6α],H-4. A 6β-methyl would be about 3.8 Å away and would lead to a very much smaller NOE at H-4. The 8β-formyl steroids are intriguing because it is known from X-ray studies (58) that 8β-methyl steroids show a distortion of their skeletal structures. 8β-Formylandrost-4-ene-3,17-dione, *57*, has NOEs at the 8β-CHO of 33% and 24% from saturation

Table 18 NOEs in Tachysterol *55*

Irradiated	Observed	NOE, %
CH_3-19	H-6	A_1[a] 45
		A_2 35
CH_3-19	H-7	B_1 18
		B_2 15
H-9	H-6	10
H-9	H-7	0

[a]Protons H-6 and H-7 form an *AB* system and NOE data is reported as area increases for each line of the quartet.

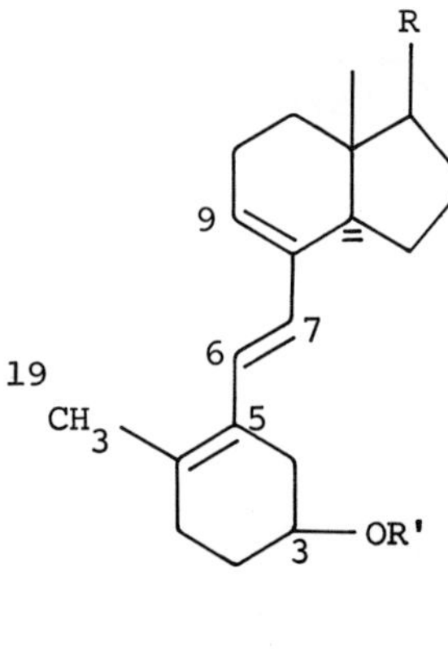

55A 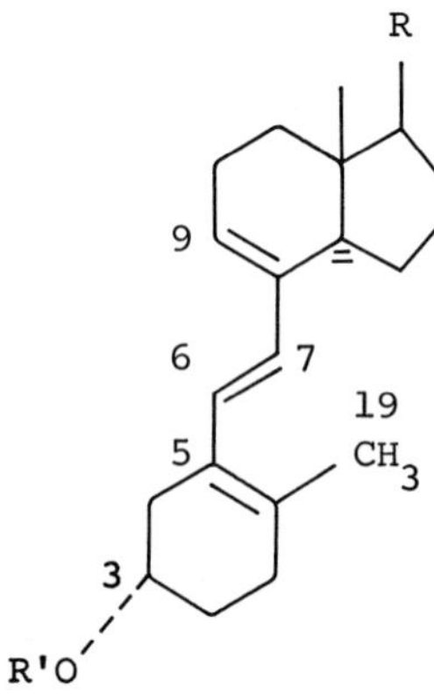 *55B*

R = C_8H_{17}
R' = 3,5-$(NO_2)_2$-4-CH_3-C_6H_3

Table 19 NOEs in Some Steroid Derivatives

	Irradiated	Observed	NOE, %
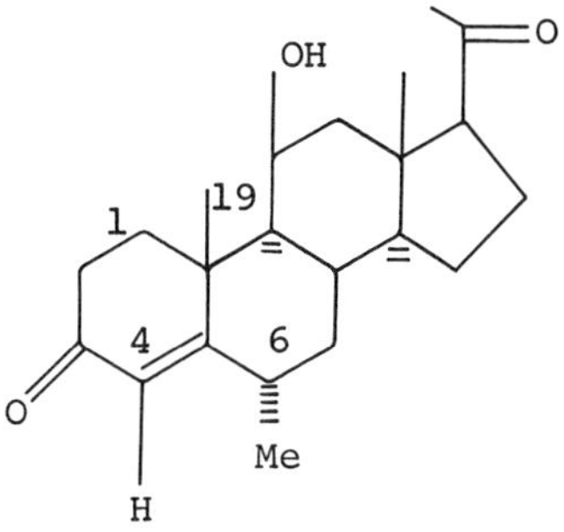 *56*	CH_3-6α	H-4	37
	CH_3-19	H-4	0
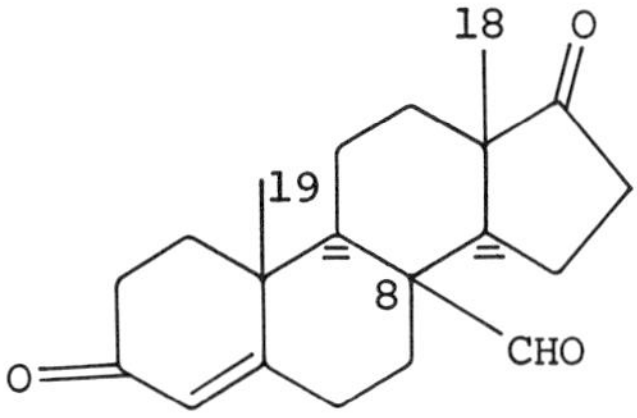 *57*	CH_3-18	H-8β	33
	CH_3-19	H-8β	24
	H-7α	H-8β	0

Table 19 Continued

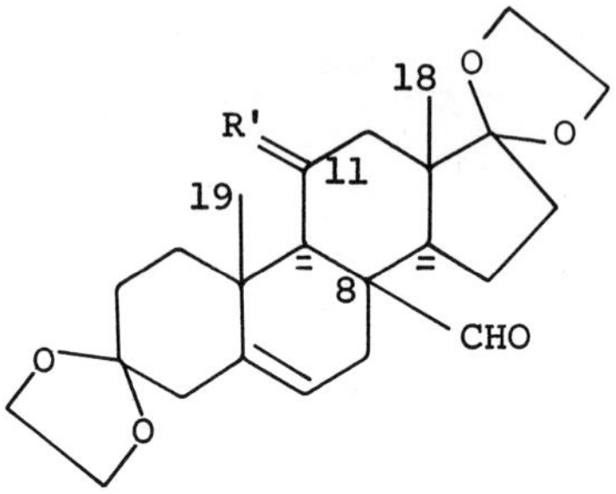

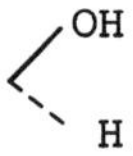

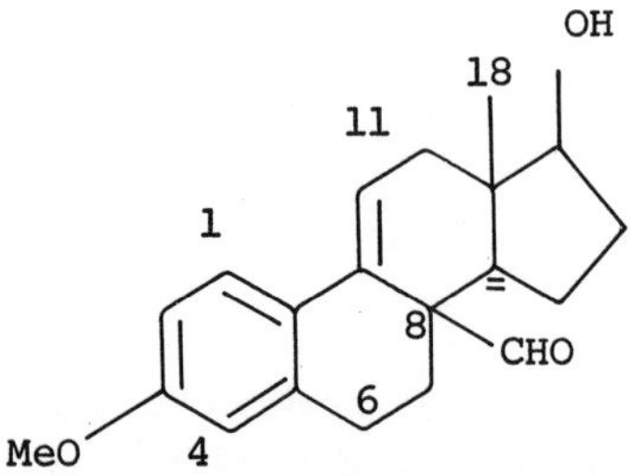

	Irradiated	Observed	NOE, %
58a: R' = O	CH_3-18	H-8β	37
	CH_3-19	H-8β	19
58b: R' = H_2	CH_3-18	H-8β	30
	CH_3-19	H-8β	16
58c: R' = OH, H	CH_3-18	H-8β	26
	CH_3-19	H-8β	14
	OH-11β	H-8β	5
59	CH_3-18	H-8β	22
	H-6α	H-4	10
	H-6β	H-4	3
	H-1	H-11	20
	H-12α+β	H-4	13
	H-11	H-1	20
	H-1	H-2	20

Table 19 Continued

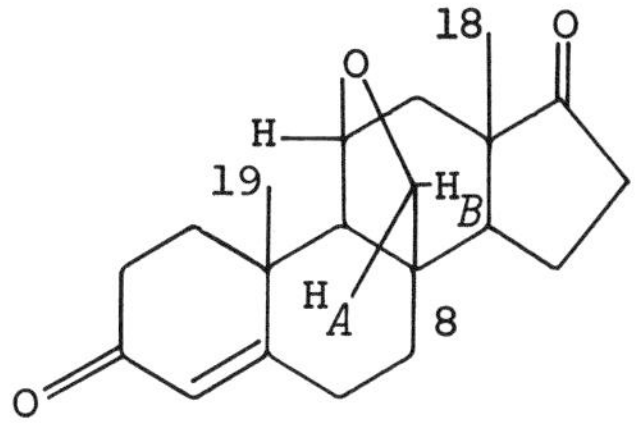

	Irradiated	Observed	NOE, %
60	CH_3-18	H-8_A	0
	CH_3-18	H-8_B	12
	CH_3-19	H-8_A	5
	CH_3-19	H-8_A	0
	CH_3-19	H-11α	10

of the 18- and 19-methyls, respectively,* showing that the formyl hydrogen is effectively sandwiched between the two methyls. Because of these strong NOEs we can conclude that the 8β-formyl group exists in only one conformation with the oxygen pointing away from the 18- and 19-methyls. The only long-range coupling observed for the formyl hydrogen, $J_{CHO,H\text{-}7\alpha}$ = 1.8 Hz, is additional compelling evidence for a single conformer. The 8β-formyl androstene ketals, *58a*, *58b*, and *58c*, have been included because they show a regular decrease in NOE at the 8β-formyl proton as we go from and 11-ketone to an 11β-alcohol. The changes can be attributed to an alteration of the skeletal distortions, coupled with the addition of a further relaxation pathway for the H-8β in *58b* and *58c*. Indeed Tori (57) reports the very interesting NOE, [OH-11β],H-8β = 5% for *58c*, which implies that the hydroxyl proton must have appreciably populated conformations where it is pointing inside the C ring.

The estrol derivative *59* has [CH_3-18],H-8β = 22%, which illustrates well the loss in steric compression attendant upon

*These two values must be in some error as their sum (57%) exceeds the maximum theoretical value of 50% for an NOE at a given proton. Probably it was not possible to uniquely saturate each of the 18- and 19-methyls independently because of their close chemical shifts (0.92 and 1.16 ppm, respectively).

removal of the 19-methyl group. The NOE of 20% between H-1 and H-11 (identical in this case in both directions) is fully consistent with the known steric interactions between H-1 and H-11 in the steroid skeleton and has been noted in the tricyclic diterpenoids already commented upon in this section. The remaining NOEs are included because they show how the NOE can be used to assign chemical shifts in a typical estrol. The example of 11β,8β-methanoepoxyandrost-4-ene-3,17-dione, *60*, is similar in that solvent-induced shifts and NOE measurements were used to locate the chemical shifts of the two 8β-methylene protons (H_A, 3.44 ppm, and H_B, 3.36 ppm). In addition, a remarkably large NOE, [CH_3-19],H-11α = 10%, was observed at H-11α. This is most unusual since a 19-methyl is normally too distant to relax an 11α-proton, and we must conclude that the methanoepoxy bridge has grossly distorted ring C and placed H-11α much closer to the 19-methyl.

In a recent total synthesis of 6-thiaestrogens (59) the intermediate ketone *61* was prepared and it was critical to determine if the ketone possessed a B/C cis ring juncture. Molecular models suggested that the 18-methyl and the C-7 axial hydrogen (H-7β) should be within 3.2 Å of each other, and this was unequivocally confirmed by the observation of an NOE [CH_3-18],H-7β = 12%. A similar kind of stereochemical problem is to be found in the protoberberine alkaloids (see Sect. III-A).

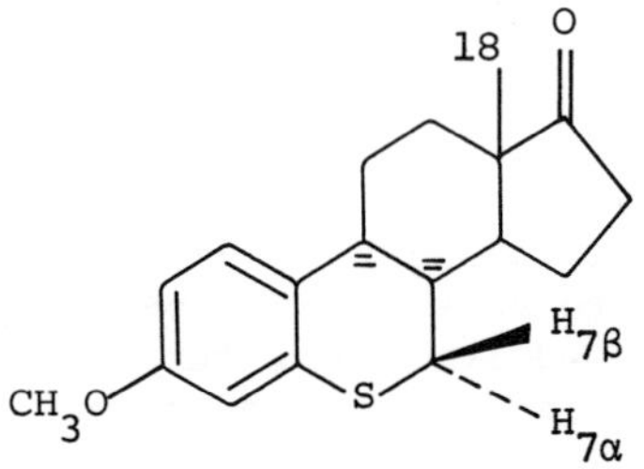

18 O

$H_{7\beta}$

CH_3O S $H_{7\alpha}$

61

C. Nucleosides and other Natural Products

The conformations adopted by mononucleosides have been the subject of extensive spectroscopic endeavors, with one of the main points of contention being the conformation about the glycosidic bond. X-ray data have shown that in the purine-base compounds (60) the conformation in the solid state is syn

(Fig. 16) with intramolecular hydrogen bonding between the OH-5

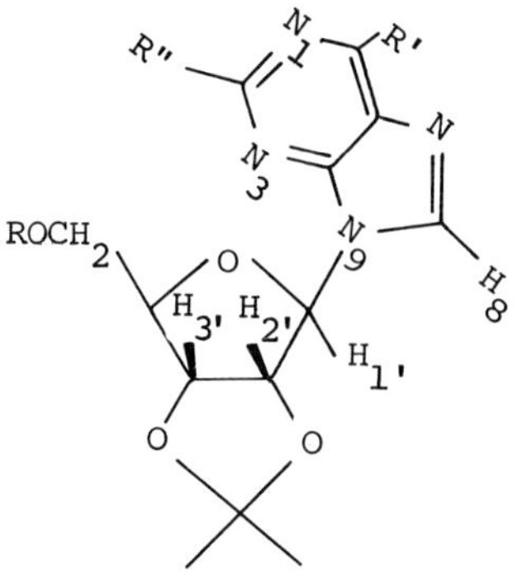

62a: R = H; R' = NH_2; R" = H
62b: R = Ac; R' = NH_2; R" = H
63: R = H; R' = OH; R" = NH_2

Fig. 16: Syn conformation of adenine and guanine nucleosides.

and N-3, whereas with a pyrimidine base the conformation is anti (61). The situation in solution is much more complex, and results derived from uv and nmr spectroscopy (62) infer that various proportions of both types of orientations are present, the proportions being dependent on substituents and conditions. Recently Hart and co-workers (63) applied NOE data in an attempt to further unravel the problem, and their results together with those of Bell and Saunders (9) are recorded in Table 20. To take into account the possibility of rapid rotation, eq. [14] (Sect. II-B) can be modified (63b) as follows: for the rotation rate $k_r \ll R_A$

$$f_A(B) = \sum_B \chi_k f_A(B,i) \qquad [18]$$

and for faster rotation rates, where $R_A < k_r < \tau_c$,

$$f_A(B) = \sum_B \frac{\langle \gamma_B \rho_{AB} \rangle}{2\langle \gamma_A R_A \rangle} - \sum_C \frac{\langle \gamma_c \rho_{AC} \rangle f_c(B)}{2\langle \gamma_A R_A \rangle} \qquad [19]$$

Table 20 NOEs in Some Adenine and Guanine Nucleosides

		NOE, %							
Irradiated	Observed	*62a*[a]	*62a*[b]	*62a*[c]	*62a*[a]	*62b*[a]	*62b*[b]	*63*[d]	*64*[e]
H-1'	H-8	20	19	17	23	27	26	12	18
H-2'	H-8	7	6	8	9	5	6		14
H-3'	H-8				3			12[f]	4
H-5' + H-5"	H-8	0	0	5	4	0		10	4
H-1'	H-2	8		6	0				10
H-2'	H-2								7
H-3'	H-2								7
H-5 + H-5"	H-2								6
OH	H-2	11	12						
H-2'	H-1'								2
H-3'	H-1'								-3
H-4'	H-1'								9
H-8	H-1'							26	16
H-1'	H-2'								1
H-8	H-2'								
H-2	H-2'								10
H-5'+H-5"	H-3'								4
H-8	H-3'								10
H-4'	H-3'								1
									-4

[a]Data recorded in pyridine-d_5 at 37°C (9,14).
[b]Pyridine-d_5 at -50°C (9,14).
[c]In DMSO-d_6/acetone-d_6 at 37°C (9,14).
[d]In DMSO-d_6 (63a).
[e]In DMSO-d_6 (63b).
[f]NOE from simultaneous irradiation of both H-2 and H-3.

Where χ_i is the fraction of molecules in conformation i, and $\langle\gamma_A R_A\rangle = \Sigma_i\ \gamma_A R_A(i)\chi_i$ and $\langle\gamma_C \rho_{AC}\rangle = \Sigma\ \gamma_C \rho_{AC}(i)\ \chi_i$, that is, an ensemble average of the NOEs, and an ensemble average of the interactions, respectively. Both eqs. [18] and [19] can be generalized to a continuous set of conformations described by an internal variable. In the following discussion, it will become clear that the NOE data cannot be interpreted in terms of a single conformation for the base ring about the glycosidic bond, but requires a Gaussian distribution between at least two conformations.

The compound 2',3'-isopropylideneadenosine in pyridine-d_5, *62a*, recorded enhancements of 20% and 7% at H-8 when H-1' and H-2', respectively, were saturated. The position defined by the distances corresponding to these NOEs places the purine ring in a syn conformation in a position similar to that obtained from the X-ray data. This infers that *62a* exists predominantly, though not necessarily exclusively, in this conformation with intramolecular hydrogen bonding being significant. The 5'-*O*-acetyl derivative *62b* gave NOEs of 27% and 5% at H-8 on saturation of H-1' and H-2', respectively. The values between H-1' and H-8 suggest that these two protons are eclipsed; however, the value between H-2' and H-8 precludes this as being the dominant conformation, and thus the molecule in solution has significant probabilities of existing in either *syn*-like or *anti*-like conformations. Cooling of either *62a* or *62b* in pyridine-d_5 caused the following changes: (*a*) all signals moved downfield uniformly as the temperature decreased; (*b*) the signal assigned to the $-NH_2$ group was transformed from a singlet at room temperature to a doublet at -30°C, and to a very broad doublet at -80°C; (*c*) in *62a* the C-5' proton absorptions, and less dramatically the C-4' proton absorptions, broadened with decrease in temperature; and (*d*) the NOEs obtained at -50°C were essentially the same as those recorded at +30°C. These data are compatible with the conclusions suggested above as to the conformations adopted by the two compounds. The results obtained for *62a* in DMSO-d_6 are indicative of both conformations being well populated. Hart and Davies (63a) also studied the guanosine derivative *63* in DMSO and suggested that this compound exists with a greater preponderance of anti conformers than in the adenosine case.

The compound 2',3'-isopropylideneinosine, *64*, studied in DMSO-d_6 gave results that were incompatible with the molecule existing in a specific conformation about the glycosidic bond. Therefore, to describe the energy versus torsion-angle profile, Hart et al. (63b) fitted a two-Gaussian distribution function to the experimental results, using various ribose ring geometries, and a least-squares criterion for the quality of fit. Equations [18] or [19] were used in the iterative procedures, with eq. [19] giving slightly better results. The computations indicated that the fits of glycosyl conformation are not highly sensitive to the ribose conformation or to the exact rate of rotation of the base ring. The conclusion reached by these authors was that *64* exists predominantly in the syn conformation ($\sim$80%) centered with τ = 355°, and about 20% of the population in the anti conformation (τ = 166°) (Fig. 17). The angle τ is defined as the dihedral angle formed by C-8, N-8, C-1', H-1', with τ being taken as 0° when the C-8,H-8 bond eclipses the C-1',H-1' bond.

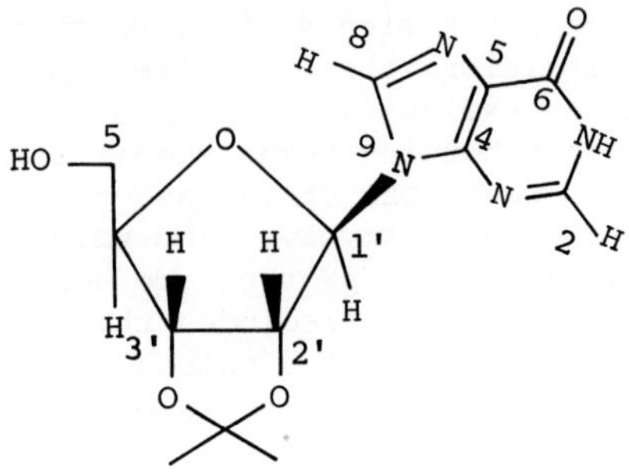

anti

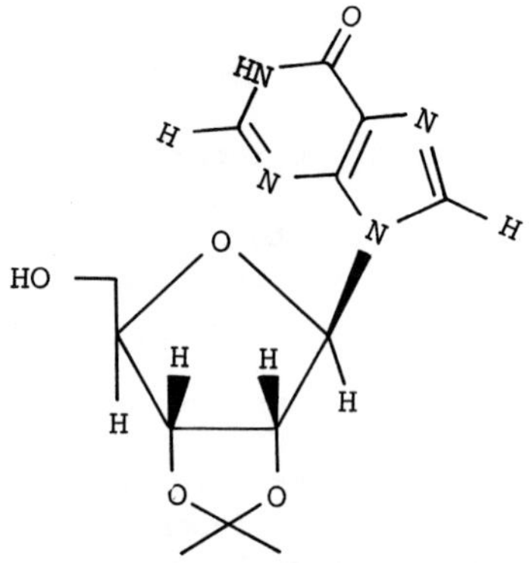

syn

Fig. 17. Syn and anti orientations for 2',3'-isopropylideneinosine, *64*.

Nucleosides that contain a pyrimidine base have been shown to exist in both *syn*- and *anti*-like orientations, the relative populations being dependent on the substituent and the conditions. The NOE results given in Table 21 for uridine, *65*, can best be explained in terms of a predominant *anti*-like orientation

Table 21 NOEs in Some Uridine and Cytidine Nucleosides[a]

		NOE, %									
Irradiated	Observed	*65*[b]	*65*[c]	*66*[b]	*66*[d]	*66*[e]	*66*[f]	*67*[g]	*67*[h]	*68*[h]	*68*[i]
H-1'	H-6	11	7	14	10	10	13	6	9	18	19
H-2'	H-6	13		4						10	10
H-3'	H-6	8		5						4	3
H-2' + H-3'	H-6		11		10	9	9	18	19		
H-5' + H-5"	H-6	0	3	0	0		5	14	4	0	0
H-5	H-6				28	26	26				
H-6	H-5				32		33				

[a]Data taken from refs. 9 and 14.
[b]In D_2O solvent.
[c]In 62% C_6D_6:38% DMSO-d_6.
[d]In pyridine-d_5 at 37°C.
[e]In pyridine-d_5 at -50°C.
[f]In 75% DMSO-d_6 and 25% acetone-d_6.
[g]In 25% pyridine-d_5 and 75% D_2O.
[h]In DMSO-d_6.
[i]In 75% DMSO-d_6 and 25% D_2O.

as the NOE between H-1' and H-6 is 7%, which is in fair agreement with the results obtained by Smith et al. (64), who studied

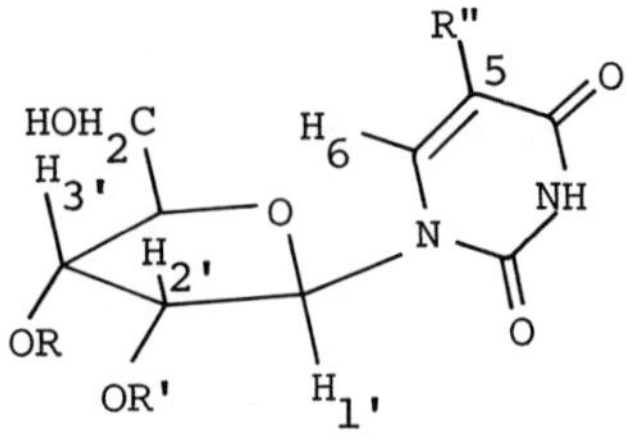

65: R = R' = R" = H

66: R" = H; R + R' = $C(CH_3)_2$

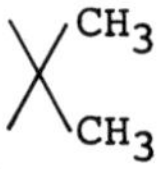

67: R = R' = H; R" = NH_2

68: R" = NH_2; R + R' = $C(CH_3)_2$

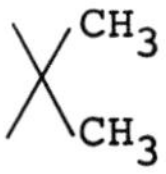

the 220 MHz spectrum and concluded that it exists solely in the anti orientation. Examination of 2',3'-*O*-isopropylideneuridine, *66*, yielded data inferring that the syn conformer is the more populated orientation. Similar results were obtained for the cytidine compounds *67* and *68*, although it appears that *68* exists almost exclusively in the *syn*-like orientation as an NOE of 19% was recorded between H-6 and H-1'. Thus the compounds can be arranged in order of increasing *syn*-like orientation as follows: *65*, *66*, *67*, and *68*. As can be seen, the choice of solvent influences the size of the NOE recorded but, because this could mirror a change in the average furanose ring conformation as well as changes about the glycosidic bond, no specific solvent effect can be ascertained. The uridine derivatives *69* and *70*, which are *R* and *S* stereoisomers at C-1" of the tetrahydropyranyl ether ring, were studied (9) in order to determine the absolute stereochemistries at C-1". The two compounds exhibited markedly different nmr and chromatographic behavior, and the NOE data for the stereoisomers are recorded in Table 22. The one major difference which appears critical for the assignment of the stereochemistry is the NOE [H-1"],H-1' = 12% for *69*, whereas for *70* this NOE was zero. The structures were assigned on the basis of a consideration of the most energetically favored conformations about the C(2')-O-C(1") bonds for both isomers. The most favored conformation for the *S* stereoisomer is one where H-1"

Table 22 NOE Results for 2'-Tetrahydro-pyranyluridine Isomers *69* and *70*[a]

		NOE, %	
Irradiated	Observed	*69*	*70*
H-1'	H-6	7	8
H-2' + H-3'	H-6	10	17
H-5' + H-5"	H-6	0	0
H-5	H-6	28	28
H-6	H-5	32	
H-1"	H-1'	12	0

[a]Data recorded in pyridine-d_5 solvent.

of the tetrahydropyranyl ring is oriented midway between H-2' and H-3' (Fig. 18). Any attempt to generate a conformation that places H-1" between H-2' and H-1' leads to severe interactions between H-2' and the equatorial proton H-2" on the tetrahydro-pyranyl ring, and any conformation that allows H-1" to reside on the underside of the ribose ring results in intolerable steric interactions between H-1" and H-1', H-4', or O-3'. It is concluded, therefore, that the *S* stereoisomer is *70* (the more polar diastereomer) and that it adopts only one conformation, where H-1" and H-1' are about 4 Å apart, and no NOE can be expected to be observed between them. The *R* stereoisomer, on the other hand, possess a conformation analogous to *70*, where a weak hydrogen bond exists between the C(1")-O and the C(3')-OH, and in addition a conformation with H-1" lying between H-2' and H-1' (Fig. 18). In the latter conformation there are no other H-2' hydrogen interactions (the C-2" equatorial H is now on the far side of the molecule) but merely a weaker H-2', C(1")-O interaction. Thus the *R* stereoisomer can exist in two conformations, one that places H-1" over 4 Å from H-1' and the other that places H-1" approximately 2.8-3.0 Å from H-1'. There will be, therefore an NOE observed between H-1' and H-1", and structure *69*, with an *R* absolute configuration at C-1", can be assigned the less polar diastereomer. Additional evidence that is consistent with the NOE data comes from the magnitude of the coupling constant $J_{1',2'}$. The value of $J_{1',2'}$ for *70* is greater than in *65*, *66*, or *69*, and it follows reasonably from the conformation of *70* depicted in Fig. 18 where the nonbonded interactions be-

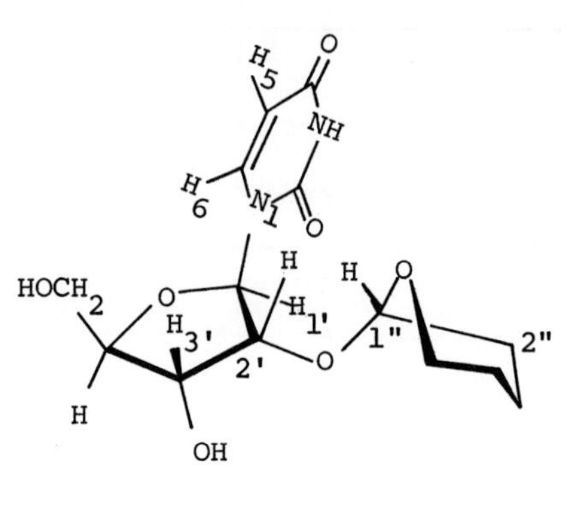

69

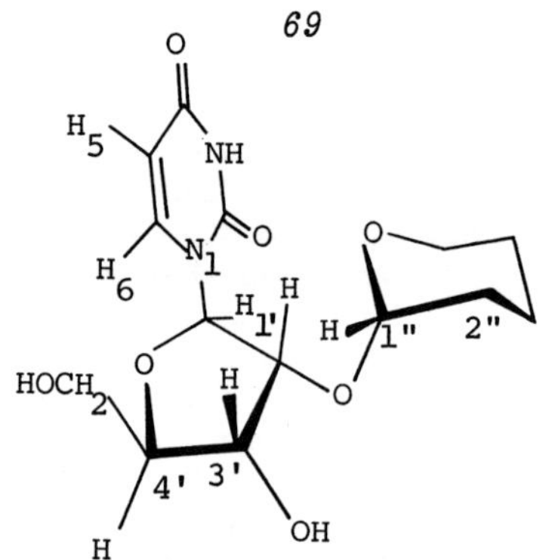

70

Fig. 18. One of the conformations adopted by C-1"(R)-2'-O-tetrahydropyranyluridine, *69*, and the only conformation adopted by C-1"(*S*)-2'-O-tetrahydropyranyluridine, *70*.

tween H-1" and H-2' and H-3' can be alleviated by forcing the ribose ring to assume a pronounced C-2' exo geometry. The resulting torsion about the C-1',C-2' bond increases the dihedral angle between H-1' and H-2', and therefore increases $J_{1',2'}$. For the diastereomer *69*, hydrogen bonding present in the conformation analogous to *70* would tend to diminish the degree of C-2' exo character, and the conformation of *69* depicted in Fig. 18

would resist any C-2' exo disposition since a strong steric interaction begins to develop between H-1" and the uracil nitrogen atom N-1. Thus the dihedral angle between H-1' and H-2' will be smaller in *69* and therefore, as is observed, $J_{1',2'}$ will be smaller. The NOE data for both isomers concerning the orientation about the glycosidic bond is best explained in terms of approximately equal populations for both orientations, the larger NOE recorded for *70* between H-2' and H-6 being a consequence of C-2' being in an exo position. Thus we can, with a fair degree of certainty, define *69* as the *R* stereoisomer and *70* as the *S*, with rapid rotation about the glycosidic bond occurring in both isomers.

Chan and Kreishman (65) used NOEs as a guide to intermolecular interactions between the nucleic acid, polyuridylic acid (poly-U), and added purine base. Irradiation of the resonances corresponding to the respective ribose protons in a solution of 0.4 *M* purine containing poly-U (0.10 *M* in uridine) resulted in an 11% NOE between the C-5' protons and H-6 of added purine and a 6% NOE between the H-3' and H-6 of the purine base. No other area enhancements were obtained. The observation of these intermolecular NOEs is important confirmation of the fact that the purine base, in the purine-intercalated complex, is preferentially oriented with either H-6 or H-8 directed at the H-3', H-5" protons of the ribose moiety.

Among the group of compounds which has been classified here as other natural products, the penicillins represent a well-studied set of synthetic and natural materials. X-ray structural studies have been performed on a number of derivatives (66), but it was of particular interest to ascertain the conformation of these molecules in solution. With two nonequivalent methyl groups and three carbon-bound protons with different chemical shifts, the system is ideally suited for NOE measurements, and these have been performed by DeMarco et al. (66b,67). The results for phenoxymethyl penicillin, *71*, the (*S*)- and (*R*)-sulfoxides, *72* and *73*, and the sulfone, *74*, are shown in Table 23. It is immediately apparent that the thiazolidine ring undergoes a dramatic change in shape in going from the sulfide *71* to the sulfoxides and sulfone. The 2β-methyl of the sulfide is genuinely axial (Fig. 19), and H-3 is disposed between the two methyl groups. The larger NOE [CH_3-2β],H-3 = 21% suggests that H-3 is closer to CH_3-2β than to CH_3-2α, and the absence of an NOE from CH_3-2α to H-5 shows that H-5 must be some 3.6-3.8 Å away from CH_3-2α. In the (*S*)-sulfoxide, *72*, on the other hand, the thiazolidine ring has inverted its shape with the 3-CO_2CH_3 group situated between the two methyl groups (Fig. 20). H-3 is now relaxed solely by CH_3-2β, and CH_3-2α is sufficiently

Table 23 NOEs in Phenoxymethyl Penicillin Derivatives

Compound	Irradiated	Observed	NOE, %[a]
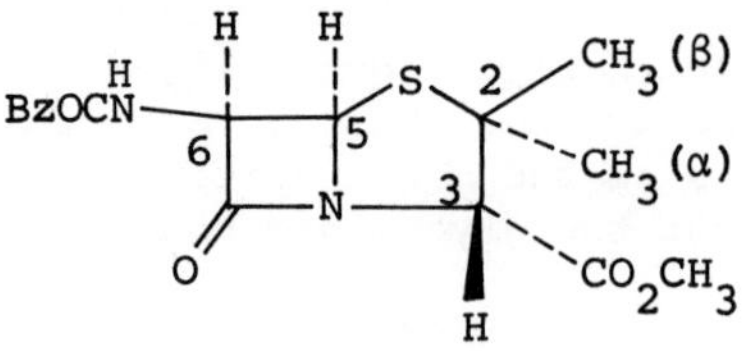 *71*	CH_3-2β	H-3	21
	CH_3-2β	H-5	0
	CH_3-2α	H-3	7
	CH_3-2α	H-5	0
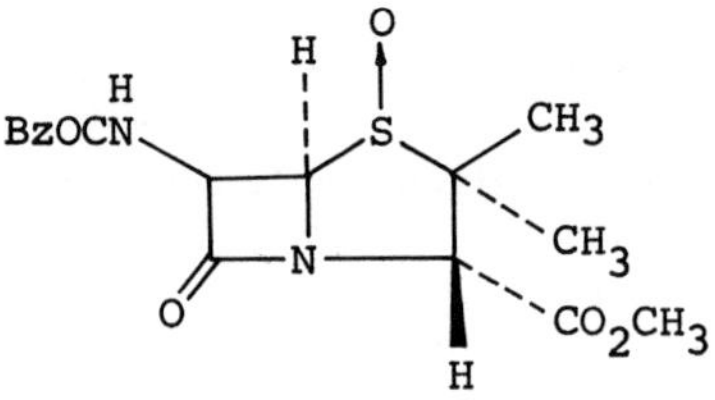 *72*	CH_3-2β	H-3	26
	CH_3-2β	H-5	0
	CH_3-2α	H-3	0
	CH_3-2α	H-5	14
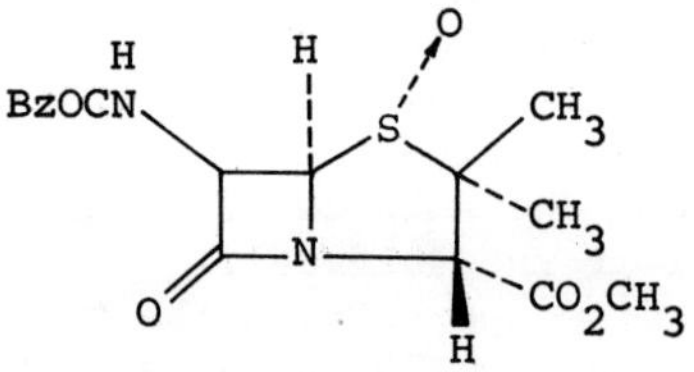 *73*	CH_3-2β	H-3	20
	CH_3-2β	H-5	0
	CH_3-2α	H-3	0
	CH_3-2α	H-5	13

Table 23 Continued

Compound	Irradiated	Observed	NOE, %[a]
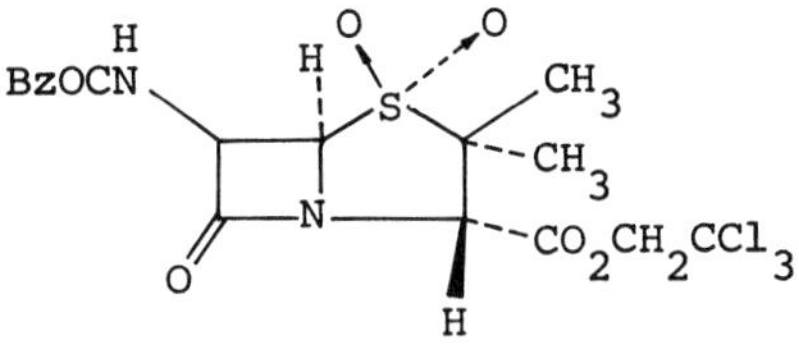 *74*	CH_3-2β	H-3	22
	CH_3-2β	H-5	0
	CH_3-2α	H-3	0
	CH_3-2α	H-5	11

[a]Spectra recorded in CCl_4 solution.

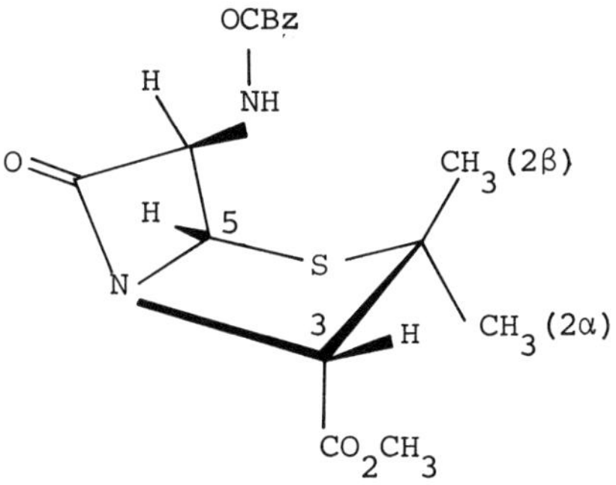

Fig. 19. Conformation adopted by penicillin sulfide *71*.

close (3.0-3.2 Å) to H-5 to affect its relaxation. In addition the NOE data shows that the (*R*)-sulfoxide, *73*, and the sulfone, *74*, both adopt conformations similar to the (*S*)-sulfoxide, and this suggests that it is not the size of the oxygen atom nor hydrogen bonding with the amide hydrogen that determines the thiazolidine ring conformation, but rather the bonding at the sulfur atom. Solvent-induced shifts are also consistent with conformations in Figures 19 and 20 (67,68), and comparison with X-ray studies shows that the thiazolidine ring possesses a

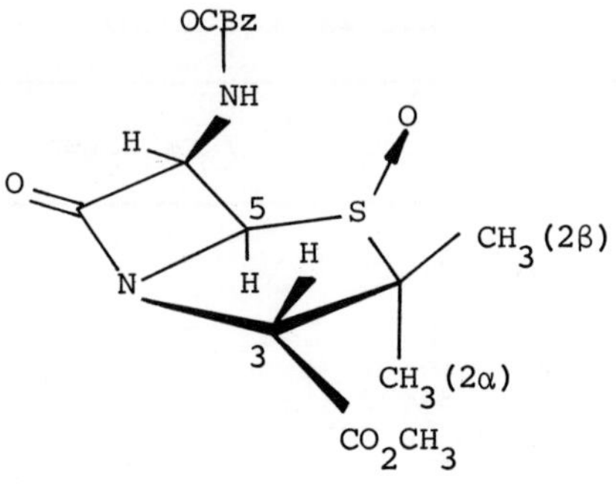

Fig. 20. Conformation adopted by penicillin sulfoxide *72*.

geometry in solution similar to that which it has in the solid state.

The coumarins are another group of natural products that are well-suited to NOE measurements because of their wide divergence in chemical shifts. A fine example of the value of extensive NOE measurements in conjunction with other spectral data in structure determination is to be found in the analysis by Tori and co-workers (69) of poncitrin. Infrared, uv, and nmr spectral evidence of the natural product itself and of a tetrahydro derivative indicated the partial structure *75*. The NOE data in Table 24 show immediately that the methoxyl is situated at position 5, between carbon 4 of the coumarin system and carbon 6 of the unsaturated ether side chain. Since there are no other aromatic protons the remaining NOEs define the positions of the 8- and 11-methyl groups and the structure of poncitrin is

Table 24 NOEs in Poncitrin, *76*

Irradiated	Observed	NOE, %
OCH_3	H-4	11
OCH_3	H-6	16
CH_3-8	H-7	22
CH_3-11	H-12	12
CH_3-11	H-13 (cis)	25

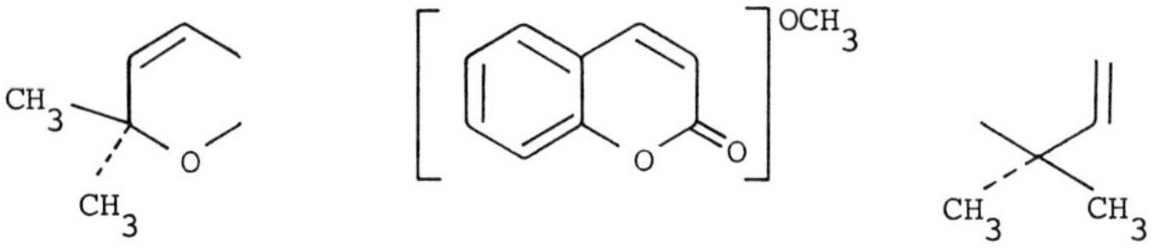

75

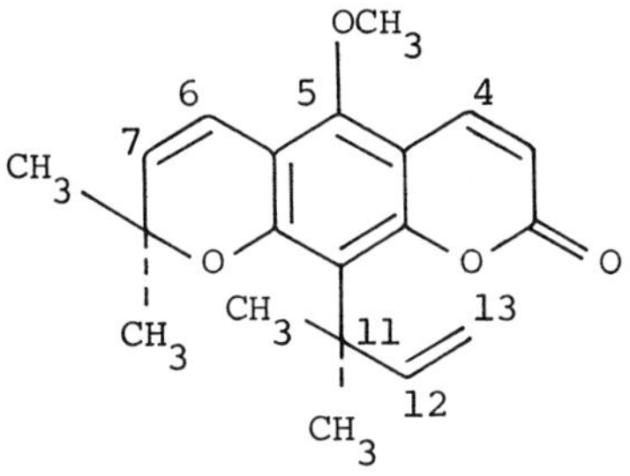

76

as shown in *76*. Another coumarin in which the NOE has been used to determine the orientation of a methoxyl group is nieshoutol (70).

There are a number of other instances where the NOE has been used in structure determinations in a confirmatory or secondary role. Frequently only single measurements that relate to the disposition of methyl or methoxyl groups are reported and because they are of a minor value from a stereochemical viewpoint, we merely list here the compounds and references for completeness. Examples of aromatic structures include aquayamycin (71), bisisodiospyrin (72), the theaflavins (73), and tolypomycinone (74). The novel polypropionate compound portentol also showed a useful NOE (75).

D. Miscellaneous Compounds

One of the first applications of the NOE in organic chemistry was that reported by Anet and Bourn (3), who studied the compounds β,β-dimethylacrylic acid, dimethylformamide, and the half-cage acetate *77*. The vinylic proton of β,β-dimethylacrylic

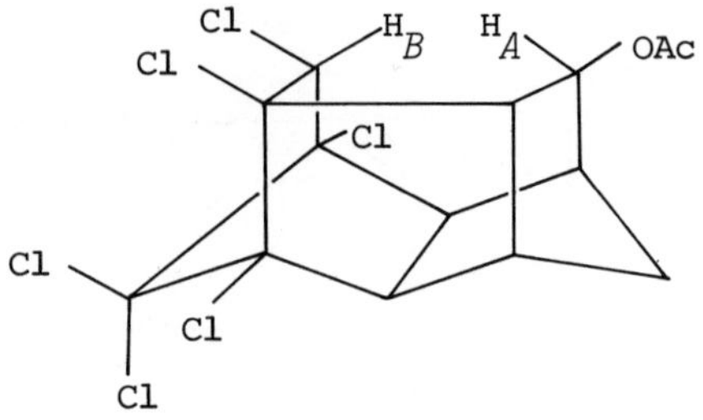

77

acid exhibited an area increase of 17% and a slight area decrease when the two methyl groups were saturated, respectively, confirming the spectral assignment of these two groups. In DMF the observance of an 18% and a -2% enhancement of the formyl proton on saturation of the high-field and low-field methyl groups, respectively, unequivocally confirmed the chemical shifts of the two methyls. The value of 18% is less than that reported subsequently (16), but no explanation of this discrepancy is available. The two protons designated as H_A and H_B in the half-cage acetate *77* gave mutual NOEs of the order of 45%, even in the presence of oxygen, illustrating the effectiveness of the dipolar interaction between the two protons forced so closely together by the cagelike configuration of the molecule.

The results obtained from a study of *t*-butylbenzenes (76) are given in Table 25. The compounds were studied in aprotic and protic solvents with both intermolecular and intramolecular NOEs being observed. The actual NOE values must be treated with some caution since a peak height rather than a peak area enhancement method was employed, and this can give erroneous results, particularly in the cases where complex multiplets were recorded. The authors obtained NOEs greater than 50% and attempted to explain these as being a result of a scalar spin-spin interaction between the *t*-butyl groups and the aromatic protons. However, such an explanation is obviously in error as a scalar spin-spin interaction would be time independent and is thus not a possible relaxation mechanism. With the understanding that the reported results are too high, it is still evident that the *t*-butyl group has a profound effect on the relaxation of the ortho protons and tends to shield those protons from solvent nuclei. It is felt that solvent-solute interactions in the more dilute solutions are less than would be indicated by the NOE results. This effect of the *t*-butyl group shielding adjacent protons from the solvent was also observed by Woods et al. (36) in their ginkolide studies.

Fraser and Schuber (15) recorded the NOE data for three

Table 25 NOE Results for Some *t*-Butylbenzenes

Compound	Irradiated	Observed, NOE, %			
		H-2	H-3	H-5	H-6
4-*t*-Butylphenol[a]	4-$(CH_3)_3C$	7	21	21	7
	4-$(CH_3)_3C$[b]	4	15	15	4
2,4-Di-*t*-butylphenol[a]	4-$(CH_3)_3C$		38	11	9
	2,4-$(CH_3)_3C$		52	23	19
	4-$(CH_3)_3C$[c]		35	20	7
2,4,6-Tri-*t*-butylphenol[a]	2,4,6-$(CH_3)_3C$		57	57	
4-*t*-Butylthiophenol[d]	4-$(CH_3)_3C$	16	16	16	16
4-*t*-Butylanisole[d]	4-$(CH_3)_3C$	7	22	22	7

[a]20% solution in CS_2.
[b]20% solution in DMSO.
[c]16% solution in CH_3CN.
[d]25% solution in CS_2.

dibenzthiepin compounds (see Table 26) in order to determine the configuration of each of the benzylic protons. Inspection of molecular models (Fig. 21) shows that in the sulfide *2*, one of each pair of diastereotopic protons lies much closer to the ortho aromatic proton and this proton (H-1) has the pro-*S* configuration in the overall *R*-biphenyl system. Thus observation of an area increase in H-ortho on saturation of the low-field benzylic resonance assigns the spectral position of H-1. Saturation of the high-field resonances of the *AB* quartet causes an area decrease in H-ortho, confirming that H-1 has the "pro-*S* in *R*" configuration. By a similar process, H-1 in the sulfone *79* (low-field proton of the *AB* quartet) also has the pro-*S* in *R* configuration. The sulfoxide *78* has two benzylic *AB* quartets because of the presence of the asymmetric sulfoxide group. When $CDCl_3$ was used as solvent no NOE was observed between H-1, or H-4, and the ortho aromatic proton. However, when the solvent was changed to DMSO-d_6 this NOE was observed. The lack of an observable NOE in $CDCl_3$ was attributed to dimerization. The results obtained (Table 26) establish a pro-*S* configuration for H-1 and a pro-*R* configuration for H-4 in the *R*-biphenyl. The configuration of H-2 and H-3 as pro-*S* and pro-*R*, respectively, was obtained by a chemical inversion of the sulfoxide group. The final stereochemical assignment, the configuration of the sulfoxide group relative to each pair of geminal benzylic pro-

Table 26 NOEs in Dibenzthiepins

		NOE, %			
Irradiated	Observed	*2*	*78*	*79*	*79*[a]
H-1	H-2	47		48	
H-1	H-ortho	23	33[b]	21	
H-2	H-1	40		30	
H-2	H-ortho	-5		0	
H-ortho	H-1	8	31[b]	10	
H-ortho	H-2	-5		-1	6
H-1	H-4		28		
H-4	H-1		24		
H-ortho	H-4		5[c]		
H-4	H-ortho		3[c]		

[a]Isotopically labeled *79* with H_1 = D.

[b]Isotopically labeled *78* with H_2,H_3,H_4 = D, in DMSO-d_6 solvent.

[c]Isotopically labeled *78* with H_1 = D_1 in DMSO-d_6 solvent.

tons, was established by a consideration of chemical shift and coupling constant data. Several interesting facts emerge from these results. Comparison of the interaction between H-2 and H-ortho when H-1 is a proton or a deuterium resulted in NOEs of -1% and +6%, respectively, when compound *79* was studied. This is a direct consequence of the interdependence of the dipolar interaction when H-1 is a proton, and the result would be predicted by the treatment of Noggle, although eq. [16] could not be used with any degree of accuracy. The result is not predicted by eq. [12] for the all-proton system, but the value of 6% obtained when H-1 is a deuterium falls within experimental error of that predicted by this equation. A further point to be noted is the inherent difficulty in obtaining accurate NOE values. Comparison of the results obtained for [H-1],H-2 and vice versa for compounds *2* and *79* shows a large discrepancy, and the value obtained at H-ortho when H-1 is saturated also varies markedly for the three compounds studied. Within a set of closely related compounds one would expect reasonable consistency in the magnitude of the NOEs, and when this does not pertain, the most probable cause is a variable but trace quantity of paramagnetic impurity in the samples themselves. In the experience of the authors, accurate and reproducible NOE values are obtainable

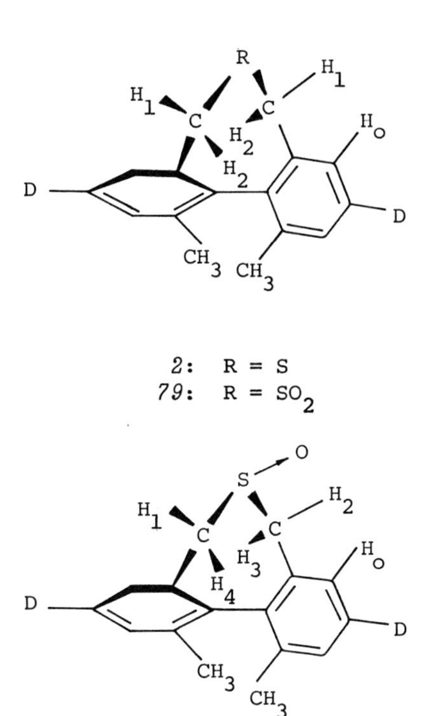

2: R = S
79: R = SO_2

78

Fig. 21. (*a*) Configuration of dibenzthiepin, *2*, and dibenzthiepin sulfone, *79*. (*b*) Configuration of dibenzthiepin sulfoxide, *78*.

only with the exercise of extreme experimental diligence and only on compounds that are in a state of strict analytical purity.

The first example of an intermolecular nuclear Overhauser effect was reported by Kaiser (2) who used a mixture of chloroform and cyclohexane to demonstrate the application of the NOE in organic chemistry. The sample contained chloroform, cyclohexane, and tetramethylsilane in the ratio 1:4.5:0.5, and a 34% NOE was observed for the $CHCl_3$ proton on saturation of the cyclohexane resonance. Although the potential for intermolecular NOEs in solvation studies would appear to be very great, only a limited number of intermolecular NOEs have been reported. The main difficulty appears to be the selection of suitable substrates, since the measurement of intermolecular NOEs is only practical when the observed proton has essentially no intramolecular relaxation pathways available to it (as in the case of $CHCl_3$).

During a study on relaxation processes, Hoffmann and Forsen (77) determined the NOEs in formic acid and acetaldehyde. The values obtained were 22.5±0.5% when observing the formyl proton and 17.5±1.3% when observing the acid proton in formic acid; for acetaldehyde, the values were 23±0.5% and 3.5±0.3% for observing the aldehydic proton and the methyl group, respectively. The method used in these examples was to study the rate of return to equilibrium when the saturating field was removed. The values of M_0^A and M_z^A were taken at t_∞ and t_0 respectively. This general method of measuring NOEs (wherein the specific values of σ_{BA} and R_B are independently measured (see Sect. II-A) does lead to more accurate NOE values, but the experimental work involved is more complex than the simple area enhancement method. In the future, however, increasing use can be expected to be made of the magnetization recovery method, since the relaxation time constants so obtained represent two additional molecular parameters. Tori (57) in his recent work on NOEs in steroids has commented on the uses of time-constant measurement in complex structures.

The 1,3-dipolar addition to the strained double bond of bicyclo[2,2,1] systems can give rise to either *80a* or *81* (Fig. 22). Oakland et al. (78) used NOEs to determine the stereochem-

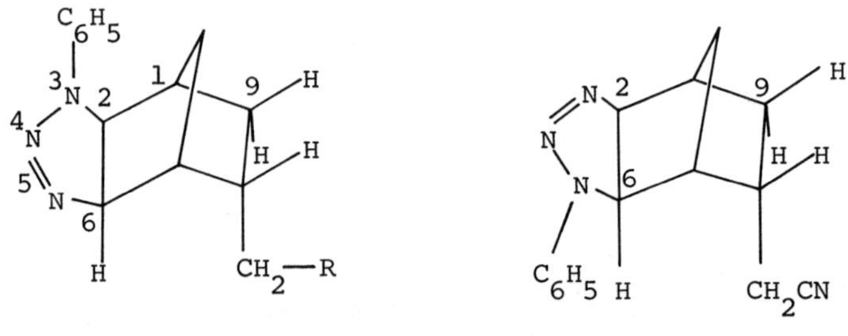

80a: R = CN
80b: R = NH_2

81

Fig. 22. Possible orientations of the triazine ring in cycloaddition products of phenyl azide and norbornenes.

istry of the product formed. NMR spectroscopic evidence suggested that the cyanomethyl group was in an endo position with the heterocyclic ring in an exo position. Saturation of the phenyl ring proton resonances caused an area increase of the signal assigned H-2, whereas saturation of the cyanomethyl protons enhanced the area of H-6 (endo) by 15%, and finally irradiation of the C-9 endo proton resonances gave rise to approximately a 10% area increase of the signal corresponding to the C-2 endo proton. By a similar technique the stereochemistry of other heterocyclic norbornanes, e.g., *80*b, was determined, and the authors point out that the application of NOE measurements could solve many problems in norbornane stereochemistry.

One problem that has plagued stereochemists for many years is the relative size of the lone pair on nitrogen. Booth and Lemieux (79) used NOEs in compound *82* (Fig. 23) in an attempt to throw more light on the subject and recorded the following results: saturation of the C-4(ax) methyl group increased the

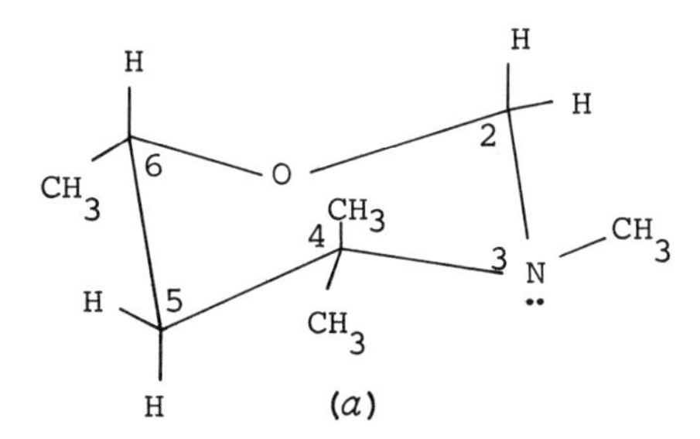

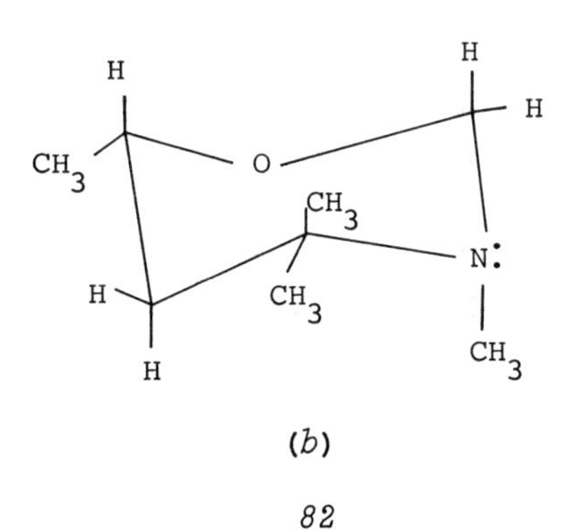

82

Fig. 23. Conformations of tetrahydro-3,4,4,6-tetramethyl-1,3-oxazine, *82*.

area of the C-2 axial proton by 9% and the C-2 equatorial proton by 2%, whereas saturation of the N-CH_3 group caused an enhancement of 6% of the axial proton at C-5. If it is assumed that the ring exists in a chair shape with staggered conformations, then the above NOE results indicate that the compound has a significant conformer population where the methyl group is axial (*82b*). Unfortunately the authors did not use degassed samples and hence the magnitude of the recorded values are only of a rough qualitative value. It should be feasible here, provided the conformations of *a* and *b* can be accurately assessed, to use NOEs obtained on stringently prepared solutions to gain a more quantitative insight into the problem.

Substituted aromatic compounds are frequently attractive candidates for NOE studies because of the wide variation in chemical shifts often encountered. An instructive example in this area is the work of Woods et al. (80) on the products of the wide variation in chemical shifts often encountered. An instructive example in this area is the work of Woods et al. (80) on the products from the bromination of 2,3-benzocycloheptenone. Two isomeric tetrabromides were obtained, *83a* and *83b*, the structures of which were shown to be stereoisomeric by a consideration of the *ABMX* system arising from the aliphatic protons, and the observance of NOEs of the same order of magnitude between H-11 and H-4 for both compounds. Dehydrobromination gave

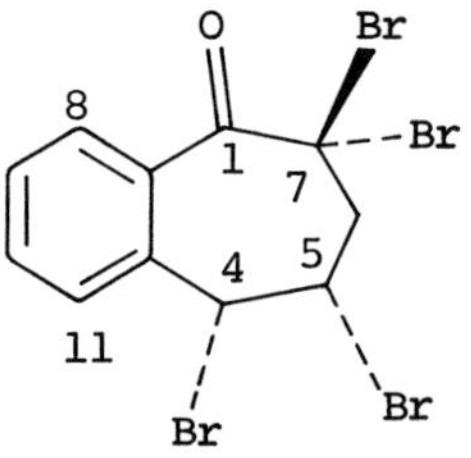

83a

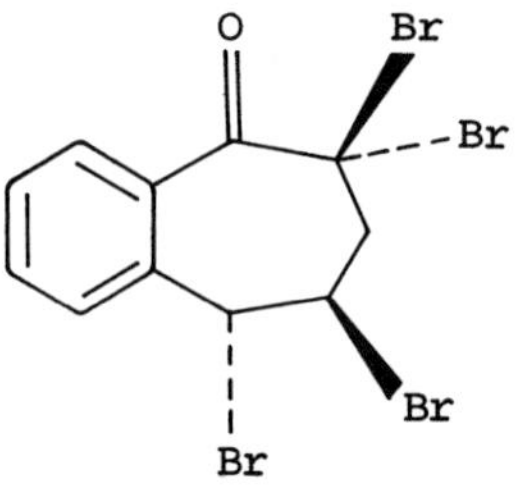

83b

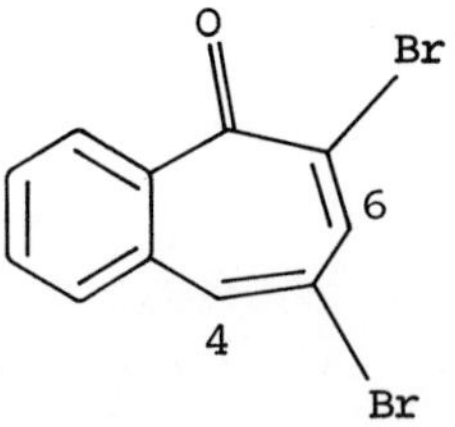

84a

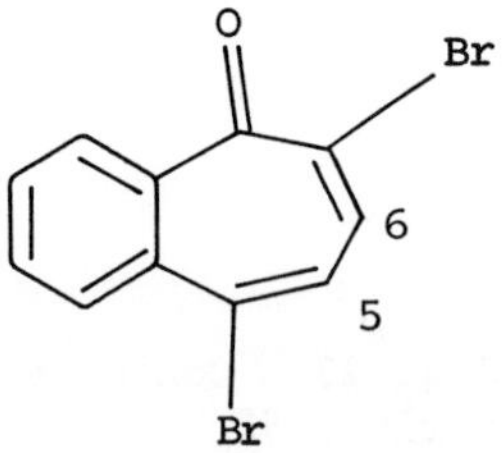

84b

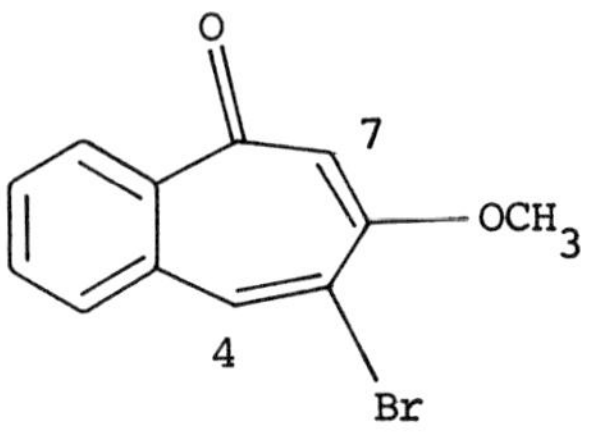

85 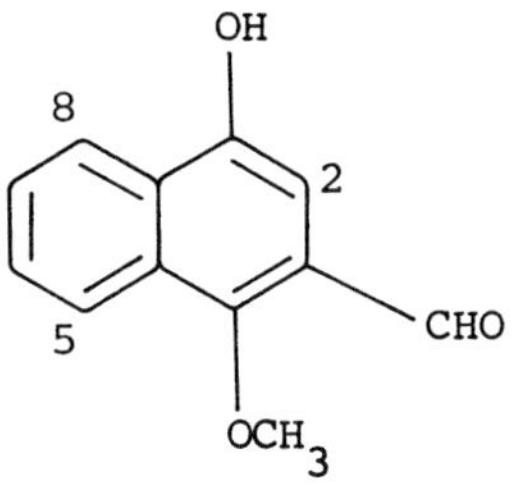 *86*

the two compounds *84a* and *84b*, the structures of which are readily analyzed by consideration of their respective nmr data. These compounds are susceptible to nucleophilic attack; for example, compound *85* is obtained by methanolysis of *84a*. Here a coupling of 0.7 Hz between the methoxyl and H-7, together with the NOE [OCH_3],H-7 = 18% confirmed the position of the methoxyl group. A side product of methanolysis is compound *86*, the structure of which is confirmed by the NOE data shown in Table 27. It is of interest to note that the peri protons, H-5 and

Table 27 NOEs in Some 2,3-Benzocycloheptenone Bromo Derivatives

Irradiated	Observed	*83a*	*83b*	*85*	*86*
H-4	H-11	16	17		
H-11	H-4	23	20		
OCH_3	H-7			18	
OH	H-2				26
CHO	H-2				8
OH	H-8				0
OCH_3	H-5				0
H-2	CHO				8
OCH_3	CHO				24

H-8, owe little of their relaxation to the adjacent methoxyl and hydroxyl groups, a form of behavior that has been noted by Woods (81) in a number of systems. Woods's results indicate that methoxyl or hydroxyl groups tend to align such that the O-H or O-C bonds are closer to the more electron-rich C=C bond. The other values recorded are consistent with the structure shown, and the value of 8% observed between the aldehydic proton and H-2, together with the large (24%) NOE between the methoxyl and the aldehydic proton, are a result of the greater steric interaction between the carbonyl and the methoxyl compared to that between the aldehydic proton and the methoxyl.

The stereochemistry of the central ring of 9-substituted 9,10-dihydroanthracenes has been the subject of much study. Brinkmann et al. (82) used the NOE results recorded for the compounds *87a* through *87e* (see Table 28) to show that in the cases

Table 28 NOE results for Some 9-Substituted 9,10-Dihydroanthracenes

R, H_M, 1, 9, 10, 4, H_A, H_B

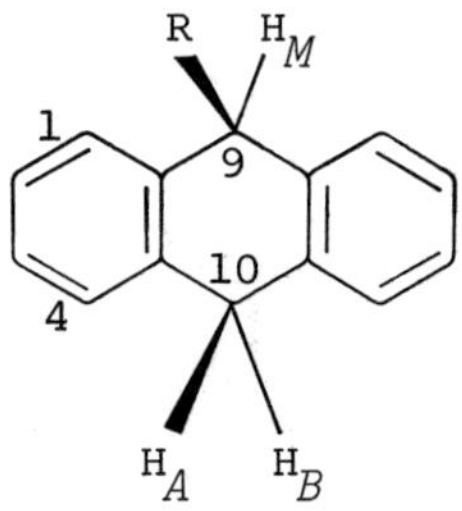

87

R	NOE,[a] %		
	H_A	H_B	H_M
87a: CH_3	—	—	15
87b: $(CH_3)_2CH$	-1	5	13
87c: $(CH_3)_3C$	-1	5	16
87d: C_6H_5	1	10	15
87e: 1,4,9-$(CH_3)_3$	0	0	7
87f: CO_2CH_3	7	11	

[a]Data recorded for the irradiation of all aryl resonances.

where the substituent at C-9 is methyl, isopropyl, or *t*-butyl, their orientation is predominantly axial, whereas if the substituent is methyl or phenyl the equilibrium contains more of the equatorial conformer, although this conformer is still the less populated one. It should be noted that in these examples all three benzylic protons are significantly relaxed by protons other than adjacent aryl protons to a greater or lesser extent. This is the reason for the differences found when observing H_B and H_M, as H_A relaxes H_B more efficiently than the R substituent relaxes H_M. Bell and Saunders (19) studied the 9-carbomethoxy derivatives, the results of which are included in Table 28. There the conformer with the substituent equatorial is well populated. However care must be exercised before quantitative comparisons can be made, as the relaxation interactions of the carbomethoxy group with H_A will be negligibly small in comparison with the relaxation effects of a C-9 alkyl or even aryl substituent. In the examples where R is alkyl or aryl the observed long-range homoallylic coupling constants were in accord with the deductions made from the NOE data.

In a very similar study of a set of 1,3,5,7-tetramethyltricyclo[5.1.0.0^{3,5}]octane derivatives, Stothers et al. (83) used NOE measurements in conjunction with long-range couplings to ascertain the stereochemistry of the 2- and 6-substituents. The results for a selection of the compounds are shown in Table 29. Although both the 2-trans and 2-cis protons show NOEs on saturation of the 1- and 3-methyl groups, the consistently larger effect shown by the cis protons is sufficient to define the stereochemistries. More interesting is the qualitative information the NOEs supply on the shape of the six-membered ring. Clearly, if NOEs are to be observed at both 2-trans and 2-cis protons, the central ring must be much flatter than is depicted in Figure 24. Indeed, the very large NOE (39%) observed for H-2 cis in *90* strongly suggests that the six-membered ring here is essentially planar. The large NOE (28%)* noted for H-2 cis of *91* is also consistent with this picture. The variation in NOEs also implies that the overall geometry of the central ring is strongly influenced by the nature of the 2- and 6-substituents.

The 7,12-dihydropleiadienes studied by Colson, Lansbury, and Saeva (84) are an analogous group of compounds where the seven-membered ring geometry can be assessed by examining across the ring interactions. The results are shown in Table 30. Compound *92* has a fixed geometry (Fig. 25) and the large NOE

*This is an exceptionally large NOE value to be observed at a geminal proton, but as the authors note (83), it may arise because of the interrelaxation effects in the closely knit tricyclic system. For a more quantitative analysis the method of Noggle should be used (5).

Table 29 NOE Results for Some Tricyclooctanes

Compound	Irradiated	Observed	NOE, %
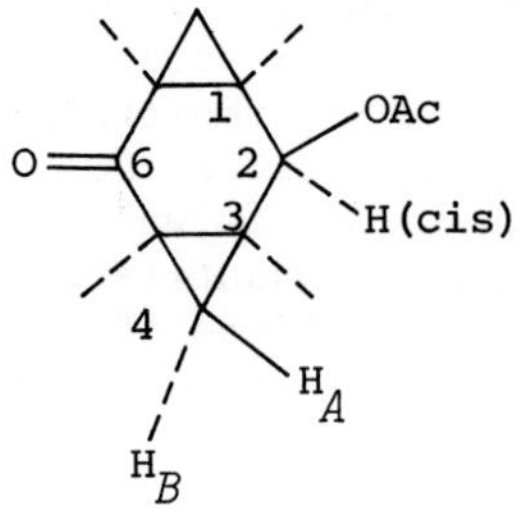 *88*	CH_3-3,1	H-2(cis)	33
	CH_3-3,1	H-B	1
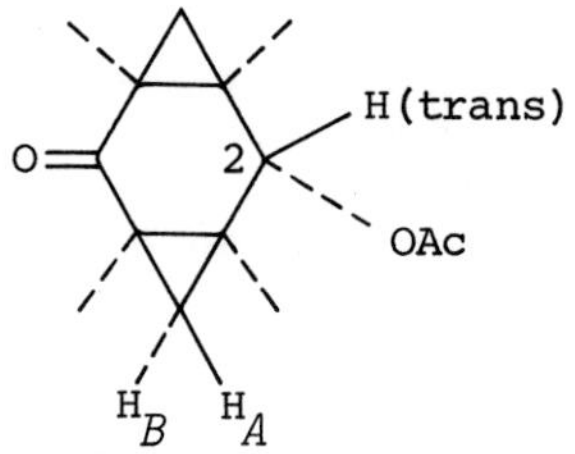 *89*	CH_3-3,1	H-2(trans)	14
	CH_3-3,1	H-B	13
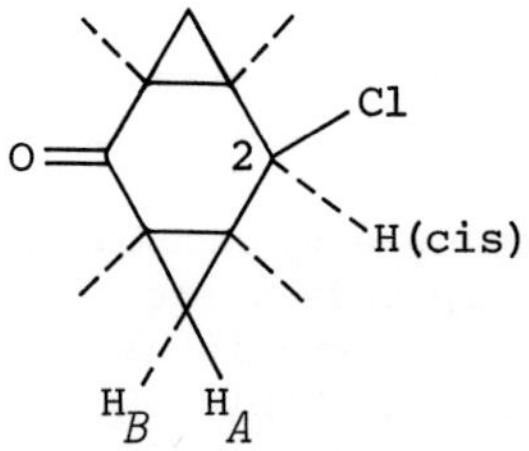 *90*	CH_3-3,1	H-2(cis)	39
	CH_3-3,1	H-B	9

Table 29 Continued

Compound	Irradiated	Observed	NOE, %
91	CH_3-3,1	H-2(cis)	28
	CH_3-3,1	H-2(trans)	6
	CH_3-3,1	H-B	15

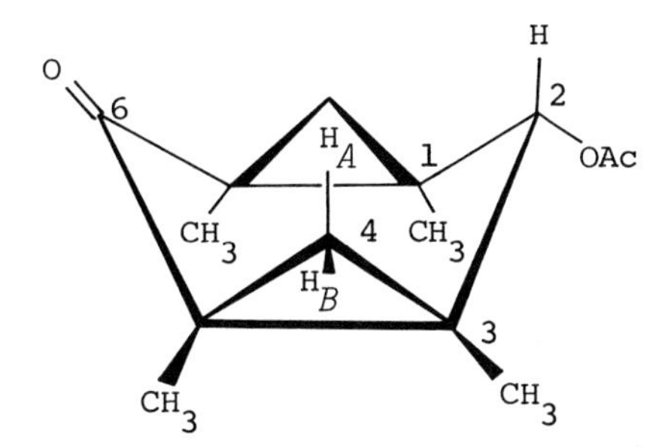

Fig. 24. Conformation adopted by 2-*cis*-acetoxy-1,3,5,7-tetramethyltricyclo[5.1.0.0^{3,5}]octan-6-one, *89*.

[CH_3-12],H-7(ax) = 26% was recorded. Compound *93* can exist in two possible forms, A and E, although at ambient temperatures it exists exclusively in the E form (both groups equatorial) since a 19% NOE between the protons at C-12 and C-7 was observed. At room temperature *94* contains both conformers, (Fig. 25) and at -20°C the interconversion rate is sufficiently slow to permit nmr observation of both conformers. The individual resonances

Table 30 NOEs in Some 7,12-Dihydropleiadienes

Compound	Irradiated	Observed	NOE, %
92	CH_3-12	H-7(ax)	26
93	H-12	H-7	19
94 (conformer E)	H-12	H-7(ax)	27
94 (conformer A)	H-12	H-7(eq)	0
96	CH_3-11	H-12	17
97b	CH_3-8	H-13	20
	CH_3-11	H-12(eq)	16

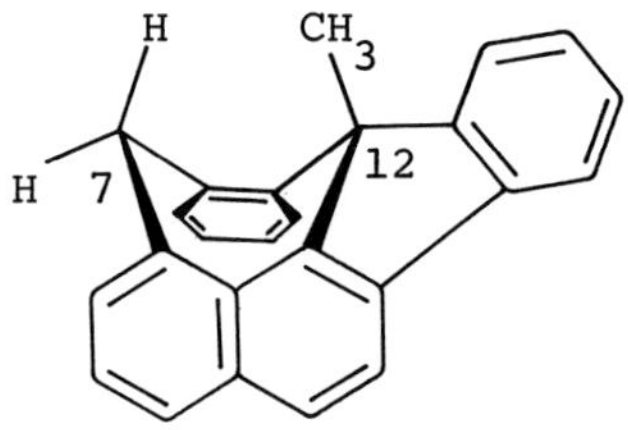

92

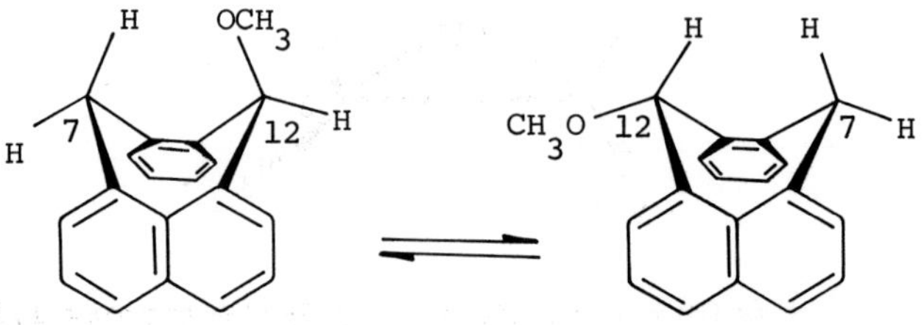

94

Fig. 25. (*a*) Fixed conformation of 12-methyl-1,12-(*O*-phenylene)-7,12-dihydropleiadiene, *92*. (*b*) Axial (A) and equatorial (E) conformers of 7-methoxy-7,12-dihydropleiadiene, *94*.

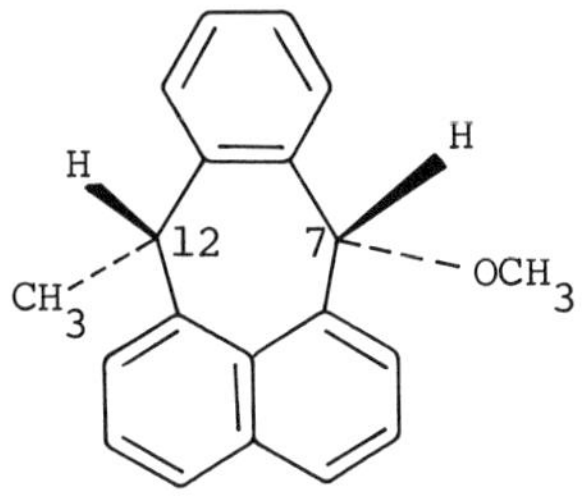

93 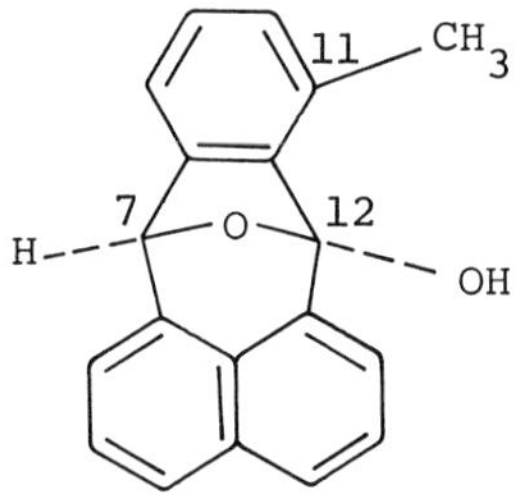 *95*

for each conformation were readily assigned, as the E form showed the NOE [H-7],H-12 = 27%, whereas conformation A showed a zero NOE. The hemiketal *95* could be converted to *96* by treatment with base, and the problem was to assign a structure to the isolated material as both compounds are tautomeric mixtures. Saturation of the aryl methyl groups in a solution containing both *95* and *96* resulted in an NOE being recorded at H-12 for *96* but not for *95*, and thus the stereochemistry was unequivocally

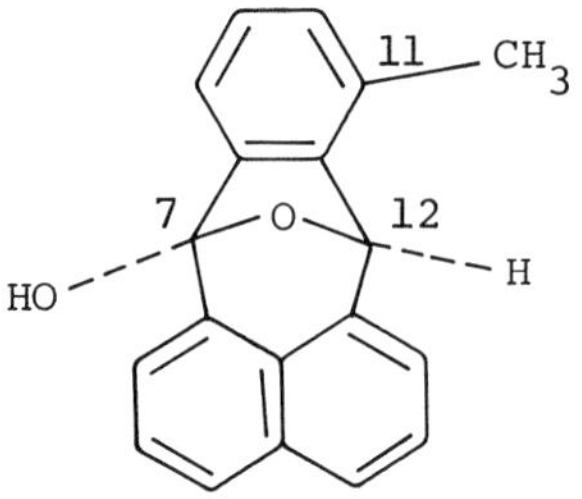

96

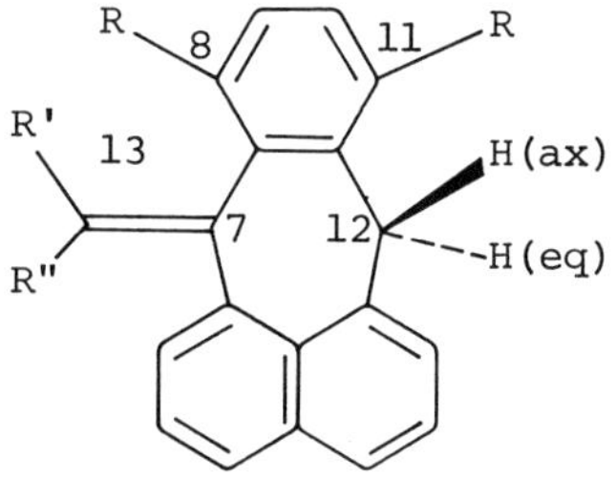

97a: R = H; R' = CH_3; R" = H
97b: R = CH_3; R' = R" = H

demonstrated. Finally, the assignment of the relative structures and spectral parameters of compounds of the type *97a* proved to be difficult until the NOEs between the C-8 methyl group and the vinyl proton H-7 and between the C-11 methyl group and H-12(eq) of *97b* were observed. The foregoing demonstrated firstly the existence of transannular NOEs between the C-7 and C-12 axial groups, and secondly how these may be employed in the determination of pleiadiene stereochemistries.

It is unfortunate that to the present time few papers have appeared using the NOE in simple monocyclic ring systems. The example of the tetrahydro-3,4,6-trimethyl-1,3-oxazine, *82*, has been commented on above. A further interesting example taken from a study by Eliel and Nader (85) on 1,3-dioxane ortho esters shows again how informative the NOE can be. The NOE data for the two ortho esters *98* and *99* are collected in Table 31. The

Table 31 NOEs in 4,4,6-Trimethyl-1,3-Dioxane Ortho Esters

Compound	Irradiated	Observed	NOE, %
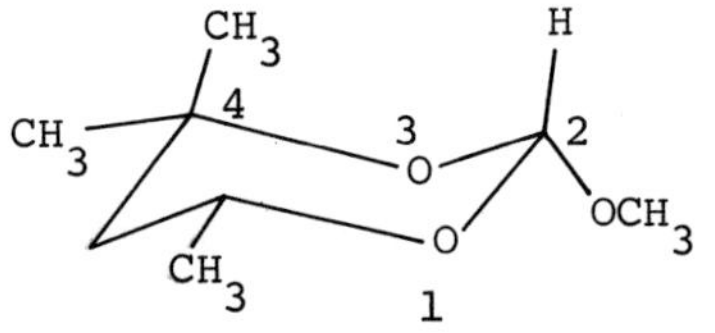 *98*	CH_3-4(ax)	H-2	12
	OCH_3	H-2	0
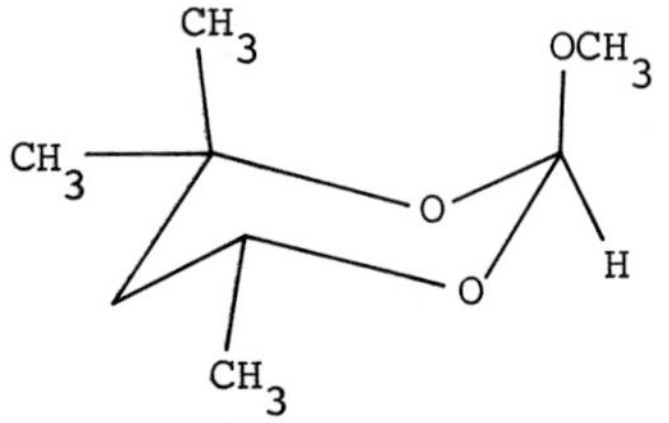 *99*	CH_3-4(ax)	H-2	0
	OCH_3	H-2	10

assignment of the equatorial methoxyl to *98* is obvious from the NOE [CH_3-4(ax)],H-2 = 12%, but more interesting are the dispositions of the methoxyl CH_3 in the two isomers. The axial isomer *99* shows a positive NOE at H-2 which means, not unreasonably, that the CH_3 groups largely points away from the ring, but the equatorial OCH_3 shows no NOE at H-2.

This unexpected result implies that the CH_3 must exist in a conformation pointing away from H-2 and lying between the 1 and 3 oxygen atoms. This conformation is the one possessing the lowest dipole moment and the fewest "rabbit-ear effects."

Other instances where the NOE has been employed in a major way but in a similar manner to the examples presented here are to be found in refs. 86, 87, and 88. Applications of the NOE as a minor adjunct to structural work are located in refs. 89 and 90. The NOE method for determining amide configurations has been extended by Bovey et al. (91), but their preliminary data is of little use since the magnitudes of the observed NOEs were not recorded.

E. Heteronuclear Nuclear Overhauser Effects

Inspection of eq. [4] (Sect. II-A) shows that the observed NOE is dependent upon the relative gyromagnetic ratios (γ) of the observed and saturated nuclei, and if we are observing the nuclei of lower γ, then the enhancement will be greater than in the homonuclear experiments. This has been of significant value in the observance of natural-abundance carbon-13 spectra, particularly prior to the advent of Fourier-transform techniques.

Nuclear Overhauser enhancements in ^{13}C spectra have been widely known since the commercial introduction of the random noise proton decoupling technique. However, the low natural abundance of ^{13}C (1%) has made quantitative measurements difficult, and in fact, only three studies have been published to date. In an early experiment, Kuhlmann and Grant (92) used ^{13}C-enriched formic acid to measure the enhancement of the ^{13}C resonance upon saturation of the proton signal. They found a value of 2.98±0.15 for the area increase, which is in exceptional agreement with the theoretical maximum value of $1/2\ \gamma_H/\gamma_{^{13}C}$ = 2.988 predicted for a pure dipolar relaxation mechanism for the ^{13}C atom. These same authors (93) examined a number of polyaromatic compounds, of which acenaphthene, *100*, is a typical

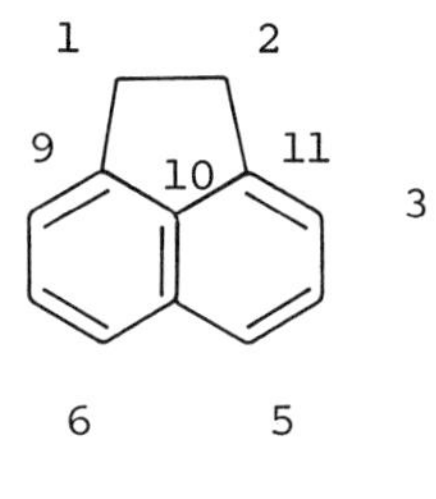

100

example, and showed that the NOE at a given carbon decreased as the distance of the carbon increased from neighboring protons. Thus carbons 3, 10, and 11 showed area ratios 2.4:1.2:1.0, respectively, when the proton signals were saturated, and the experiment is an elegant demonstration of how the Overhauser effect can be employed for the assignment of ^{13}C resonances. In a more recent and particularly lucid presentation of the current Overhauser effect theory (94), Kuhlmann, Grant, and Harris have deduced that all carbon atoms bearing directly bonded protons should show identical NOEs irrespective of the number of protons involved. This conclusion was experimentally verified by measurements of the NOEs obtained for adamantane where the area enhancements of the CH_2 and CH carbons were found to be identical. A similar result has been noted by Allerhand (95) who used the Fourier-transform technique.

Overhauser effects between protons and ^{15}N have been studied by Lichter and Roberts (96) and the results show an intriguing facet of the overall NOE problem. ^{15}N has a very small *negative* magnetic moment (= -2.830 nuclear magnetons) which means that Overhauser effects observed on saturating proton resonances will be negative; i.e., emission lines will be observed instead of normal absorption curves. For pure dipolar relaxation of the ^{15}N nuclei, the Overhauser "enhancement" factor will be $1/2\ \gamma_H/\gamma_{15_N} = -4.93$, and we should expect to see lines of the same order of intensity as ^{13}C. However, the experimental results for $^{15}NH_4Cl$ at pH 0 show only a -2.22 enhancement factor, and this has been interpreted by Roberts (96) as indicating the presence of other spin relaxation mechanisms, possibly spin rotation. A selection of aliphatic and cyclic amines was also examined with rather divergent results, and it appears that pH (controlling proton-exchange rates at the nitrogen), sample purity, and molecular size are all factors which must be taken into consideration. There is no doubt that more work is required on this fascinating and potentially useful nucleus.

The NOE between protons and fluorine was, in fact, the first NOE recorded (97), although the original work and several subsequent publications used the technique as an aid in relaxation studies. The first application of ^{1}H-^{19}F NOEs in organic chemistry was that of Bell and Saunders (98) who studied a series of fluoroaromatic compounds, the results of which are shown in Table 32. The relaxation processes of fluorine are more complex than that of the proton, as interactions other than dipole-dipole can be important. However, a study of the effect of temperature and field strength on T_1 (99) showed that, with the exception of *o*-fluorotoluene where spin rotation is important, the dominant relaxation process is the dipole-dipole interaction.

Table 32 NOE Results for Some Fluoroaromatics

Compound	Irradiated	Observed	NOE, %
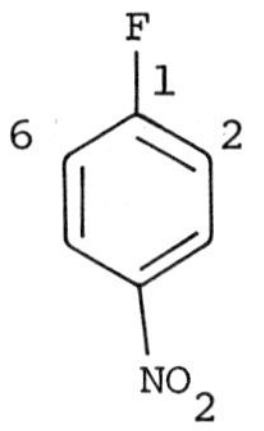 *101*	H-2		
	H-6	F-1	32
	F-1	H-2	10
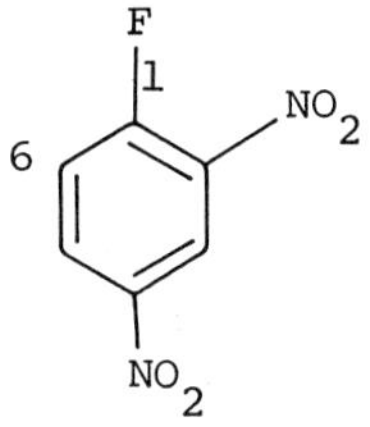 *102*	H-6	F-1	22
	F-1	H-6	10
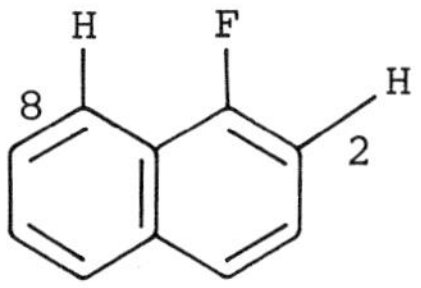 *103*	H-2 + H-8	F-1	50

Table 32 Continued

Compound	Irradiated	Observed	NOE, %
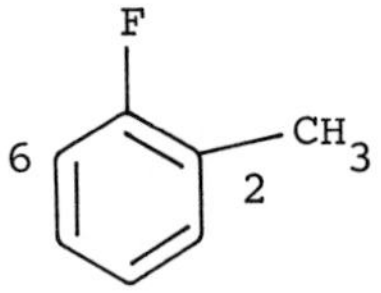			
104	H-6	F-1	18
	CH_3-2	F-1	11
	F-1	H-6	8

This conclusion was further substantiated by the 32% NOE observed at the fluorine of 4-nitrofluorobenzene, *101*, when the ortho protons were saturated, since the experimental NOE is close to the value predicted by equation [12]. The value recorded for the area enhancement of the fluorine when the ortho proton was saturated in compound *102* is indicative of a distortion where the C-F bond is bent away from the nitro group toward the proton. For 1-fluoronaphthalene, *103*, the NOE value of 50% is close to the predicted maximum, but as the 2- and 8-aromatic protons could not be saturated independently, no information on the distance dependence of the NOE nor of the peri interaction was available.

The heterocyclic compounds, *105*, *106*, and *107* represent

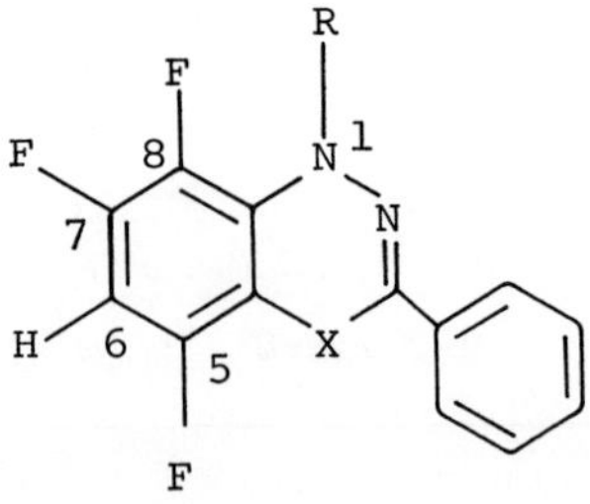

105: X = S; R = H
106: X = S; R = Ac
107: X = O; R = Ac

typical examples of how heteronuclear NOEs can be used to determine the position of the proton relative to the nitrogen. The NOEs recorded (14) in Table 33 show unequivocally that in all four compounds, the proton is meta to the nitrogen. The

Table 33 NOE Results for Some 4-Oxo and 4-Thiadiazoles

Irradiated	Observed	*105*	*106*	*107*
H-6	F-8	0	0	0
	F-7	24	23	22
	F-5	34	32	30
N-R	F-8	34	12	12
F-8	F-7	8	11	10
F-7	F-8	10	9	
F-7	H-6	20	19	22
F-5	H-6	27	26	26

magnitude of the NOEs indicate a distortion of ring A, as the values recorded at H-6 are nonsymmetrical with respect to F-5 and F-7. The actual values imply a "normal" (consistent with eq. [12]) distance H-6,F-7, and a steric or polar interaction causing a decrease in the distance between F-5 and H-6. The NOE recorded between N-H and F-8 is larger than that predicted on the basis of eq. [12], and this again suggests an F-F repulsion which forces F-8 closer to the nitrogen atom. The homonuclear NOE [F-7],F-8 = 12%, is less than predicted and appears to indicate the presence of out-of-plane distortions. However, the latter results may be in error as the signals are widely separated and difficulty was experienced in obtaining complete saturation when performing homonuclear saturation. Overall, one can conclude that ring A is distorted such that the distances F-5,H-6 and F-8,N-H have decreased, but one cannot be certain with respect to the distortion, if any, at C-7.

REFERENCES

1. J. D. Baldeschwieler and E. W. Randall, *Chem. Rev.*, *63*, 81 (1963).
R. A. Hoffman and S. Fórsén, *Progr. Magn. Resonance*, *1*, 15 (1966).

2. F. A. L. Anet and A. R. J. Bourn, *J. Amer. Chem. Soc.*, *87*, 5250 (1965).
3. R. Kaiser, *J. Chem. Phys.*, *42*, 1838 (1965).
4. R. A. Bell and J. K. Saunders, *Can. J. Chem.*, *48*, 1114 (1970).
5. R. E. Schirmer, J. H. Noggle, J. P. Davis, and P. A. Hart, *J. Amer. Chem. Soc.*, *92*, 3266 (1970).
6. I. Solomon, *Phys. Rev.*, *99*, 559 (1955).
7. A. Abragam, *The Principles of Nuclear Magnetism*, Oxford University Press, New York, 1961, Chap. VIII.
8. H. G. Hertz, *Progr. Magn. Resonance*, *3*, 159 (1967).
9. R. A. Bell and J. K. Saunders, unpublished results.
10. J. W. ApSimon, W. G. Craig, A. Demayo, and A. A. Raffler, *Can. J. Chem.*, *46*, 809 (1968); J. W. ApSimon, W. G. Craig, P. V. Demarco, D. W. Mathieson, L. Saunders, and M. Whalley, *Tetrahedron*, *23*, 2357 (1970).
11. J. H. Noggle and R. E. Schirmer, *The Nuclear Overhauser Effect: Chemical Applications*, Academic Press, New York, 1971.
12. J. H. Noggle, *Abstracts of the 3rd Annual NMR Conference*, Gainesville, Florida, 1970.
13. R. A. Bell and J. K. Saunders, *Can. J. Chem.*, *46*, 3421 (1968).
14. J. K. Saunders, Thesis, McMaster University, Hamilton, Ontario, Canada.
15. R. R. Fraser and F. J. Schuber, *Can. J. Chem.*, *48*, 633 (1970).
16. R. A. Bell and J. K. Saunders, *Can. J. Chem.*, *48*, 512 (1970).
17. S. Combrisson, B. Roques, P. Rigny, and J. J. Basselier, *Can.. J. Chem.*, *49*, 904 (1971).
18. (a) R. A. Bell and J. K. Saunders, paper in preparation; (b) R. E. Schirmer, private communication.
19. R. A. Bell and J. K. Saunders, paper presented at the 52nd C.I.C. Conference, Montreal, Quebec, May 1969.
20. T. Fukumi, Y. Arata, and S. Fujiwara, *J. Mol. Spectry.*, *27*, 443 (1968).
21. J. K. Saunders, R. A. Bell, C.-Y. Chen, D. B. MacLean, and R. H. F. Manske, *Can. J. Chem.*, *46*, 2876 (1968).
22. M. Castillo, J. K. Saunders, D. B. MacLean, N. M. Mollov, and G. I. Yakimoo, *Can. J. Chem.*, *49*, 139 (1971).
23. R. H. F. Manske, R. G. A. Rodrigo, D. B. MacLean, D. E. F. Gracey, and J. K. Saunders, *Can. J. Chem.*, *47*, 3585 (1969).
24. C. K. Yu, J. K. Saunders, D. B. MacLean, and R. H. F. Manske, *Can. J. Chem.*, *49*, 3020 (1971).
25. R. H. F. Manske, R. G. A. Rodrigo, D. B. MacLean, D. E. F. Gracey, and J. K. Saunders, *Can. J. Chem.*, *47*, 3589 (1969).
26. S. McLean, M.-S. Lin, A. C. MacDonald, and J. Trotter,

Tetrahedron Lett., *1964*, 3819; S. McLean, M.-S. Lin, and R. H. F. Manske, *Can. J. Chem.*, *44*, 2449 (1966).
27. S. McLean, M.-S. Lin, and J. Whelan, *Tetrahedron Lett.*, *1968*, 2425.
28. K. Nakanishi, *Infrared Absorption Spectroscopy*, Holden-Day, San Francisco, 1954, p. 174.
29. C.-Y. Chen and D. B. MacLean, *Can. J. Chem.*, *46*, 2501 (1968).
30. D. B. MacLean, C. K. Yu, R. Rodrigo, and R. H. F. Manske, *Can. J. Chem.*, *48*, 3673 (1970).
31. D. B. MacLean, D. E. F. Gracey, J. K. Saunders, R. Rodrigo, and R. H. F. Manske, *Can. J. Chem.*, *47*, 1951 (1969).
32. J. Slavik, L. Doleyis, V. Hannus, and A. D. Cross, *Coll. Czech. Chem. Commun.*, *33*, 1619 (1968).
33. G. R. Clark, R. H. F. Manske, C. J. Palnuk, R. Rodrigo, D. B. MacLean, L. Baczynskyj, D. E. F. Gracey, and J. K. Saunders, *J. Amer. Chem. Soc.*, *92*, 4998 (1970).
34. T. J. Batterham, K. H. Bell, and U. Weiss, *Australian J. Chem.*, *18*, 1799 (1965).
35. G. R. Newkome and N. S. Bhacca, *Chem. Commun.*, *1969*, 385.
36. M. C. Woods, I. Miura, Y. Nakadaira, A. Terahara, M. Maruyama, and K. Nakanishi, *Tetrahedron Lett.*, *1967*, 321 and preceding papers.
37. R. T. Major, *Science*, *157*, 1270 (1967).
38. K. Nakanishi, K. Habaguchi, Y. Nakadaira, M. C. Woods, M. Maruyama, R. T. Major, M. Alauddin, A. R. Patel, K. Weinges, and W. Bähr, *J. Amer. Chem. Soc.*, *93*, 3544 (1971).
39. M. Shiro, T. Sato, H. Koyama, Y. Maki, K. Nakanishi, and S. Uyeo, *Chem. Commun.*, *1966*, 98.
40. M. C. Woods, H.-C. Chiang, Y. Nakadaira, and K. Nakanishi, *J. Amer. Chem. Soc.*, *90*, 522 (1968).
41. R. C. Cambie, G. R. Clark, D. R. Crump, and T. N. Waters, *Chem. Commun.*, *1968*, 183.
42. R. A. Bell and E. N. C. Osakwe, *Chem. Commun.*, *1968*, 1093.
43. R. H. Bible and R. R. Burtner, *J. Org. Chem.*, *26*, 1174 (1961).
44. R. A. Bell and M. B. Gravestock, *Can. J. Chem.*, *47*, 3661 (1969).
45. S. Itô, M. Kodama, M. Sunagawa, H. Honma, Y. Hayashi, S. Takahashi, H. Ona, T. Sakan, and T. Takahashi, *Tetrahedron Lett.*, *1969*, 2951.
46. R. A. Bell, M. B. Gravestock, J. K. Saunders, and V. Taguchi, unpublished results.
47. K. Tori, M. Ohtsuru, I. Horibe, and K. Takeda, *Chem. Commun.*, *1968*, 943.
48. Y. Ishiaki, Y. Tanahashi, T. Takahashi, and K. Tori, *Chem.*

Commun., *1969*, 551.
49. J. T. Pinhey and S. Sternhell, *Tetrahedron Lett.*, *1963*, 275.
50. K. Tori and I. Horibe, *Tetrahedron Lett.*, *1970*, 2881.
51. K. Takeda, K. Tori, I. Horibe, M. Ohtsuru, and H. Minato, *J. Chem. Soc.*, *1970*, 2697.
52. N. S. Bhacca and N. H. Fischer, *Chem. Commun.*, *1969*, 68.
53. W. von E. Doering and W. Roth, *Angew. Chem. Intern. Ed. Engl.*, *2*, 115 (1963).
54. N. H. Fischer, T. J. Mabry, and H. B. Kagan, *Tetrahedron*, *24*, 4091 (1968).
55. H. Hikino, C. Konno, T. Takemoto, K. Tori, M. Ohtsuru, and I. Horibe, *Chem. Commun.*, *1969*, 662.
56. J. Lugtenburg and E. Havinga, *Tetrahedron Lett.*, *1969*, 2391.
57. K. Tori, lecture presented at the *3rd International Congress on Hormonal Steroids*, Hamburg, September 1970.
58. H. Koyama, M. Shiro, T. Sato, Y. Tsukuda, H. Itazaki, and W. Nagata, *Chem. Commun.*, *1967*, 812.
59. W. N. Speckamp, J. G. Westra, and H. O. Huisman, *Tetrahedron*, *26*, 2353 (1970).
60. J. T. Rao, M. Sundaralingam, *J. Amer. Chem. Soc.*, *92*, 4963 (1970).
61. D. W. Miles, S. J. Hahn, R. K. Robins, M. J. Robins, and H. Eyring, *J. Phys. Chem.*, *72*, 1483 (1968), and references cited therein.
62. P. O. P. Ts'o, in *Fine Structure of Proteins and Nucleic Acids*, G. D. Fasman and S. N. Timasheff, Eds., Dekker, New York, 1970, p. 49; M. P. Schweizer, J. T. Witowski, and R. K. Robins, *J. Amer. Chem. Soc.*, *93*, 277 (1971); D. W. Miles, W. H. Inskeep, M. J. Winkley, R. K. Robins, and H. Eyring, *J. Amer. Chem. Soc.*, *92*, 3866 (1970); W. H. Inskeep, D. W. Miles, and H. Eyring, *J. Amer. Chem. Soc.*, *92*, 3866 (1970).
63. (a) P. A. Hart and J. P. Davis, *J. Amer. Chem. Soc.*, *93*, 753 (1971); (b) R. E. Schirmer, J. P. Davis, J. H. Noggle, and P. A. Hart, submitted for publication--private communication.
64. B. J. Blackburn, A. A. Grey, I. C. P. Smith, and F. E. Hruska, *Can. J. Chem.*, *48*, 2866 (1970).
65. S. I. Chan and G. P. Kreishman, *J. Amer. Chem. Soc.*, *92*, 1102 (1970).
66. (a) D. Crowfoot, C. W. Bunn, B. W. Rodgers-Low, and A. Turner-Jones, *Chemistry of Penicillin*, Princeton University Press, Princeton, N. J., 1949, p. 310; (b) R. D. G. Cooper, P. V. DeMarco, J. C. Cheng, and N. D. Jones, *J. Amer. Chem. Soc.*, *91*, 1408 (1969).

67. R. A. Archer and P. V. DeMarco, *J. Amer. Chem. Soc., 91*, 1530 (1969).
68. D. H. R. Barton, F. Comer, and P. G. Sammes, *J. Amer. Chem. Soc., 91*, 1529 (1969).
69. T. Tomimatsu, M. Hashimoto, T. Shingu, and K. Tori, *Chem. Commun., 1969*, 168.
70. R. D. H. Murray and M. M. Ballantyne, *Tetrahedron, 26*, 4473 (1970).
71. M. Sezaki, S. Kondo, K. Maeda, H. Umezawa, and M. Ohno, *Tetrahedron, 26*, 5171 (1970).
72. K. Yoshihira, M. Tezuka, and S. Natori, *Tetrahedron Lett., 1970*, 7.
73. T. Bryce, P. D. Collier, I. Fowlis, P. E. Thomas, D. Frost, and C. K. Wilkins, *Tetrahedron Lett., 1970*, 2789.
74. T. Kishi, M. Asai, M. Muroi, S. Harada, E. Mizuta, S. Terao, T. Miki, and K. Mizuno, *Tetrahedron Lett., 1969*, 91.
75. D. J. Aberhart and K. H. Overton, *Chem. Commun., 1969*, 162.
76. E. Lippma, M. Alla, and A. Sugis, *Izv. Akad. Nauk Est. SSR, Ser. Tekhn. Fiz. Mat. Nauk, 16*, 385 (1967).
77. R. A. Hoffman and S. Fórsén, *J. Chem. Phys., 44*, 2049 (1966).
78. F. Shunman, D. Barraclough, and J. S. Oakland, *Chem. Commun., 1970*, 714.
79. H. Booth and R. A. Lemieux, *Can. J. Chem., 49*, 777 (1971).
80. M. C. Woods, S. Ebine, M. Hoshino, K. Takahashi, and I. Miura, *Tetrahedron Lett., 1969*, 2879.
81. M. C. Woods and I. Miura, *Abstracts 7th NMR Symposium*, Nagoya, Japan, 1968, p. 145.
82. A. W. Brinkmann, M. Gordon, R. G. Harvey, P. W. Rabideau, J. B. Stothers, and A. L. Ternay, Jr., *J. Amer. Chem. Soc., 92*, 5912 (1970).
83. M. Gordon, W. C. Howell, C. H. Jackson, and J. B. Stothers, *Can. J. Chem., 49*, 143 (1971).
84. J. G. Colson, P. T. Lansbury, and F. D. Saeva, *J. Amer. Chem. Soc., 89*, 4987 (1967).
85. E. L. Eliel and F. Nader, *J. Amer. Chem. Soc., 91*, 536 (1969).
86. J. Lugtenburg and E. Havinga, *Tetrahedron Lett., 1969*, 1505.
87. R. H. Martin and J. C. Nouls, *Tetrahedron Lett., 1968*, 2727.
88. J. L. C. Wright, A. G. McInnes, D. G. Smith, and L. C. Vining, *Can. J. Chem., 48*, 2702 (1970).
89. N. S. Bhacca, L. J. Luskus, and K. N. Houk, *Chem. Commun., 1971*, 109.
90. Y. Nakadaira and H. Sakurai, *Tetrahedron Lett., 1971*, 1183.
91. A. H. Lewin, M. Frucht, and F. A. Bovey, *Tetrahedron Lett., 1970*, 1083, 3707.
92. K. F. Kuhlmann and D. M. Grant, *J. Amer. Chem. Soc., 90*,

7355 (1968).
93. A. J. Jones, D. M. Grant, and K. F. Kuhlman, *J. Amer. Chem. Soc.*, *91*, 5013 (1969).
94. K. F. Kuhlmann, D. M. Grant, and R. K. Harris, *J. Chem. Phys.*, *52*, 3439 (1970).
95. A. Allerhand, Lecture presented at *Symposium on Nuclear Magnetic Resonance*, University of Waterloo, Ontario, Canada, June 7-11, 1971.
96. R. L. Lichter and J. D. Roberts, *J. Amer. Chem. Soc.*, *93*, 3200 (1971).
97. I. Solomon and N. Bloembergen, *J. Chem. Phys.*, *25*, 261 (1956).
98. R. A. Bell and J. K. Saunders, *Chem. Commun.*, *1970*, 1078.
99. D. K. Green and J. G. Powles, *Proc. Phys. Soc.*, *85*, 87 (1965).

STEREOCHEMICAL ASPECTS OF THE SYNTHESIS OF 1,2-EPOXIDES

GIANCARLO BERTI

Istituto di Chimica Organica,
Università di Pisa, Pisa, Italy

I. INTRODUCTION

1,2-Epoxides (oxiranes) are among the most useful intermediates in organic synthesis. The ease with which they can be prepared, their versatile reactivity, and the interest they have acquired in recent years in biosynthesis have fully justified the attention which organic chemists have been devoting to them in the more than hundred years that have elapsed since the discovery of ethylene oxide by Würtz in 1859 (1). The volume of literature dealing with compounds belonging to this class has been increasing year by year, and although several extensive and excellent reviews on the topic are available (2-5), none of them covers the work done in the last eight years, nor deals specifically with stereochemistry, with the exception of a recent review on stereoselective epoxide cleavage reactions (6).

This chapter deals with the stereochemical aspects of the synthesis of epoxides. Only in recent years, with the advancement achieved by physical methods, has it become possible and usually easy to analyze completely mixtures of diastereoisomers. Much of the older work of pre-glpc and pre-nmr days, though often entirely valid from a preparative point of view, was deficient as to the quantitative stereochemical characterization. The largest part of the work that is discussed in this article was carried out during the 1960s, though earlier results are included when relevant to the discussion. This review does not attempt to be a complete survey of stereochemical results in epoxide synthesis. Well over a thousand papers dealing with epoxides have been published since 1960, and most of them contain some stereochemical implications. Our purpose is rather to select material that has particular significance for the topics under discussion, or practical interest for organic synthesis. The literature has been surveyed up to the beginning of

1971. Space limitations do not allow treatment of the very interesting problems connected with the steric course of epoxide reactions, which are in part discussed in the review by Buchanan and Sable (6).

II. STERIC COURSE OF THE SYNTHESIS OF 1,2-EPOXIDES

The oxidation of alkenes with peroxyacids and the dehydrohalogenation of halohydrins are still by far the most widely used methods for the laboratory synthesis of epoxides, but some recent techniques, such as the reactions of ketones and aldehydes with sulfur ylides, have considerably enlarged the preparative possibilities in this field. A good knowledge of the steric course of the formation of epoxides by the different available methods is a prerequisite for the complete exploitation of the synthetic applications of these compounds.

A. Oxidation of Alkenes with Peroxyacids*

1. *Mechanism*

The reaction of olefins with peroxyacids has been extensively surveyed by Swern (2), Rosowsky (4), and Malinovskii (5); the topic has also been briefly reviewed recently by Fahey (7) and, as applied to steroids, by Kirk and Hartshorn (8).

Epoxidation with peroxyacids is usually considered as an electrophilic addition, since it is facilitated by electron donation to the double bond (9,10) and electron withdrawal from the peroxyacid CO_3H group (11). It differs, however, from most other reactions of this class, since the electrophilic attack is not followed by a reaction with an external nucleophile: the peroxidic oxygen acts both as the electrophile and as the nucleophile. The reaction could be assumed to proceed first by attack of OH^+ (not necessarily as a free ion) to give a hydroxy carbonium ion *1*, followed by nucleophilic ring closure to the oxiranium ion *2* and deprotonation. The main objection to such a stepwise mechanism is the fact that, in contrast to most other electrophilic additions, epoxidation is absolutely *syn*-stereospecific, even in cases where electron-donating substituents could stabilize the carbonium ion *1*. Thus, for instance, oleic

*The following abbreviations will be used to designate the most commonly used peroxyacids: PAA = peroxyacetic acid, PBA = peroxybenzoic acid, CPBA = *m*-chloroperoxybenzoic acid, NPBA = *p*-nitroperoxybenzoic acid, MPA = monoperoxyphthalic acid, PLA = peroxylauric acid, and PTFA = peroxytrifluoroacetic acid.

$$RCO_3H + \;>C{=}C< \;\longrightarrow\; >\overset{OH}{\underset{}{C}}-\overset{+}{C}< \;\longrightarrow\; >C\overset{\overset{+}{OH}}{\frown}C< \;\longrightarrow\; >C\overset{O}{\frown}C<$$

1 2

$$\downarrow RCO_2^-$$

$$>\overset{OH}{C}-\overset{OCOR}{C}<$$

3

and elaidic acid are converted, respectively, into exclusively *cis*- and *trans*-9,10-epoxystearic acid (12). Similarly, *cis*- and *trans*-stilbene give exclusively the *cis*- and *trans*-epoxides (11). There has not yet been reported a single instance, among the very large number of epoxidations of known steric course, of a *cis*-olefin giving some *threo*-epoxide, or *trans*-olefin giving *erythro*-epoxide. Although the isolation of glycol monoesters *3* in the reactions with peroxyacids derived from strong acids and in the presence of electron-donating groups on the alkene has sometimes been attributed to a direct attack of an anion on *1*, no direct evidence has so far supported this assumption. Kinetic data concerning the glycol monobenzoates obtained in the reaction of substituted stilbenes with PBA point to the formation of the epoxide, followed by a slower opening of the ring by benzoic acid (13). Also in favor of the latter hypothesis is the fact that the mixtures of formic esters *5* obtained in the reaction of 1-phenyl- and 1-methylcyclohexene (*4*) with peroxyformic acid have the same composition as the reaction products of the corresponding epoxides *6* with formic acid (14,15).

R; HCO_3H / HCO_2H →; R, OH, OCHO; ← HCO_2H; R, O

R = Me,Ph *5* *6*

4

The "butterfly" mechanism first proposed by Bartlett (16),

involving a cyclic transition state such as *7*, accounts well for the stereospecificity and is still largely accepted, even though a variation in which the formation of the two C-O bonds is not entirely simultaneous could explain the experimental results equally well, provided the rate of ring closure is at least two orders of magnitude greater than the rate of rotation around the

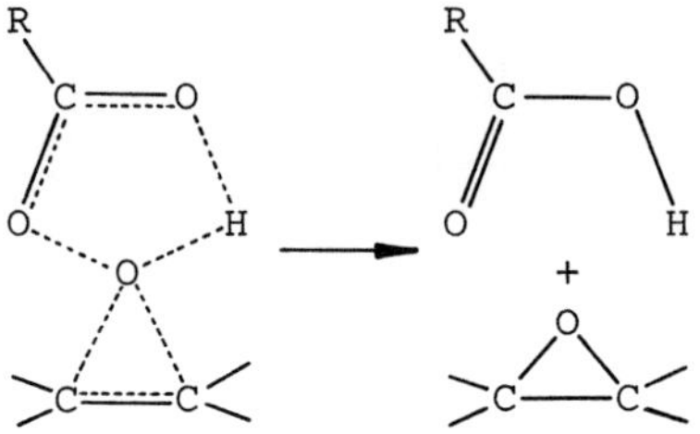

7 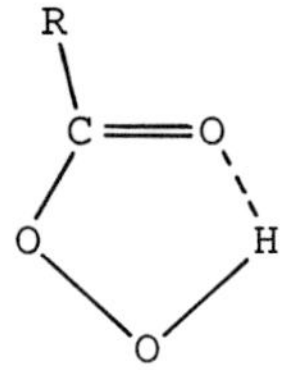 *8*

C-C bond of *1*. It may be pointed out that the oxiranium ion *2* is formally similar to a halonium ion, and that the opening of epoxides under acidic conditions, e.g., the reaction with hydrogen bromide, which most probably involves ions of type *2*, often has the same steric course as electrophilic additions to the corresponding alkenes, e.g., brominations (17). The same arguments that are used in favor of or against the intermediacy of halonium ions in halogen additions to olefins (18) can apply to that of oxiranium ions in epoxidations.

A transition state resembling *7* is also in agreement with the finding from infrared evidence that peroxyacids are present in solution as monomeric intramolecularly hydrogen-bonded forms *8* (19,20), and that basic solvents considerably depress the rate of epoxidation (21), probably because of interference with the intramolecular hydrogen bonding. This effect increases with increasing base strength of the solvent (22). On the other hand, in nonbonding solvents a rough parallelism between rate and dielectric constant can be observed, in accordance with the transition state being more polar than the reactants (23).

The question of whether epoxidation is subject to acid catalysis is still unsettled, since contradictory reports exist on this point. Although Lynch and Pausacker (11) and Campbell and co-workers (13) obtained clean second-order kinetics in epoxidations with PBA, indicating the absence of catalysis by the benzoic acid formed in the reaction, a definite increase in rate was observed when stilbenes and stilbenecarboxylic acids were epoxidized in the presence of about equimolar amounts of trichloroacetic acid (24,25). Possibly only strong acids can exert the rate-accelerating effect, but they must be present in sufficiently high amounts, since they are rapidly consumed by reaction with the formed epoxide. Very definite evidence for catalysis by strong acids during the epoxidation of allyl chloride with PAA in solvents of low basicity was also found by Sapunov and Lebedev (26). According to a proposal first made by Swern (2), the acid catalyst could facilitate the nucleophilic attack of the peroxyacid on the double bond by the protonation of one of the peroxidic oxygens, which would make the O-O bond cleavage easier, as shown in *9*. The enhancement in the rate of epoxidation by an allylic hydroxyl group, capable of hydrogen bonding with the peroxyacid (see Sect. II-A-9), could also involve intramolecular acid catalysis. The lack of

9

catalysis by trifluoroacetic acid in the epoxidation of ethyl crotonate (23) could be due to competition by a deactivating effect on the double bond due to protonation of the ester grouping. The matter clearly requires further investigation.

An alternative mechanism of epoxidation, recently proposed by Kwart (27), considers it as a 1,3-dipolar addition of the peroxyacid in the form *10* to the double bond, to give an adduct *11*, which decomposes to epoxide and acid. Although there are some similarities in the reactivity parameters and general

10 *11*

kinetic characteristics of epoxidation and 1,3-dipolar additions, in other instances such as agreement does not exist (28). Furthermore, compounds of type *11* have been prepared and found to be rather stable under the conditions of epoxidation. The mechanistic hypothesis has therefore been modified (29) by assuming that *11* is not actually formed as an intermediate, but that the reaction goes through a transition state *12*, resembling *11*. Although calculations by the Pariser-Parr SCF-MO

R—C ... O ... O ... O—H → epoxide + RCOOH

12

method do not reveal any relevant differences between the total energies of the intramolecularly hydrogen bonded and dipolar forms of peroxyacids (30), it can be noted that formulation *12* has so far not found much support, and that it might be quite difficult, if not impossible, to bring decisive proof in favor of or against either of the two mechanisms.

Although the complete *syn*-stereospecificity in the formation of epoxides from alkenes does not leave any doubt as to the relative configurations at the oxirane ring carbons once the configuration of the starting alkene is known, if the substrate contains one or more chiral centers, mixtures of diastereoisomers are usually formed. Their composition can depend on several factors of steric or polar nature, on the type of peroxyacid and of solvent used, etc. The next sections analyze these aspects.

2. *Acyclic Alkenes*

Very little is known about stereoselectivity in the epoxidation of alkenes with double bonds in noncyclic positions, and chiral atoms in the molecule. One would not expect much asymmetric induction, except in very special cases, such as that of allylic alcohols, which will be discussed in Sect. II-A-9. 3,4-Dimethyl-1-pentene (*13*) is converted by MPA into an equimolar mixture of the diastereoisomeric epoxides (31); the same is observed in the epoxidation of 24,25 double bonds in the side chain of triterpenoids *14* (32), and even the vinylbicyclo[2.2.1]-hexane derivative *15*, in which a rather crowded quaternary chiral center α to the vinyl group would be expected to favor stereoselectivity, gives with PBA a 55:45 mixture of the two epoxides

(33). It is therefore surprising that the MPA epoxidation of

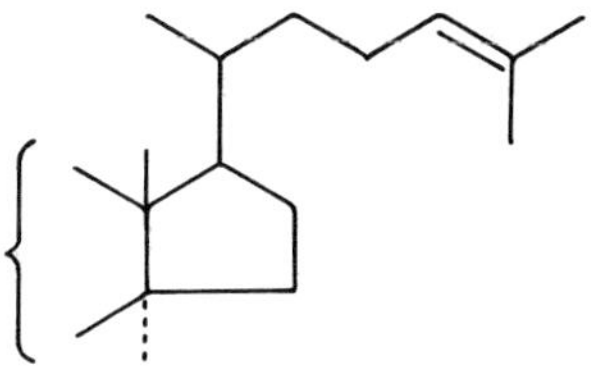

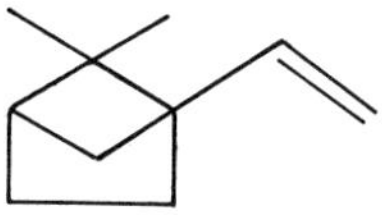

$Me_2CHCH(Me)CH{=}CH_2$

13 *14* *15*

trans- (*16*) and *cis*-chrisanthemic acid (*17*) is highly selective (34). Whereas *16* forms a single epoxide, *17* is resistant to peroxyacid attack under the same conditions. These results have been interpreted in terms of different shielding of the double bond by the carboxyl group in *16* and *17*, and the configuration of the epoxide should therefore be *18*, although this has not been directly proved.

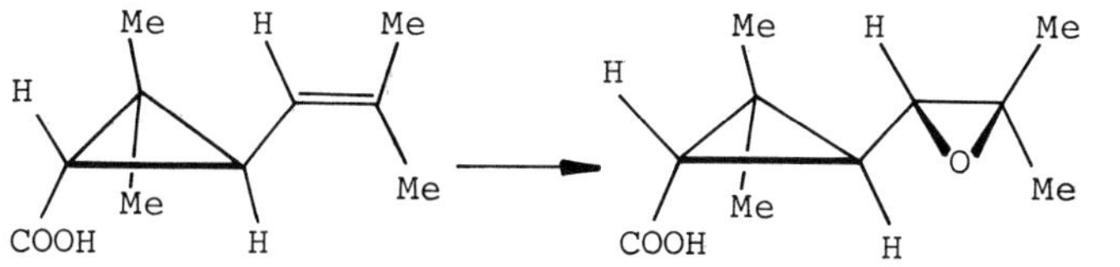

16 *18*

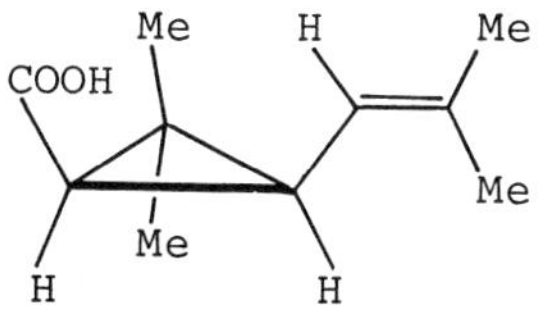

17

3. *Monocyclic Alkenes*

Relatively few reliable data are available on the steric course of the epoxidation of alkylmonocycloalkenes (Table 1). Alkyl substituents have a limited effect on the direction of attack if they are in an equatorial allylic or homoallylic

Table 1 Epoxidation of Alkylcycloalkenes

Olefin	% *cis*-Epoxide[a]	% *trans*-Epoxide[a]	Peroxyacid	Solvent	Refs.
19	46.4	53.6	CPBA	Et_2O	35
20	60.5	39.5	CPBA	Et_2O	35
21	10	90	CPBA	CH_2Cl_2	38
22	13	87	CPBA	Et_2O	35
23	50	50	PBA	$CHCl_3$	39
23	60	40	PBA	C_6H_6	40
24	50	50	PBA	$CHCl_3$	41-43
25	50	50	PBA	$CHCl_3$	39
25	40	60	PBA	$CHCl_3$	40
26	55	45	PBA	$CHCl_3$	44
26	60	40	PBA	$CHCl_3$	15
27	25	75	PBA	$CHCl_3$	45
27	17	83	PBA	$CHCl_3$	46
28	64	36	CPBA	AcOEt	47
29	27	73	PLA	$(CH_2)_5$	48
29	26	74	PLA	MeCN	48

[a]Cis and trans refer to the relation between the epoxide function and the larger of the alkyl groups not attached to the oxirane ring.

position, a more definite one if they are in an axial one. These effects have been explained as being largely or entirely of steric and conformational, rather than of electronic, nature, mainly on the basis of their insensitivity to changes in solvents (35,36).

Most of the treatments of steric effects in epoxidations are based on the reasonable assumption, first made by Rickborn and Lwo (35), that the conformation in the transition state must be very similar to that of the starting alkene, since there is only little change in geometry in passing from olefin to epoxide. It therefore should not constitute a violation of the Curtin-Hammett principle (37) to infer the steric course of the reaction from the ground-state conformational population. This assumption has found support from a rather accurate kinetic and (in part) steric analysis of the epoxidation of several substituted cyclohexenes (35), which has shown that an axial 4-methyl group retards peroxyacid attack syn to itself. In order to explain the slight preference for anti attack in 4-methylcyclohexene (*19*) the rate of epoxidation was dissected as shown in *30*, and it was assumed that $k_e(s) = k_e(a) = k_a(a) > k_a(s)$. The

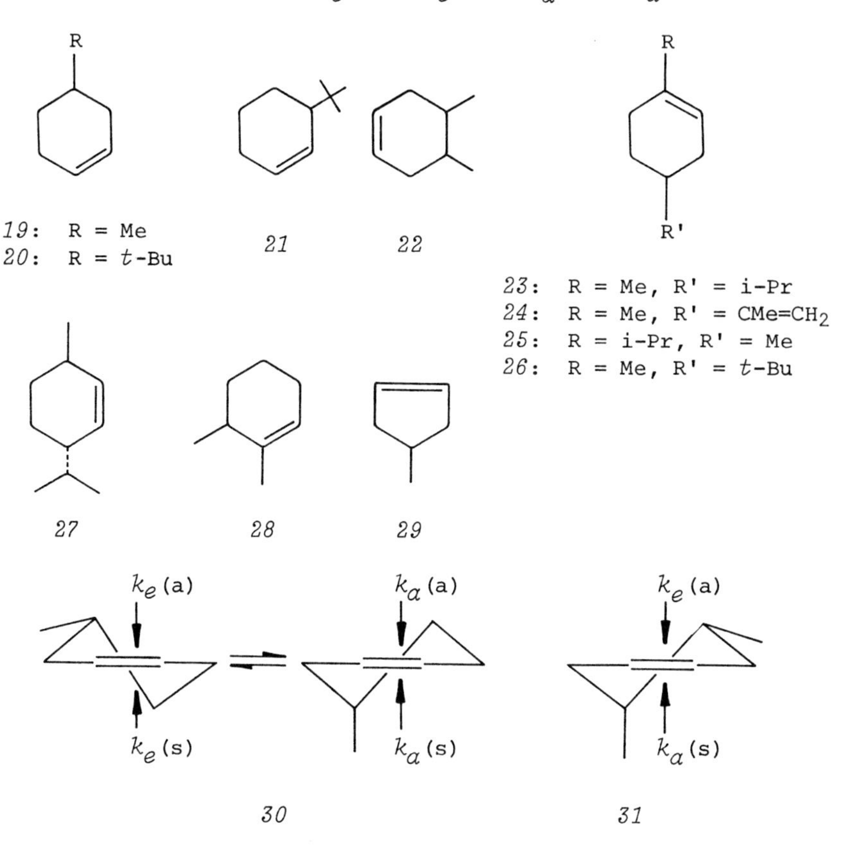

ratio of $k_e(a)$ to $k_a(s)$ was deduced from the epoxidation of *cis*-4,5-dimethylcyclohexene (*22*), in which one of the methyl groups has to be axial (*31*), and which gives an 87:13 ratio of *trans*- (*32*) to *cis*-epoxide (*33*). Although these assumptions must be taken as rather approximative because they neglect such factors as deformations in the cyclohexene conformation by substituents and torsional effects, they provide a reasonable value of 85±5% equatorial methyl conformational preference in 4-methylcyclohexene, on the basis of 53.6:46.4 trans-cis ratio obtained in the epoxidation of *19*.

The 50:50 product ratio obtained in the epoxidation of 1-menthene (*23*) (39) and limonene (*24*) (41-43) can be explained by the greater bulk of the isopropyl group which decreases the population of the axial conformer, but 3-menthene (*25*) would be

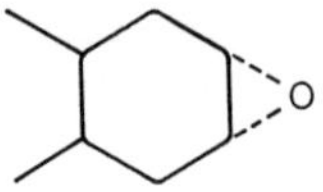

32 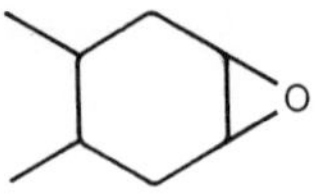 *33*

expected to behave as 4-methylcyclohexene does. It must, however, be stressed that the data reported for the latter compound (39,40) probably are not as accurate as the 50:50 value for *24*, which has been repeatedly confirmed on the basis of glpc and nmr measurements.

The above treatment fails completely in the case of 4-*t*-butylcyclohexene (*20*) and 1-methyl-4-*t*-butylcyclohexene (*26*), since it could in no way explain the excess of syn attack by the peroxyacid (15,35,44). A distortion from the chair form, caused by the presence of the bulky substituent, can best account for the results. Calculations by Altona and Sundaralingam (49) indicate that in the most stable conformation of *t*-butylcyclohexane the axial hydrogen α to the *t*-butyl group is bent toward the center of the ring by about 5°; this increases the shielding on the side of the ring trans to the substituent and could account at least in part for the formation of a slight excess of *cis*-epoxide from *20* and *26*.

In the case of 3-*t*-butylcyclohexene (*21*) a much stronger direct steric effect favors the *trans*-epoxide by about 9:1 (38). In any of the rotameric positions at least one of the methyls of the *t*-butyl group shields the cis but not the trans side (*34*). A similar situation holds for *trans*-2-menthene (*27*), where two alkyl groups are present on the allylic carbons, but attack by the peroxyacid takes place preferentially anti (*35*) to the bulkier isopropyl group (45,46).

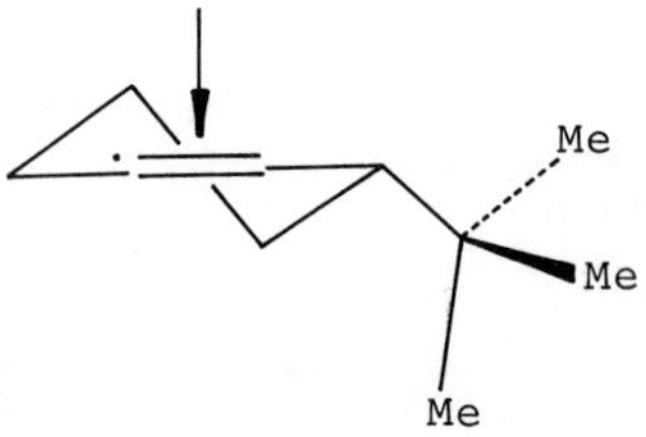

Me Me Me Me i-Pr

34 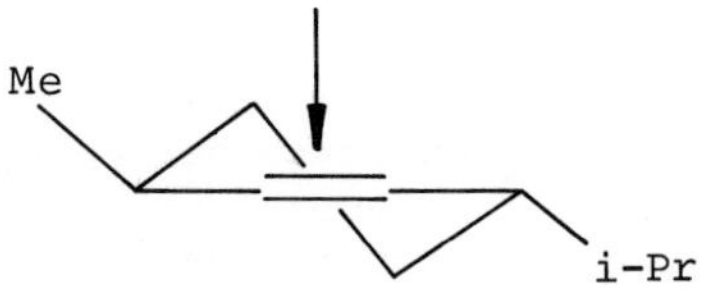 *35*

Also an alkoxy group in the allylic position can induce

preferential anti attack; this effect must be steric at least in part, since with the methoxy derivative *36* the ratio of *trans*- to *cis*-epoxide is 3:1, with the *t*-butoxy analogue (*37*) 9:1 (50).

36: R = Me
37: R = *t*-Bu

An apparent anomaly is found in the epoxidation of 1,6-dimethylcyclohexene (*28*), which gives an excess of *cis*-epoxide. Torsional stain can explain the prevalence of syn over anti attack: whereas the latter increases the eclipsing between the two methyl groups, the former decreases it, as shown in *38* and *39* (47).

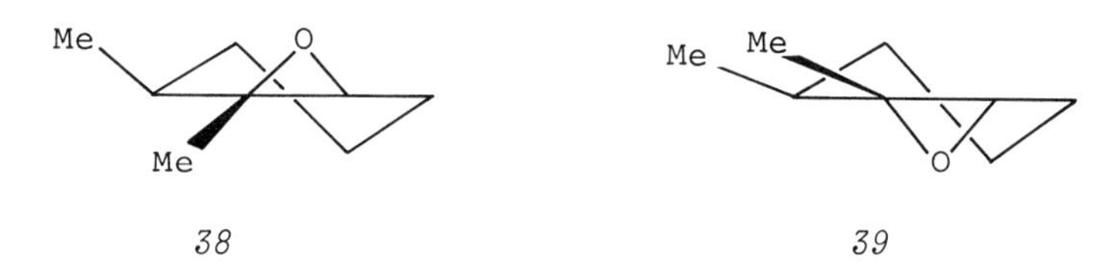

38 *39*

A purely steric effect can account for the 73:27 trans-cis ratio in the epoxidation products of 4-methylcyclopentene (*29*) with PLA, since this ratio is practically unaffected by a change in solvent from cyclopentane to acetonitrile (48). Such solvent independence is not found in a comparison of the rates of epoxidation of cyclopentene and 4,4-dimethylcyclopentene, which are in ratios ranging from 3.9 in cyclohexane to 5.7 in chloroform (51). Solvation of the CO_3H group by the polar halogenated solvents, increasing its effective bulk and depressing the rate of attack on the more hindered substrate, can explain the increase in the rate ratio.

4. *Bicyclic Alkenes*

In conformationally rigid polycyclic systems it is usually relatively easy to predict gross asymmetric induction effects in epoxidations on the basis of the shielding by substituents on each of the two sides of the double bond: for instance, attack by the peroxyacid will preferentially take place anti to a pseudoaxial allylic or axial homoallylic alkyl group. This is exemplified by the epoxidation of the octalin derivative *40* with NPBA, which gives exclusively the α-epoxide *41*, because the angular methyl group prevents approach of the peroxyacid from the β side (52).*

40 *41*

More subtle conformational or torsional effects are less easy to assess. Thus the reason for the fact that the hexahydroindene *42* and the octalin *43* yield with NPBA the epoxides *44* and *45*, in ratios of 60:40 (53) and 55:45 (54), respectively, is not obvious on the basis of undistorted half-chair models for the cyclohexene ring. Although these stereoselectivities are rather small and unimportant from a practical point of view,

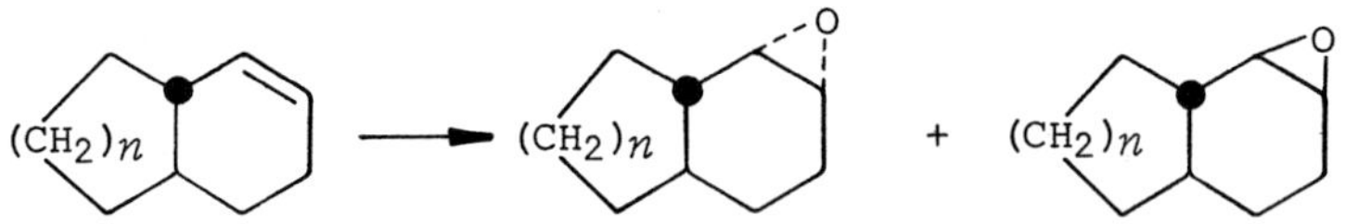

42: $n = 1$
43: $n = 2$

44 *45*

it is worthwhile to examine possible explanations in some detail, since in other cases they may influence the steric course of epoxidations in a more profound manner.

A first explanation can be sought in the fact that trans fusion of a second ring to the 3- and 4-positions of cyclohexene decreases the dihedral angle between the equatorial bonds involved in the ring formation (55), and consequently pushes the pseudoaxial and axial hydrogen atoms at the ring junctures a little toward the double bond. This can make attack by the reagent slightly more hindered syn than anti to the nearer ring

*To simplify the discussion, α and β will be used to indicate substitution below and above the "plane" of the ring system, written as shown in the formulas according to the steroid nomenclature, even for simpler bi- and polycyclic systems.

juncture. Some support of this interpretation is provided by

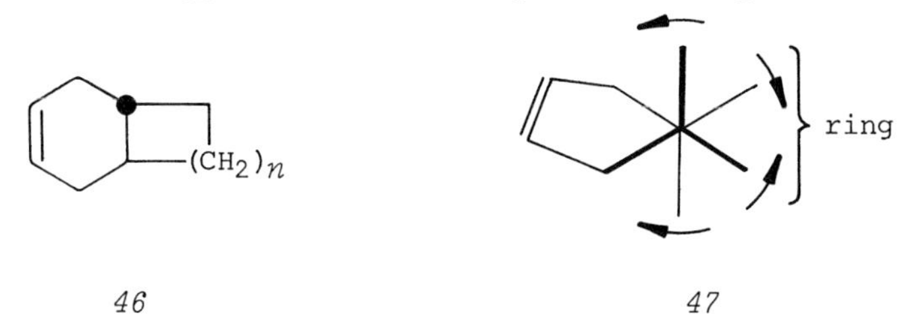

kinetic data on the more symmetric olefins *46*, which can produce only one epoxide; their rates of reaction with NPBA decrease from 1.84×10^{-3} for $n = 3$ to 0.81×10^{-3} for $n = 1$ (56), in accordance with the fact that the reduction in the dihedral angle and the consequent push of the axial hydrogens toward the double bond is greater, the smaller the fused ring, as shown in *47*.

A second, and probably more important, effect of a torsional nature is indicated in *48*, *49*, and *50*, which are Newman projections relative to the $C_{(1)}$-$C_{(9)}$ bond of *43*, *45*, and *44* ($n = 2$). The dihedral angle θ, which is about 35° in the olefin *43* (55), decreases to about 15° on epoxidation syn to $H_{(9)}$ (*49*), but increases to about 55° on anti epoxidation (*50*). Torsional strain therefore increases in the former and decreases in the latter case. This effect is in part counterbalanced by an

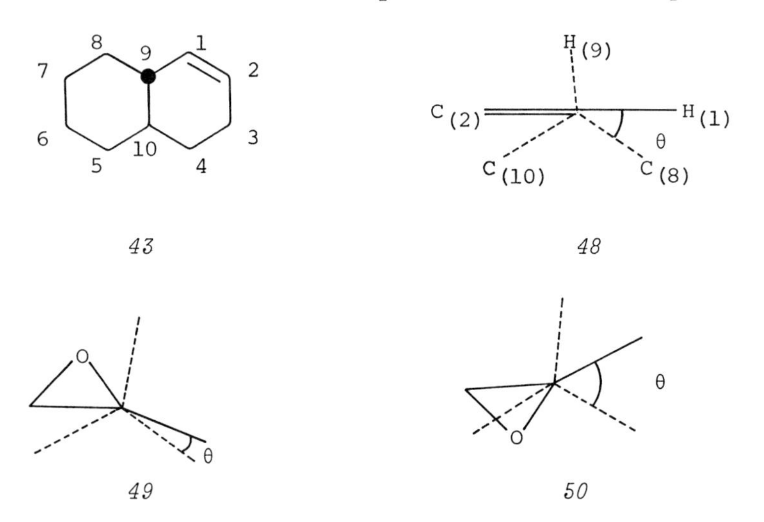

opposite trend in the torsional strain between the C-H bonds in 3 and 4, which should be less important, however, since it involves two C-H, instead of one C-H and one C-C bond, and also because the dihedral angle $H_{(2)},C_{(2)},C_{(3)},H_{(3e')}$ in the olefin is larger than θ (about 45°) (55), and should decrease and increase to 25° and 65°, respectively, in *50* and *49*. A precise evaluation of the relative importance of the steric and torsional effects discussed above would require accurate structural data, which are as yet not available.

In *cis*-fused bicyclic systems a preference is usually observed for epoxidation syn to the substituents at the ring junctures, which is understandable on the basis of the geometry of these systems. The second ring, being joined to the unsaturated ring through one of its axial bonds, normally provides a stronger hindrance to the approach of the peroxyacid than does the axial hydrogen or alkyl at the ring juncture; this is valid for both the half-chair and chair conformations of *cis*-Δ^1-octalin (*51*) and (*52*). For instance, compounds *53* (57), *54* (58), and *55* (59) have been reported to give exclusively the β-epoxides, in spite of the presence of one or more alkyl substituents on the β side. Also the glucoside *56* is preferentially attacked by CPBA on the β side of the cyclopentene ring (60), and the heterocyclic derivative *57* is epoxidized predominantly syn to the angular methyl group (61).

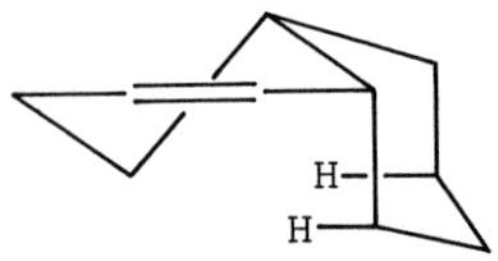

H— H— H— H— CO_2Me

51  *52* 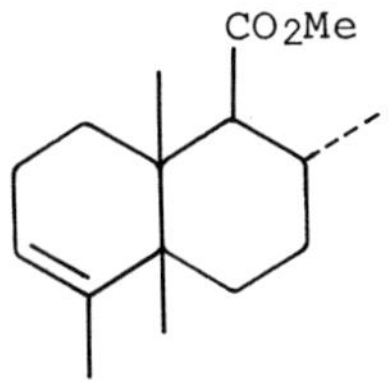 *53*

OH

CH_2OH O

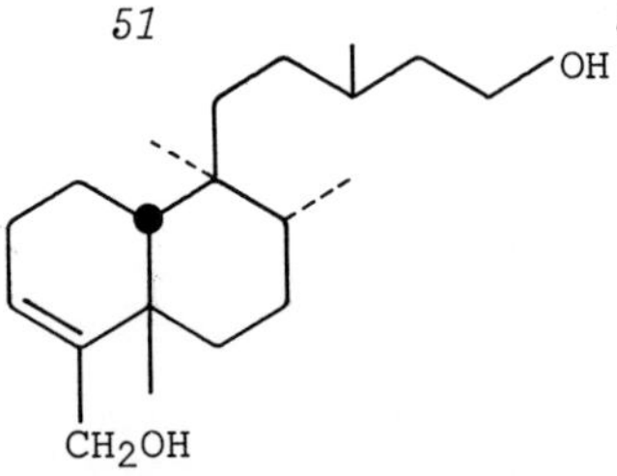

54 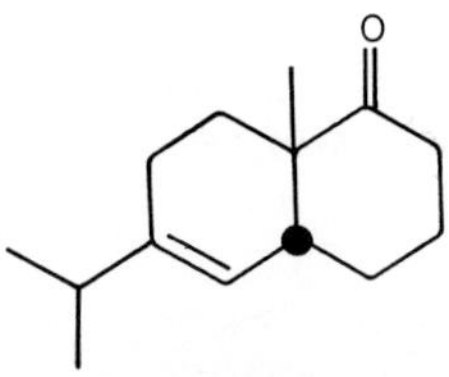 *55*

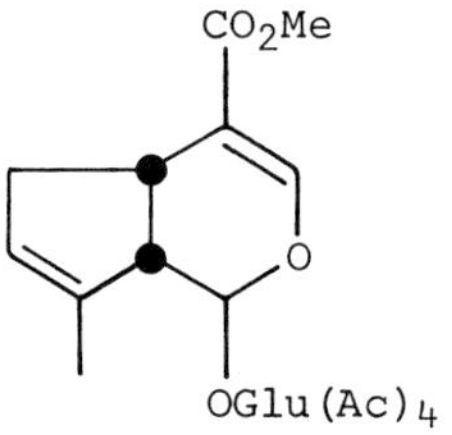

CO_2Me CH_2OH O N $OGlu(Ac)_4$

56 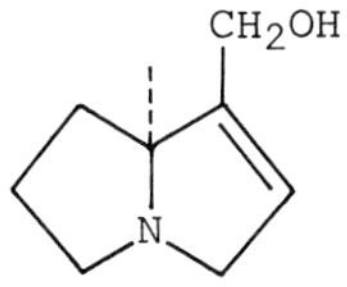 *57*

The limited stereoselectivity observed in the epoxidation by NPBA of the *cis*-hexahydroindene derivative *58* (58:42 ratio of β- to α-epoxide) (62) has been explained by the fact that, of the two boat conformations, *59* is more stable than *60*. With the *cis*-bicyclo[3.3.0]octene *61* the ratio of β- to α-epoxide is higher (87:13), since the molecule has less conformational liberty, and α attack is inhibited by its "U shape" (63).

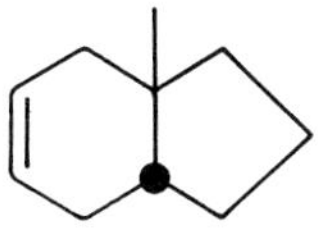

58 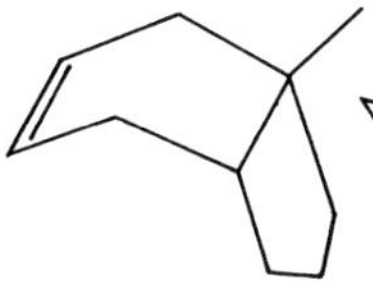 *59* 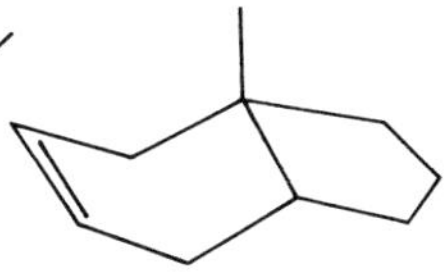 *60* 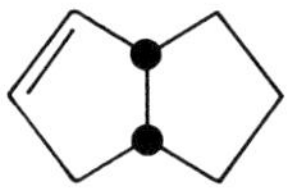 *61*

When the double bond is exocyclic to one of the rings (*62*) it is more difficult to predict the steric outcome of the reaction, because the unsaturated ring enjoys a certain conformational mobility, and epoxidation, involving one of the ring junctures, leads to two diastereoisomeric products of different stability, one with *cis*- (*63*) and the other with *trans*-fused (*64*) carbocyclic rings. If the reaction is at least partly influenced by product development control, one would expect, in the absence of other factors, the formation of an excess of *trans*-epoxide in the decalin series, and possibly of the cis isomer in smaller ring systems. However, the fact that the dodecahydrophenanthrene *65* gives with PBA the epoxides *66* and *67* in a ratio of 55:45 (64), which is an excess of the diastereoisomer having two *cis*-fused decalin systems, indicates that the type of ring junction in the product should not be particularly important in determining the steric course of the epoxidation of olefins of type *62*, at least in the decalin series.

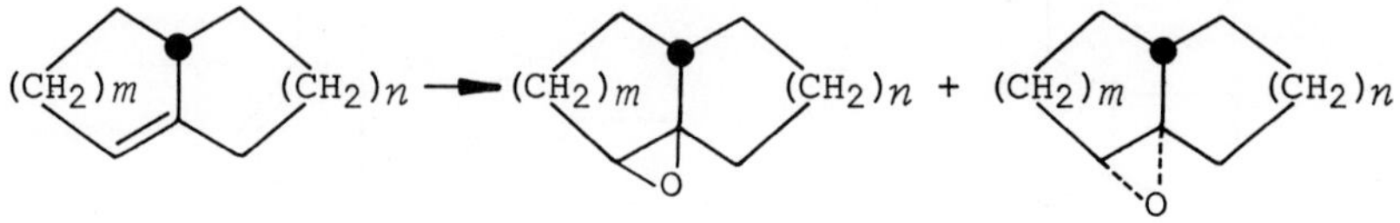

$(CH_2)m$ $(CH_2)n$ → $(CH_2)m$ $(CH_2)n$ + $(CH_2)m$ $(CH_2)n$

62 *63* *64*

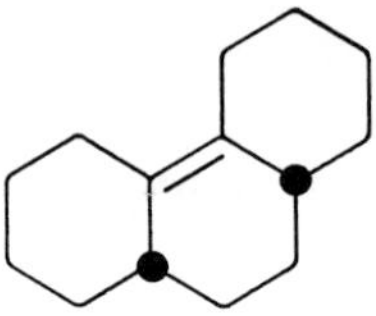

65 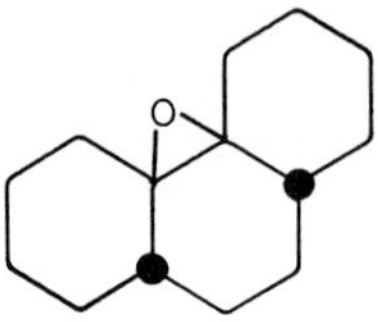 *66* 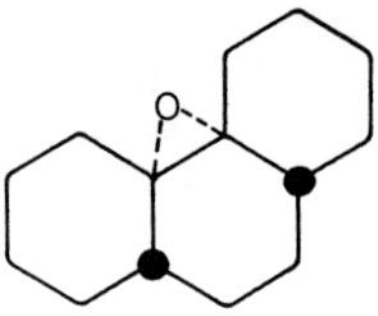 *67*

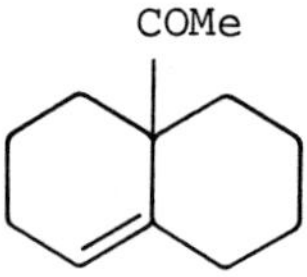

COMe

68 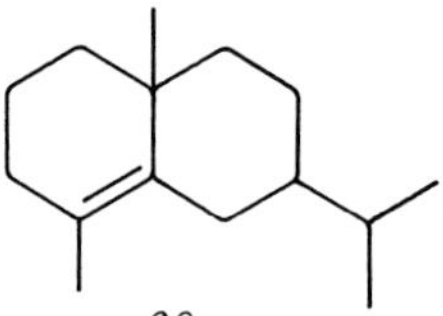 *69* 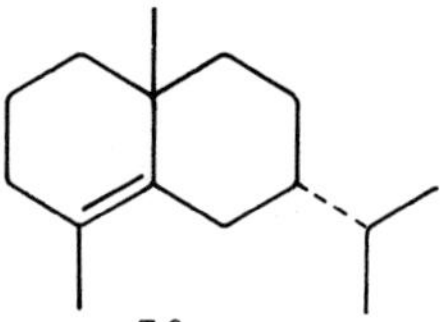 *70*

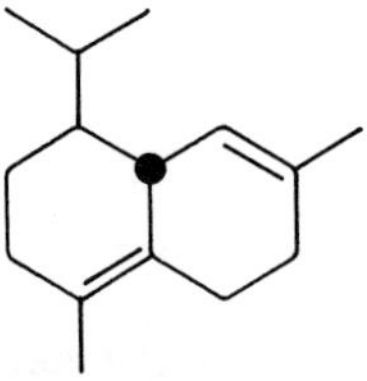

71 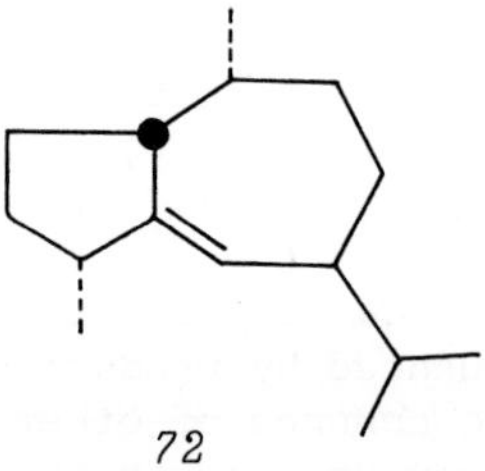 *72*

The octalin derivative *68* was reported to give only the α-oxide with PBA (65), and *69* about equal amounts of the two epoxides (66). Only the β-epoxide, however, was obtained from the isomer *70* (66,67), in which the isopropyl group is axial. The preferential epoxidation of δ-cadinene (*71*) on the β side of the tetrasubstituted double bond (68), and of γ-gurjunene (*72*) on the α side (69), further illustrate the unpredictability of the steric outcome of epoxidations of this type of compounds.

In the bicyclo[4.1.0]heptane series both 2- (*73*) and 3-carene (*74*) are epoxidized exclusively anti to the cyclopropane ring. For *73* this is easily predicted, since in both its half-chair conformations the cyclopropane ring exerts a considerable shielding to syn approach (70); even in the presence of a 4α substituent, as in *75* and *76*, the α-epoxide is the only product (71). The situation is less obvious in *74* since of its two boat conformations, *77* and *78*, only the latter presents a very strong hindrance to syn approach. However, nmr studies by Acharya (72) indicate a high conformational preference for *78* over *77*. This can justify the exclusive formation of the α-oxide (73,74) without the necessity of assuming cyclopropyl participation (75).

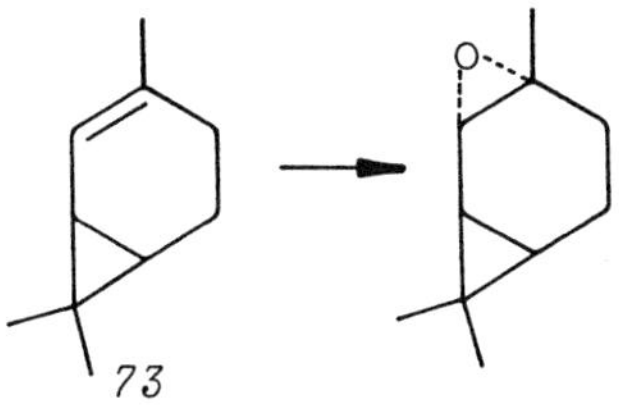

73 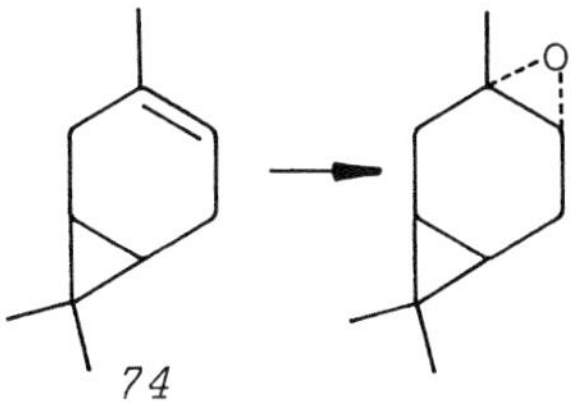 *74*

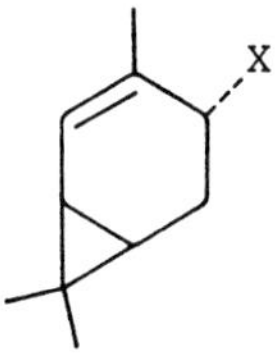

75: X = CH_2OAc
76: X = COMe

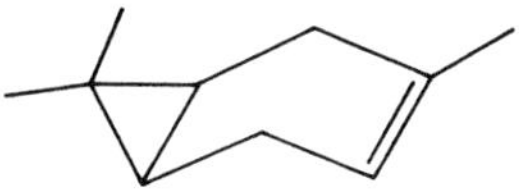

77 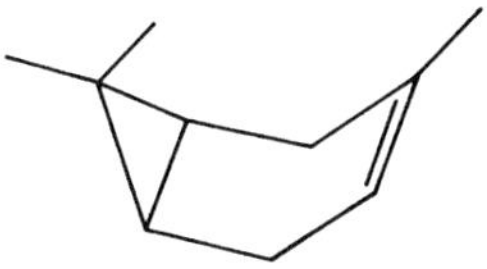 *78*

5. *Polycyclic Alkenes*

Among tricyclic compounds the formation with MPA of a large excess (ca. 9:1) of the β-oxide from *79* (76) and of the α-oxides from *80* and *81* (77) can clearly be attributed to the presence of the axial methyl groups, which shield from syn attack. This effect is particularly evident in the steroid series, where the methyl groups at $C_{(10)}$ and $C_{(13)}$, which are located on allylic or homoallylic carbons with respect to double bonds being in any of the possible endocyclic positions, decrease the rate of β attack with respect to that of α attack: "rule of α attack" (78). Although it is not possible within the limits of this chapter to survey the extremely abundant literature in this field, reference is made to the extensive, if not complete,

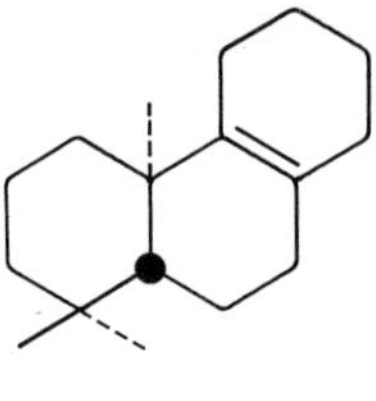

OMe OMe

79 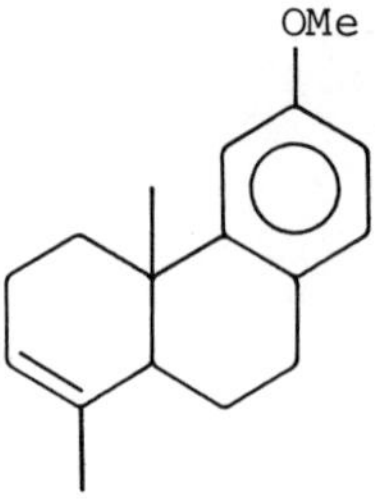 *80* 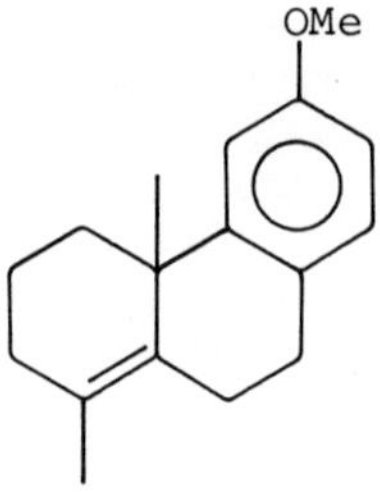 *81*

reviews by Swern (2,79) and Djerassi (80). We shall limit our discussion to cases that deviate from the rule of α attack.

That shielding by the angular methyl groups is not always the sole cause of the preference for α epoxidation is illustrated by the fact that, although cholesteryl acetate (*82*) produces the α and β oxides in a ratio of about 7:3 (81), the 19-nor steroid *83*, which, because of the absence of one of the angular methyl groups, would be expected to give rise to a higher percentage of β-oxide, actually is epoxidized more selectively from the α side (90% α-epoxide) (82). In such a complicated system as that present in steroids, other factors that are rather difficult to evaluate accurately and completely, such as torsional effects in the transition state and van der Waals interactions in the final products, may play a major role. From the results obtained with *82* and *83*, one may conclude that α attack is favored in the Δ^5 derivatives independently of the presence of the substituent at $C_{(10)}$. A methyl group at $C_{(10)}$ diminishes the preference for α attack, even if it increases crowding on the β side. One possible reason for this can be sought in the 1,2-diplanar form (55) assumed by ring B in the β-epoxide (*84*), in which the hydrogen or alkyl at $C_{(10)}$ is much farther from $H_{(8)}$ than in the monoplanar ring B of the α-epoxide (*85*). Although this decrease in *syn*-diaxial nonbonding interaction is relatively unimportant in *83*, it is considerably greater in the normal steroid derivative *82*, and this may explain the formation of more β-epoxide in the latter case.

Axial substituents on the α side of the steroid system can reverse the preference for α attack. Thus, whereas 5-cholestene and several of its 3β substitution derivatives, like *82*, are converted by PLA in benzene into mixtures of epoxides containing 70-80% of the α isomer, 3α-methoxy-5-cholestene gives under the same conditions a large excess (8:2) of the β-epoxide (81). In this case, besides the steric effect of the axial α-methoxy group, a polar effect (see Sect. II-A-11) could also contribute to favoring β attack. An ethylenedioxy group in position 3 can

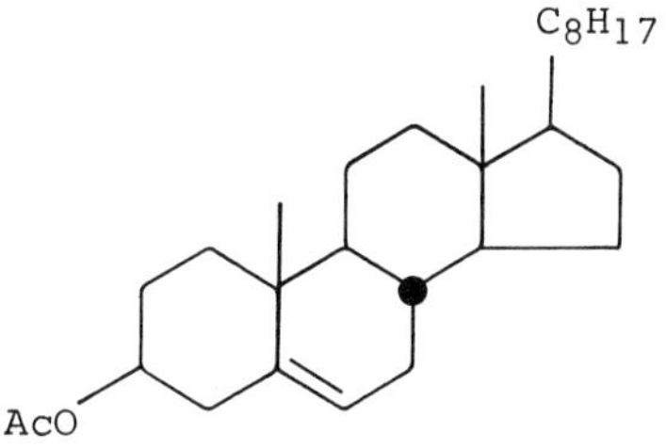

82 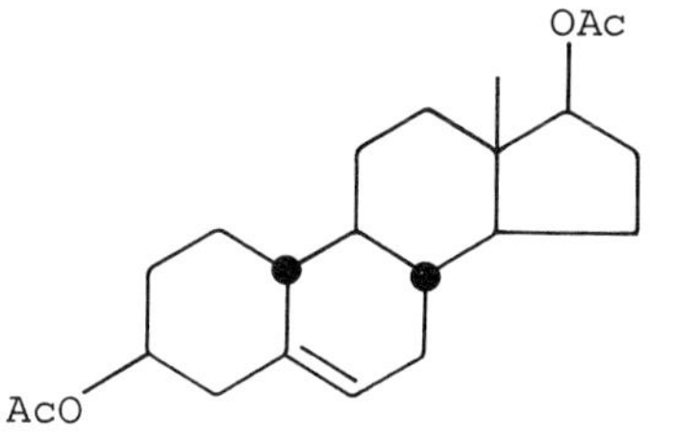 *83*

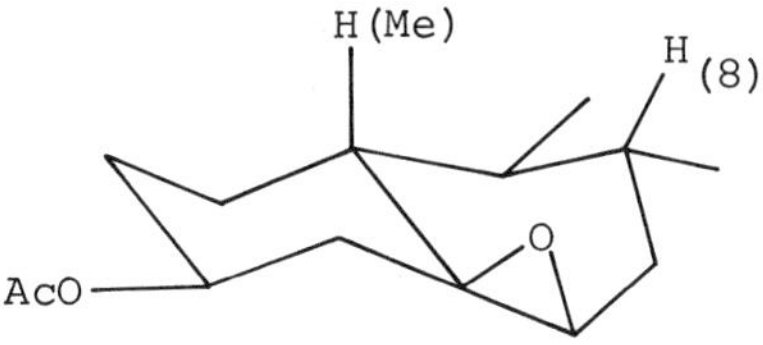

84 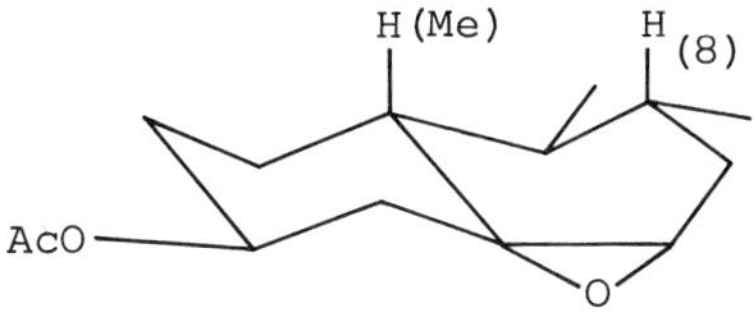 *85*

produce a similar effect, because of the presence of the axial 3α oxygen, but the situation is complicated by a dependence of the diastereoisomer ratio on other structural features: 3,3-ethylenedioxy-5-cholestene (*86*) gives the α- and β-epoxides in a ratio of 37:63 (81), and the corresponding 6-methyl derivative *87* in one of about 50:50 (83), whereas an excess of α-epoxides is reported in the epoxidation of several 3,3-ethylenedioxy derivatives of pregnane (84) and of 11β-hydroxy (85) and 11-oxo steroids (86-88). The diene *88* produces good yields of the 5,6β-epoxide and of the 5,6β,8,11α-diepoxide, respectively, with one and two equivalents of PBA (85). No obvious explanations can be provided for these differences, but it must be stressed that most of the data mentioned above refer to actually isolated compounds, since no complete product analyses were carried out.

Further exceptions to the rule of α attack are found in 5βH steroids containing a double bond in ring A. In the stable conformation of these compounds (*89* for 5β-cholest-2-ene), the 19-methyl group is equatorial with respect to ring A, and a more severe shielding to peroxyacid attack is caused by the axial 7αH and 9αH, which considerably reduces the percentage of α attack. Whereas *90* gives only α-oxide, the 5β epimer *91* and the spirostane derivative *92* are converted exclusively into the β-oxides (89,90).

In 5β,Δ^6-derivatives (*93*) the situation is different, since the 19-methyl group is axial with respect to ring B, and prevents

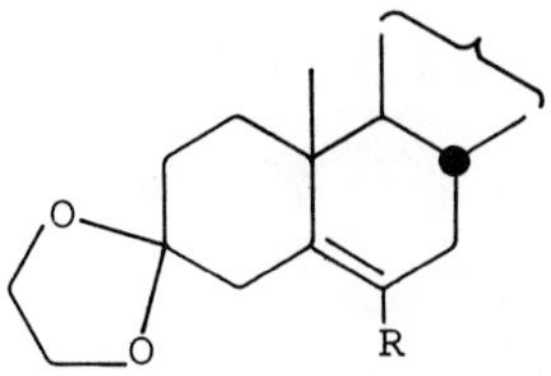

86: R = H
87: R = Me

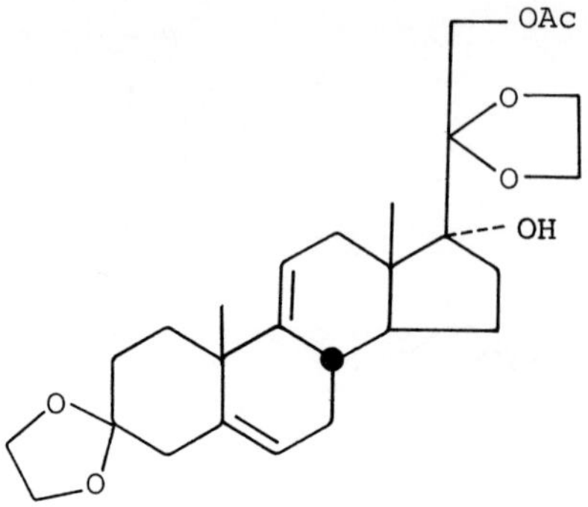

88

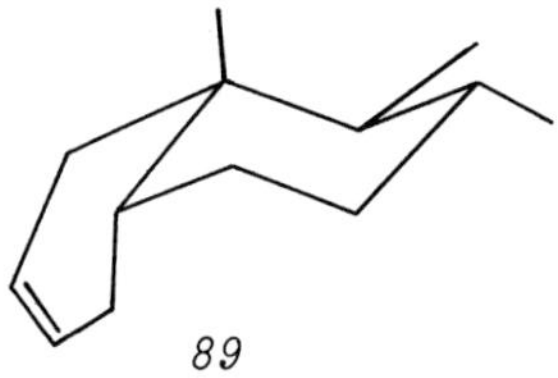

89

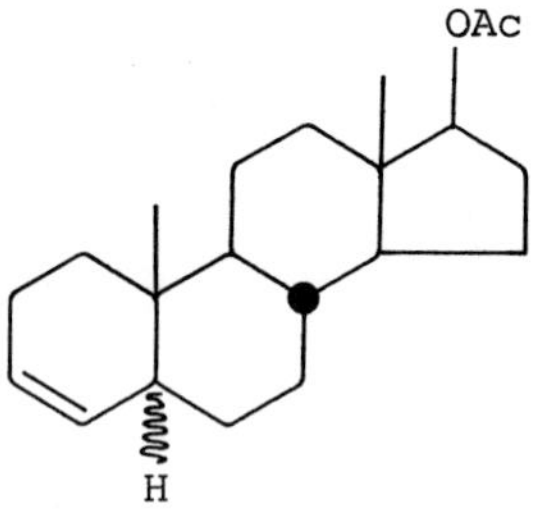

90: 5αH
91: 5βH

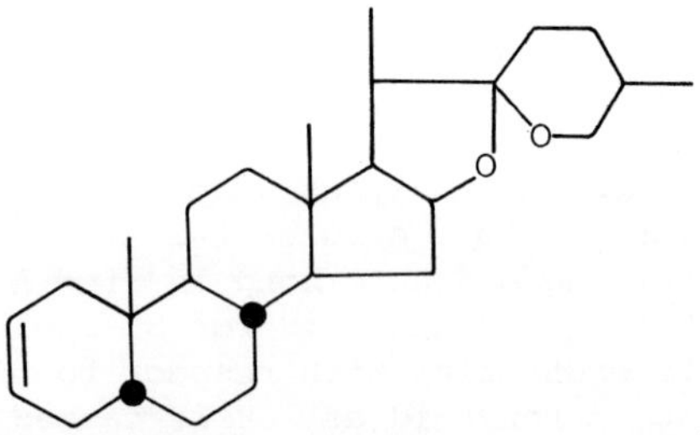

92

syn attack more than does the pseudoaxial methylene $C_{(4)}$. Therefore, the bile acid derivative *94* gives 80% α-epoxide; interestingly enough, however, osmium tetroxide, which should be much bulkier than PBA, produces a high yield of the β,β-diol (91). This fact, which emphasizes the shortcomings of reasoning based only on a rough assessment of primary steric effects, will be discussed in Section II-A-6.

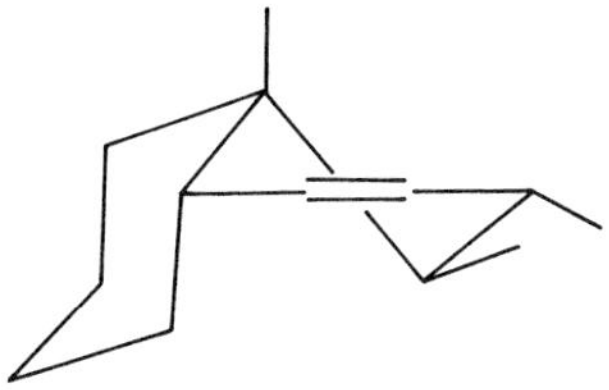

CO_2Me

93 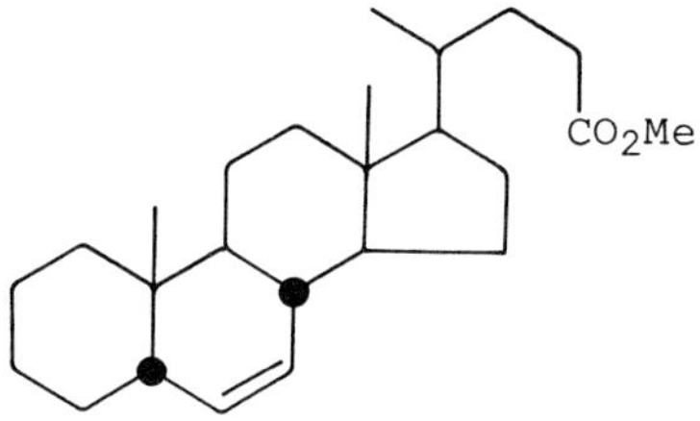 *94*

The type of C/D ring juncture can affect the steric course of epoxidations: the *trans*-joined enol acetate *95* gives the α-epoxide (92), but the *cis*-joined one *96* yields only the β-epoxide (93). The β epoxidation that is observed with Δ^{15}-17-

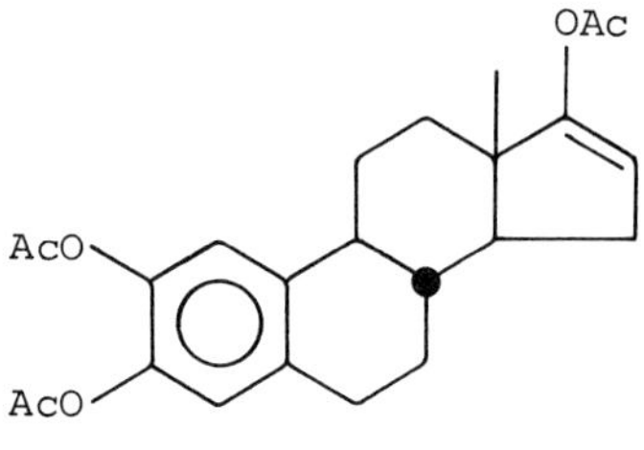

OAc AcO AcO OAc AcO

95 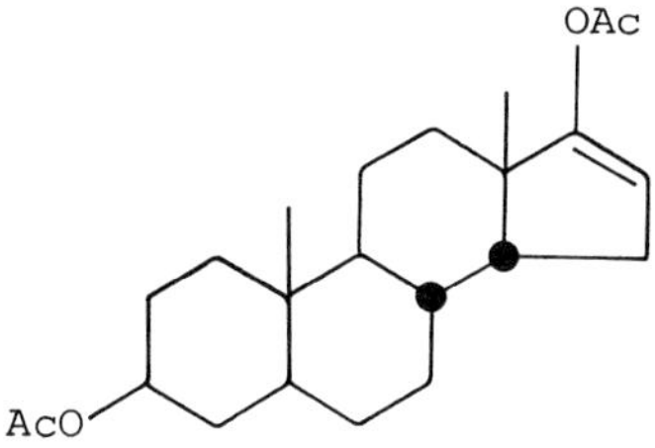 *96*

oxosteroids (94), such as *97*, can be attributed to the formation of the more stable C/D ring juncture, but if a 17β substituent is present (*98*) the α-epoxide is the main product (95,96).

The exclusive formation of the β-epoxide from anhydrohokagenin (*99*) (97), in contrast with the behavior of most Δ^4 steroids, is probably due to a preference for conformation *100* of ring A, in which two of the acetyl groups and the 19-methyl are pseudoequatorial or equatorial, and the β face is relatively unhindered.

4-Methyl and 4,4-dimethylsteroids and tetracyclic triterpenoids behave similarly to steroids in their preference for α attack. The same holds for rings A and B of most pentacyclic

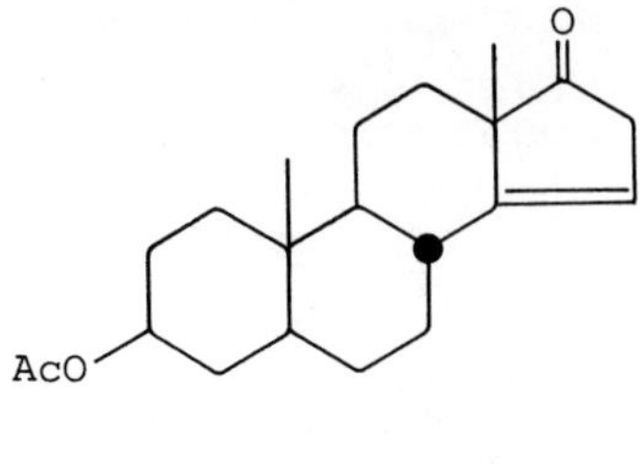

AcO O

97

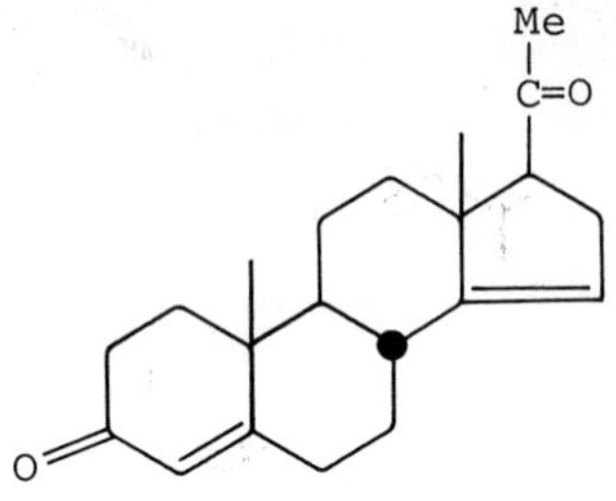

Me C=O O

98

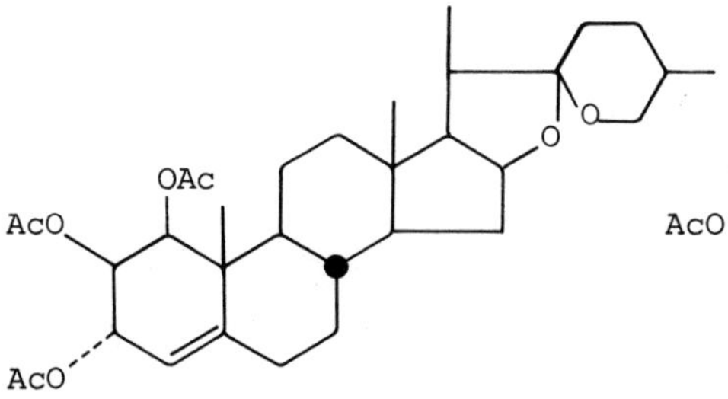

AcO OAc AcO O O

99

OAc AcO AcO

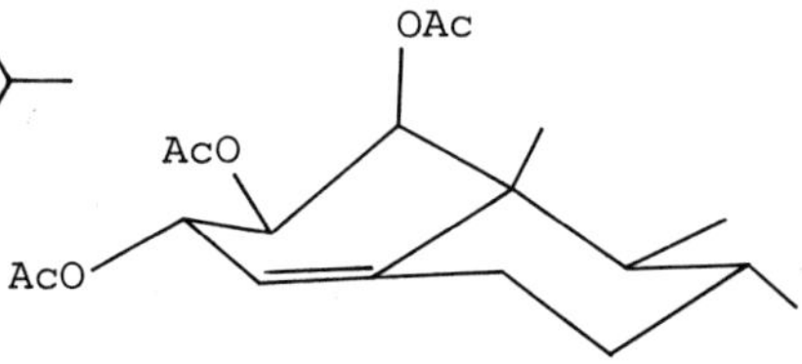

100

triterpenoids. Thus, for instance, 2-lupene (98), 4,4-dimetnyl-2-cholestene (99,100), and anhydroglycyrrhetic acid (101), all of which have structure *101* for ring A, give only α-epoxides, in agreement with the fact that the additional pseudoaxial 4β-methyl group increases the shielding of the β face. However, the latter group is apparently less efficient than the axial 19-methyl in preventing β approach, since 5α-cholest-2-ene (*102*) forms exclusively the α-epoxide (102), but 4,4-dimethyl-2-estrene (*103*) is converted into a 7:3 mixture of α- and β-epoxides by NPBA (103).

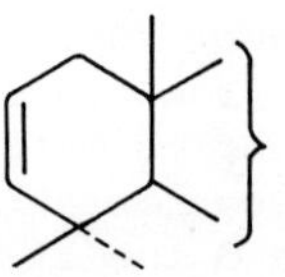

101 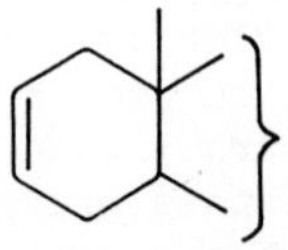 *102* 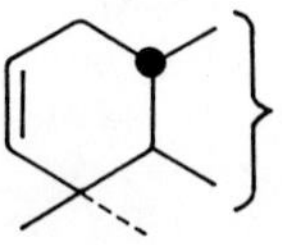 *103*

The presence of an additional methyl group at $C_{(14)}$ and differences in the configurations at $C_{(13)}$ and $C_{(17)}$ in the tetracyclic triterpenoids can also affect the steric course of epoxidation. Thus, whereas Δ^8 steroids usually form only the α-epoxide, the tirucallane derivative *104* produces a mixture of α- and β-epoxides (104).

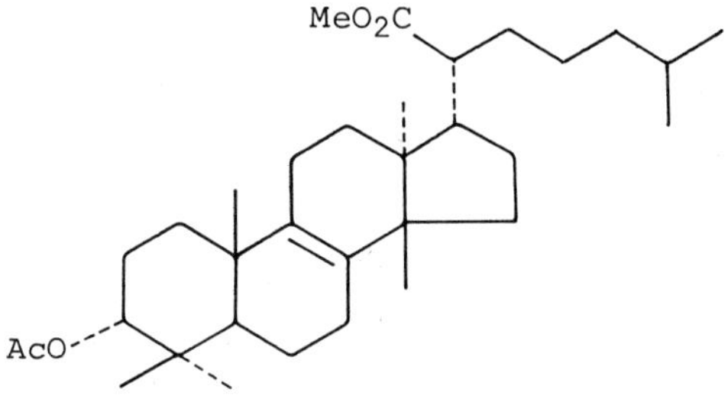

104

Another interesting example of difference in steric course between epoxidation and reaction with osmium tetroxide is given by the A-norsteroids *105-107* (105). Table 2 shows the high stereoselectivity of these reactions, epoxidation taking place anti to, osmate ester formation syn to, the hydrogen or substituent at $C_{(10)}$. The fact that there is little change in the

Table 2 Comparison between Epoxidation and Reaction with OsO_4

Olefin	α/β ratios	
	NPBA in pentane (epoxides)	OsO_4 (*cis*-diols)
105	97:3	10:90
106	82:18	5:95
107	9:91	99:1

α/β ratio in passing from the 10-methyl to the 10-nor compounds, and that the bulkier reagent is the one that preferentially attacks from the syn side, clearly indicates that the shielding effect of the methyl groups plays only a very secondary role in this case, and that the tentative explanation (105) based on a steric effect in the olefin in the case of epoxidation, and on the difference in stability of the diastereoisomeric osmic esters in the case of hydroxylation, is not entirely satisfactory.

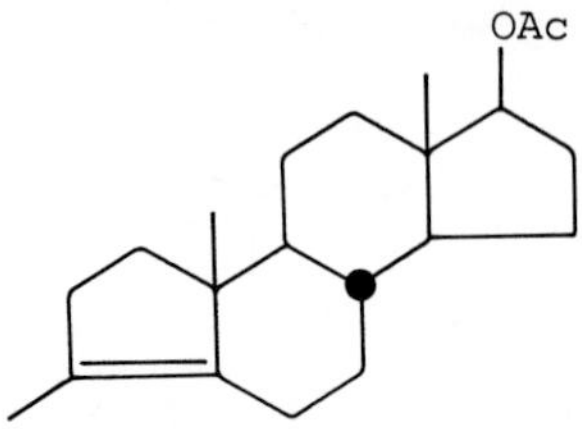

OAc

OAc

105

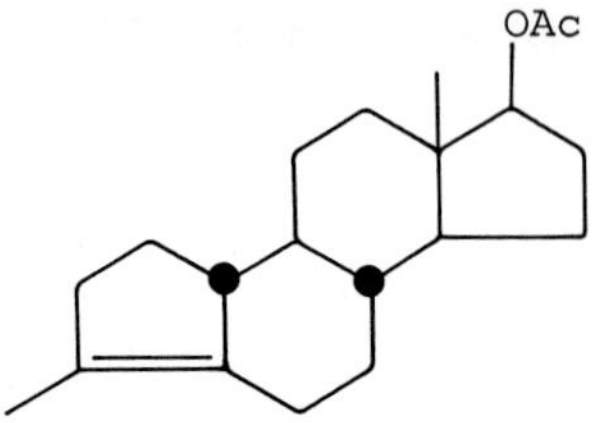

106

O

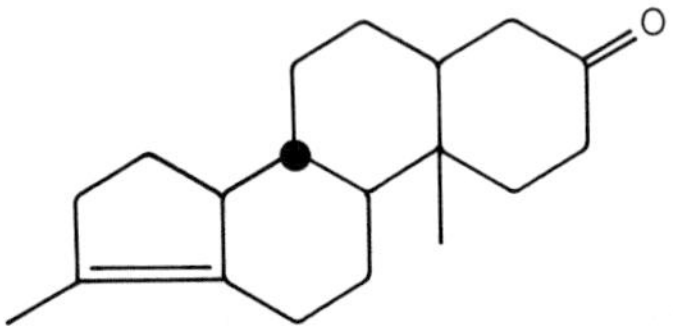

107

Compounds *105-107* can be considered as analogues of methylene-cyclohexane derivatives, and behave similarly to them, since they undergo "axial attack" by peroxyacid and "equatorial attack" by osmium tetroxide. Possible explanations for this behavior will be discussed in the next section.

6. *Exocyclic Alkenes*

Unhindered methylenecyclohexane derivatives are attacked by peroxyacids preferentially from the axial side; for instance, the olefins *108-110* give an excess of the epoxides derived by attack from the side indicated by the arrows. The mole fraction of the axial epoxides varies from 0.6 to 0.85, depending on the substrate, peroxyacid, and solvent (Table 3) (51,106). This behavior is analogous to the nucleophilic attack on the structurally rather similar unhindered cyclohexanones, which also involves a preference for axial attack by the nucleophile with formation of the equatorial alcohol (107). In both cases an increase in the steric crowding around the reaction site, or of the effective size of the reagent, can reverse the steric course. The "steric approach control"-"product development control" dichotomy first postulated by Dauben and co-workers (108) to

Table 3 Epoxidation of Exocyclic Alkenes

Compound	Peroxyacid	Solvent	% Axial Attack	Ref.
108	CPBA	CH_2Cl_2	69	106
	MPA	Et_2O	79	106
	PLA	$CHCl_3$	79	51
109	CPBA	CH_2Cl_2	59	106
	CPBA	MeOH	68	106
	MPA	Et_2O	69	106
110	CPBA	CH_2Cl_2	65	106
	MPA	Et_2O	76	106
114	CPBA	CH_2Cl_2	2	106
	MPA	Et_2O	17	106
115	PLA	$CHCl_3$	17	51
	PLA	Et_2O	35	51
116	PBA	$CHCl_3$	<25	77
117	CPBA	$CHCl_3$	10	116
118	PBA	$CHCl_3$	10	117
119	MPA	Et_2O	0	118
120	MPA	Et_2O	80	118
	CPBA	Et_2O	85	119
121	MPA	Et_2O	60	118
122	CPBA	Et_2O	63	119
123	CPBA	Et_2O	82	119
125	MPA	Et_2O	66	121

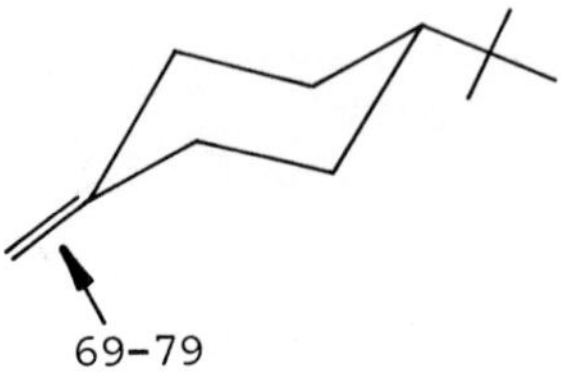

69-79

108

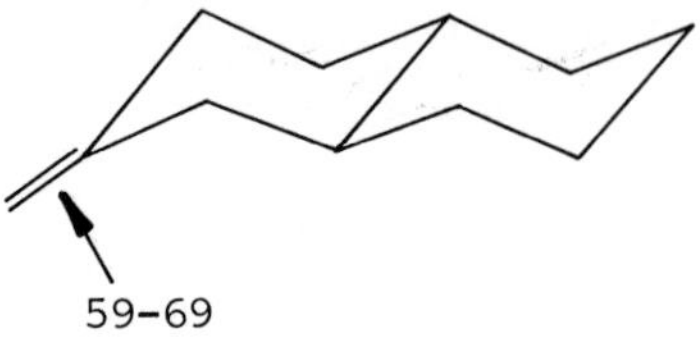

59-69

109

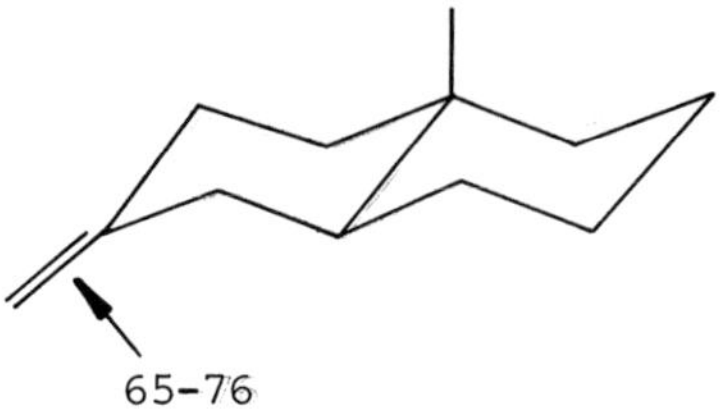

65-76

110

interpret the ketone reactions has been repeatedly criticized, and alternative explanations based on steric (109,110) and torsional effects (111,112) have been proposed to account for the axial attack in unhindered systems. The matter is far from being completely settled, however, since recent evidence has been provided for the fact that in nucleophilic additions to ketones the position of the transition state along the reaction coordinate depends upon the type of nucleophile and can be reactantlike or productlike (113). However, it is rather unlikely that the axial preference in epoxidations is due to product development control, i.e., to the higher stability of the epoxide with an equatorial CH_2 group. The $-\Delta G$ value for the equilibrium *111* $\rightleftharpoons$ *112* has been estimated as 0.15 (114) to 0.27 kcal/mole (115), and would not be high enough to account for the excess of axial epoxidation product.

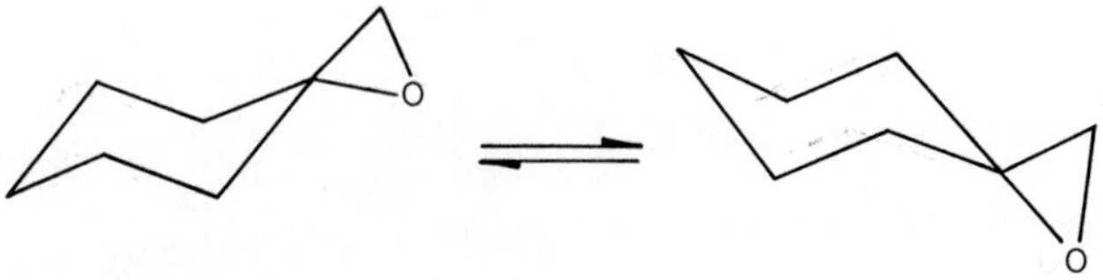

O O

111 *112*

Provided that the transition state is more reactantlike, there should not be much difference between the steric factors involved in the electrophilic attack by peroxyacid on the C=C double bond and in the nucleophilic attack on carbonyl. This would, however, not be true for product development control, since the final geometry is quite different in the two cases. As an alternative to product development control, the preference for axial attack on unhindered cyclohexanones has been explained as being due to a steric interaction of the reagent with the axial α hydrogens (109,110), which can outweigh the steric effects of the axial β ones, provided that the attacking species is very near to the carbonyl group in the transition state. A similar model can be valid for the epoxidation of methylenecyclohexanes; measurements along a line perpendicular to the plane of the double bond at the center show that the distances of less than 1.8 Å from the double bond the axial hydrogens on the α carbon are closer than the ones on the β carbon (*113*) (106). A more recent interpretation (111), which attributes the slower rate of equatorial attack on cyclohexanones to torsional

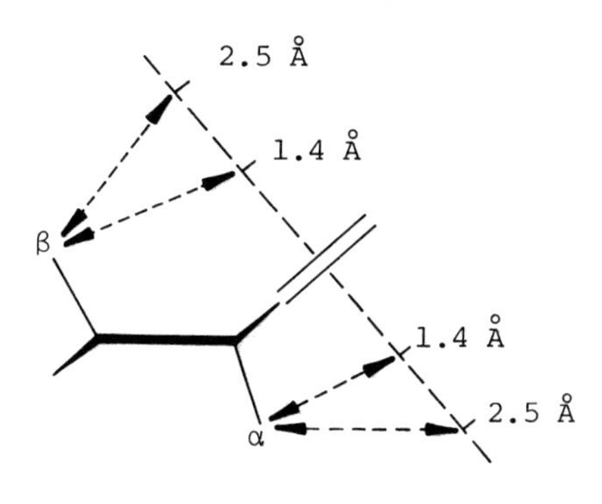

113

strain in the transition state, involving the partially formed bond and the axial α hydrogens, which almost eclipse each other, could also apply to epoxidations.

Axial substituents on the homoallylic carbon reverse the situation and induce preferential equatorial epoxidation, by increasing the steric screening of the axial side. This is exemplified by compounds *114* (106), *115* (107), *116* (77), *117* (116), *118* (117), and *119* (118). On the other hand, an axial methyl group on the γ carbon slightly increases the percentage of axial attack by shielding the equatorial side, as seen from a comparison between *109*, *110*, and *120* (118,119). Even an equatorial substituent on the allylic carbon seems to exert some influence, as shown by the decrease in the axial/equatorial ratio in going from *120* to *121* (118) and *122* (119); the axial 14α-hydrogen atom of *121* and the analogously placed hydrogens of the 2α-methyl group of *122* can be responsible for this slight increase in shielding of the axial side. Finally, an axial methyl group on the allylic carbon produces the expected increase in the axial/equatorial ratio; however, its influence is smaller than that

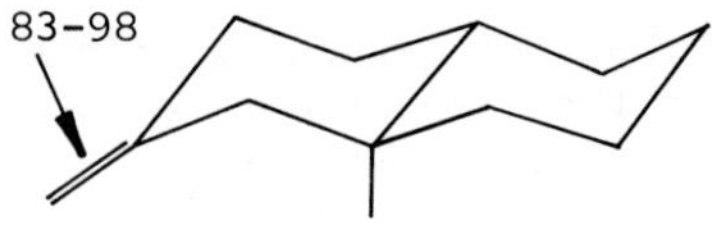

83-98

114

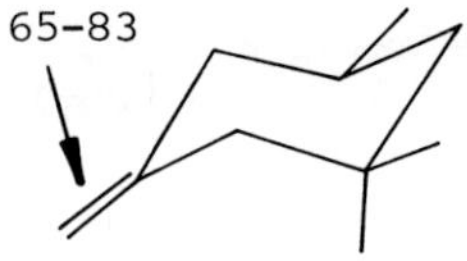

65-83

115

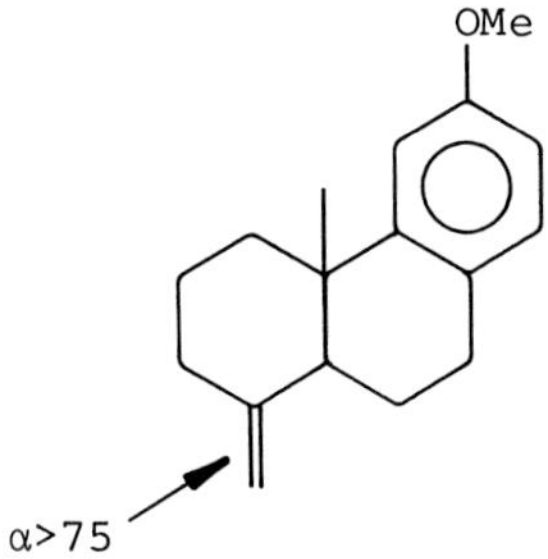

OMe

α>75

116

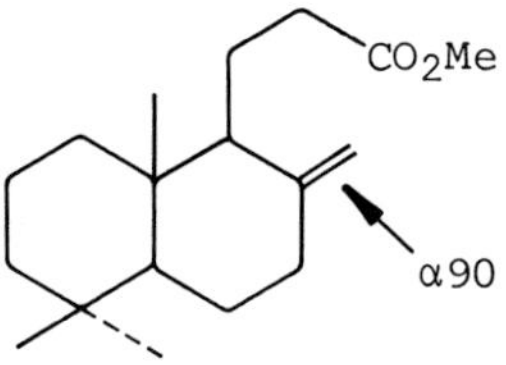

CO_2Me

α90

117

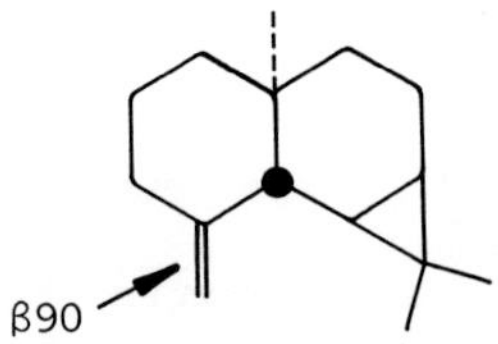

β90

118

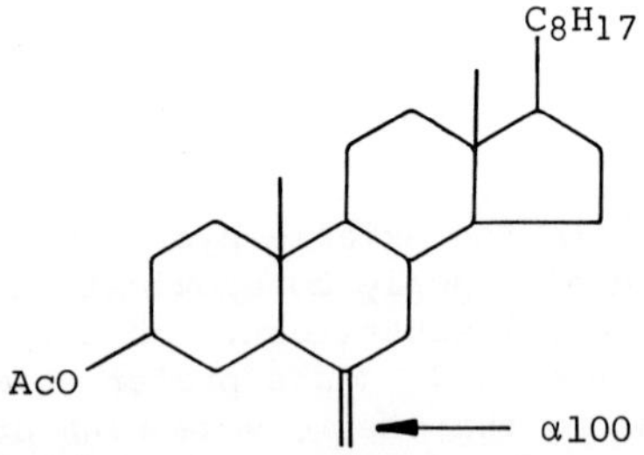

C_8H_{17}

AcO

α100

119

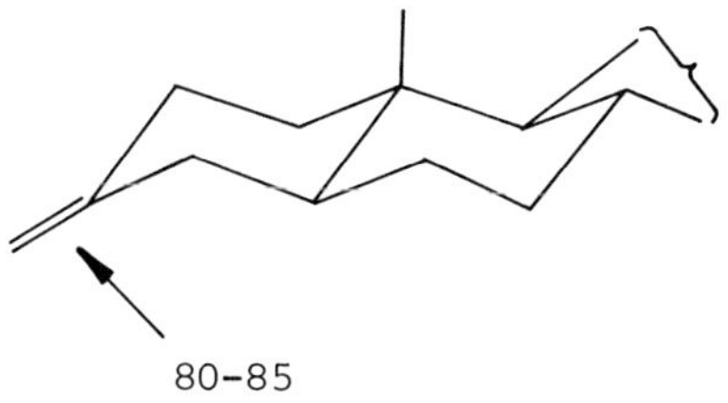

80-85

120

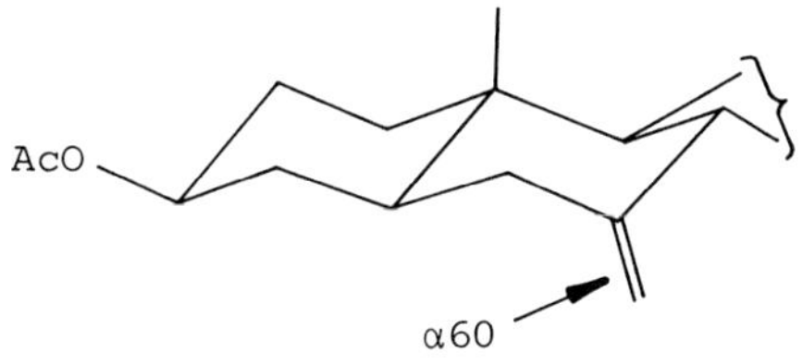

AcO

α60

121

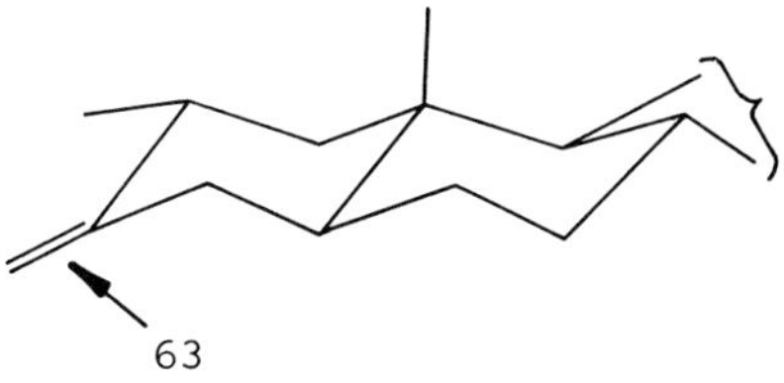

63

122

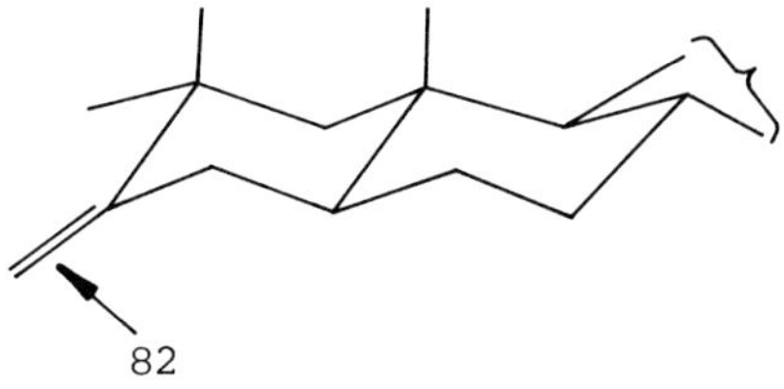

82

123

of an axial homoallylic methyl in decreasing the same ratio, as shown by a comparison between *122* and *123* on one side, and *109* and *114* on the other. A similar observation was also made for the reduction of cyclohexanones with lithium tri-*t*-butoxy-aluminohydride (120), and is in agreement with the fact that an increase in size causes a greater increase in shielding for an axial β substituent than for an α one.

Osmium tetroxide attacks exocyclic double bonds from the equatorial side even in the case of unhindered compounds, probably because the reagent has larger steric requirements and is farther away from the double bond in the transition state than the peroxyacid, so that interaction with the axial homoallylic hydrogens becomes more important than that with the axial allylic ones. Thus the reaction of *108* with osmium tetroxide gives the osmate ester corresponding to the diol *124* (36). This is true even in the presence of an axial methyl group on the allylic carbon, as shown by the fact that 12-methylenetigogenin (*125*), which gives a 2:1 mixture of α- and β-epoxide (prevalent axial attack), undergoes attack exclusively from the equatorial side with osmium tetroxide (121). The difference between the steric course of epoxidation and osmium tetroxide reactions has already been mentioned for compounds *94*, *105*, *106*, and *107*, and can be explained in similar terms.

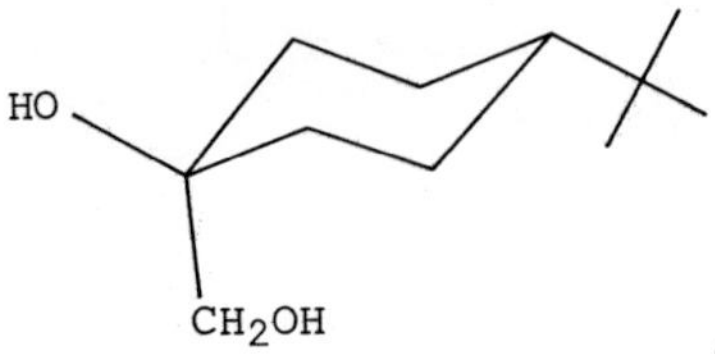

HO

CH_2OH

AcO

124 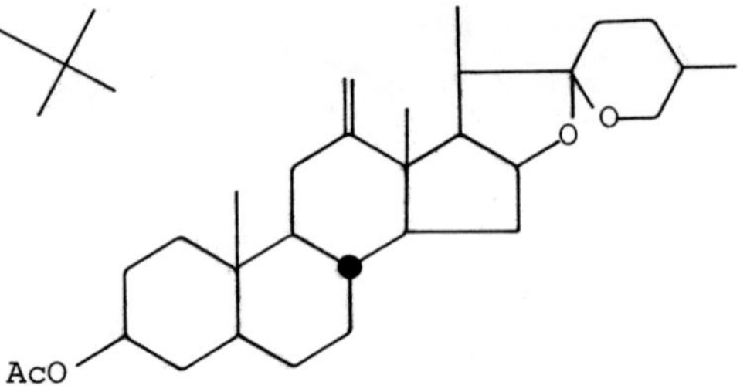 *125*

Other types of conformational effects can influence the steric course of epoxidations of exocyclic double bonds. For instance, in the case of the diastereoisomers *126* and *127*, reaction with CPBA gives an α/β ratio of 20:1 for *126* and 4:1 for *127* (122). In the former, interaction between the two phenyl groups forces the one that is attached to the ring into an axial position and causes a high preference for attack anti to it.

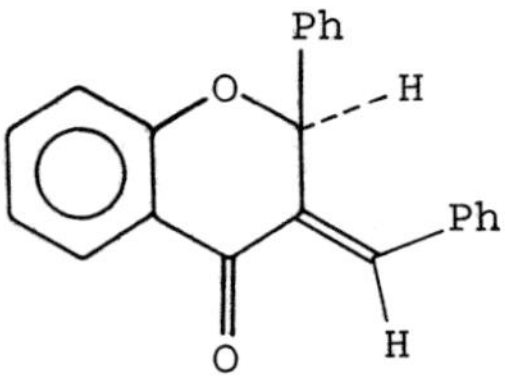

126 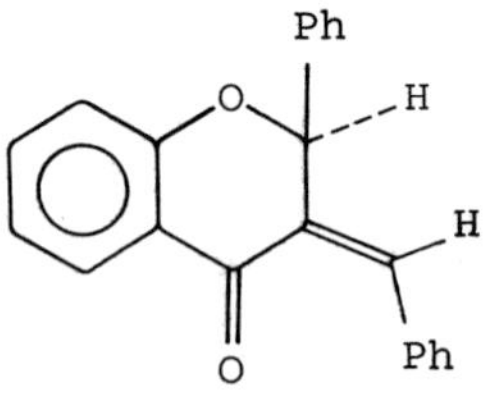 *127*

7. Bridged Cycloalkenes

The epoxidation of norbornene (*128*) has been repeatedly investigated, and found to produce mainly the *exo*-epoxide (*129*). Although the concomitant formation of some of the endo isomer (up to 6%) has been deduced from indirect evidence (123,124), a recent very accurate analysis of the CPBA epoxidation product of *128* revealed that *129* is formed in about 99% yield, with no evidence for the presence of *endo*-epoxide (125). The preference for exo attack, which is quite general in additions to *128*, is due in part to the shielding on the endo side by the endo hydrogens at $C_{(5)}$ and $C_{(6)}$, which can, however, hardly account for the high stereoselectivity that is observed. Schleyer (126) has proposed that torsional strain may be an important contributing factor: the quasi-eclipsing between the C-H bonds in positions 1 and 2 (and 3 and 4) of *128* is relieved in the transition state for exo attack, in which the dihedral angle between these bonds is increased, whereas the opposite holds for endo attack. Exclusive formation of the *exo*-oxides is also observed with compounds *130* (127) and *131* (128).

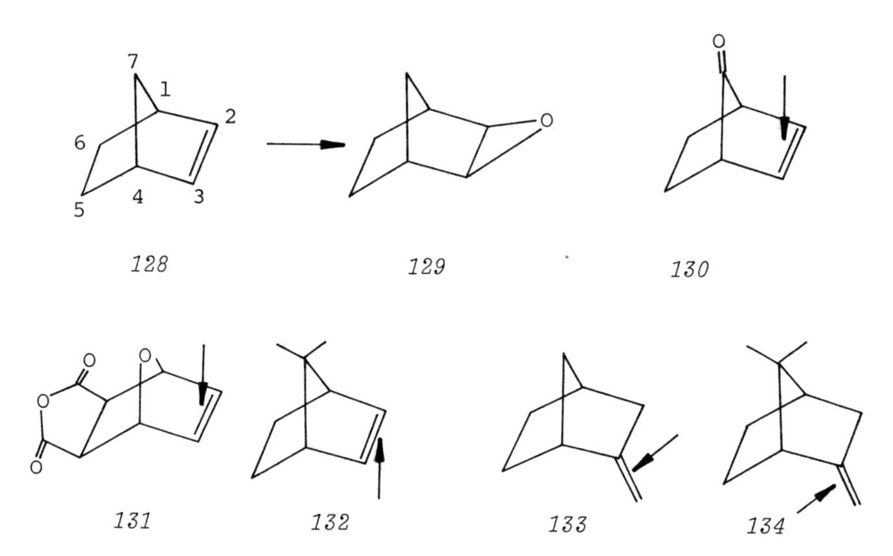

128 *129* *130*

131 *132* *133* *134*

With the 7,7-dimethylnorbornene (*132*) the rate of epoxidation with CPBA in methylene chloride decreases to about 1/100 of that of *128*, and a ratio of 90:10 of endo to exo isomer is obtained. The *syn*-7-methyl group provides such a strong shielding to exo attack that the unfavorable endo attack becomes predominant. A similar, even if considerably smaller, effect is observed in the exocyclic olefins *133* and *134*, which form, respectively, 86% *exo*- and 84% *endo*-oxide (125).

Exclusive formation of the *exo*-epoxides is also expected, and actually found in compound *135* (63) and in cedrene (*136*) (129). Although the epoxidation of α-pinene (*137*) occurs exclusively anti to the bridge (130), it takes place syn to it in *138* (131), which can be considered as an analogue of *137* with an exocyclic double bond.

Preference for exo attack is also observed in the azabicyclo[3.2.1]octane series; the trifluoroacetate of 2-tropene (*139*) is epoxidized by PTFA anti to the CH_2-CH_2 bridge. Peroxyacids derived from weaker acids, such as PBA, preferentially attack *139* on nitrogen, with formation of tbe *N*-oxide (132).

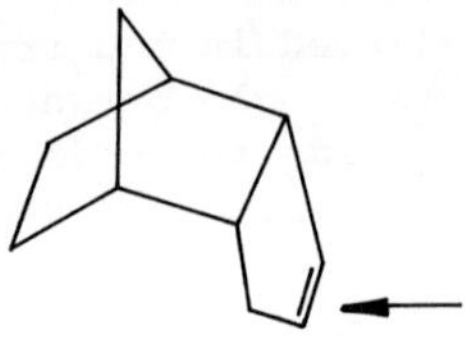

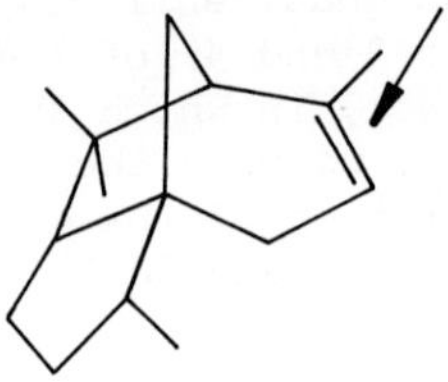

135 *136*

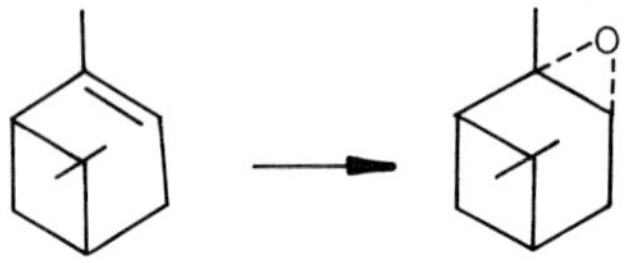

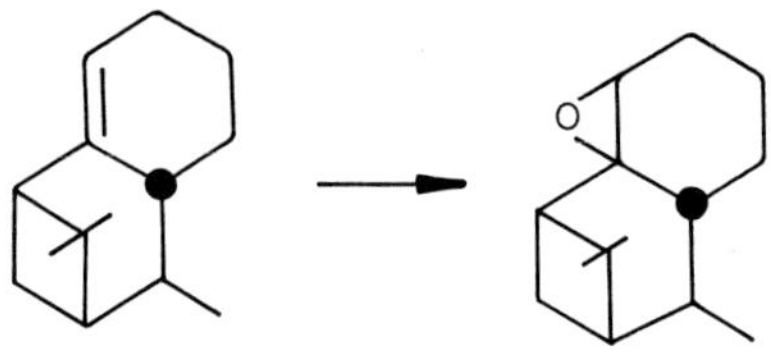

137 *138*

In the bicyclo[2.2.2]octane series the diene *140* is epoxidized by PBA with formation of an excess of *141* (133); the smaller effective bulk of the unsaturated bridge can account for this selectivity of attack syn to it.

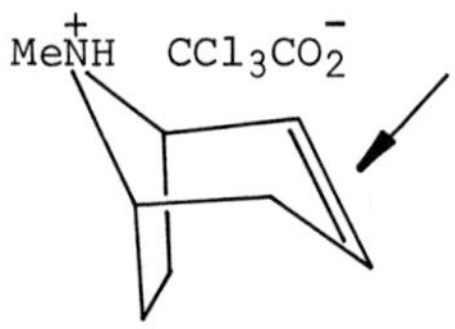

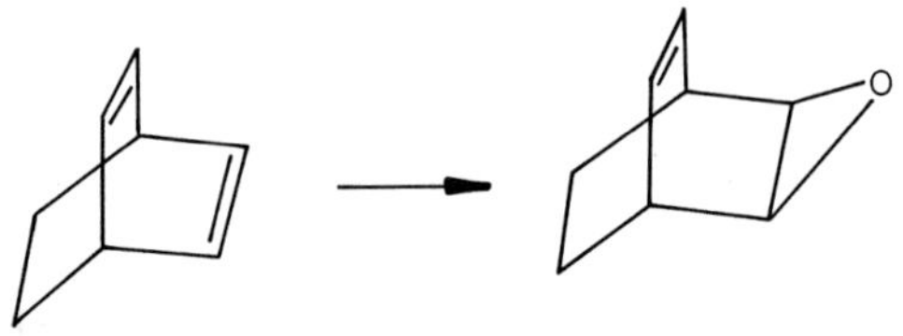

139 *140* *141*

8. *Polyenes*

When two or more nonequivalent double bonds are present in a compound, it is often possible to epoxidize selectively one of them, provided their nucleophilic reactivities are sufficiently different and a slowly reacting system, such as PPA in ether, is employed. This is usually the case when the double bonds carry different numbers of substituents, the most highly substituted one reacting the fastest (9). Thus *142* (134), *143* (42), and *144* (68) give monoepoxides by attack at the positions indicated by the arrows. In some cases steric factors can outweigh electronic ones, and a trisubstituted double bond can be epoxidized at a much faster rate than a tetrasubstituted one; this is the case of lanostadiene (*145*), in which the double bond in the side chain is much more reactive than the endocyclic one (32).

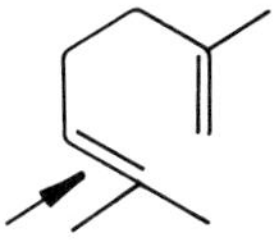

142 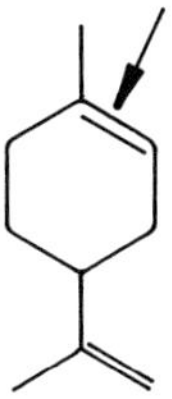 *143* 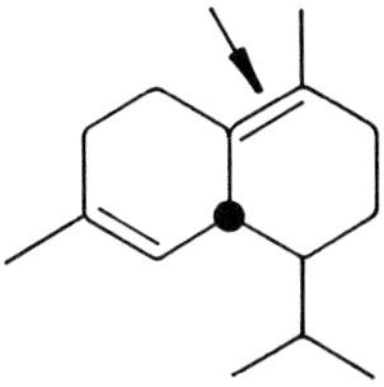 *144*

In *cis*,*trans*-1,5-cyclodecadiene (*146*) preferential epoxidation of the trans double bond is observed, because of its more strained nature (135).

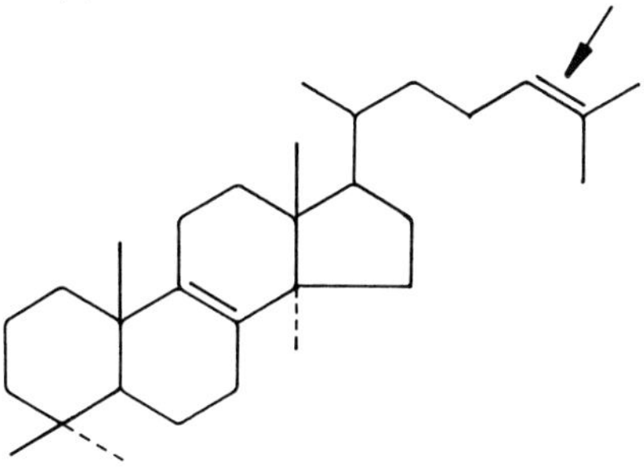

145

146

1,4-Cyclohexadiene (*147*), when treated with either an excess of PBA in chloroform or of MPA in ether, gives only the *trans*-diepoxide, whereas some of the cis isomer is also formed with PAA or CPBA in chloroform (136). However, the tetramethyl derivative *148* is converted by PAA exclusively into the *cis*-diepoxide *149* (137). The monoepoxide is forced to assume one of the two flattened boat conformations *150* and *151*, of which the latter should be favored by more than 2 kcal/mole, because it has two Me-H and two O-H eclipsings less than *150*. Since attack on *151* from the trans side is severely hindered by the "flagpole-bowsprit" methyl groups, formation of *149* is the preferred course.

Also *cis*,*cis*-1,5-cyclooctadiene (*152*) forms only the *cis*-diepoxide on treatment with either PAA or PBA (138), in accordance with an attack by the peroxyacid on the less hindered side of the double bond in a "tub" form of the monoepoxide (*153*); reaction on the rigid chair conformation *154* would be expected to yield the *trans*-diepoxide. Potential energy calculations and IR data seem to indicate that the diene *152* is more stable in the tub than in the chair form (139). It may therefore be assumed that also for the monoepoxide *153* is the preferred conformer. On the other hand, only *trans*-diepoxides have been obtained from 1,3- and 1,4-cyclooctadiene with NPBA, and this, too, can be justified on the basis of ease of approach to the more stable conformations of the monoepoxides (140).

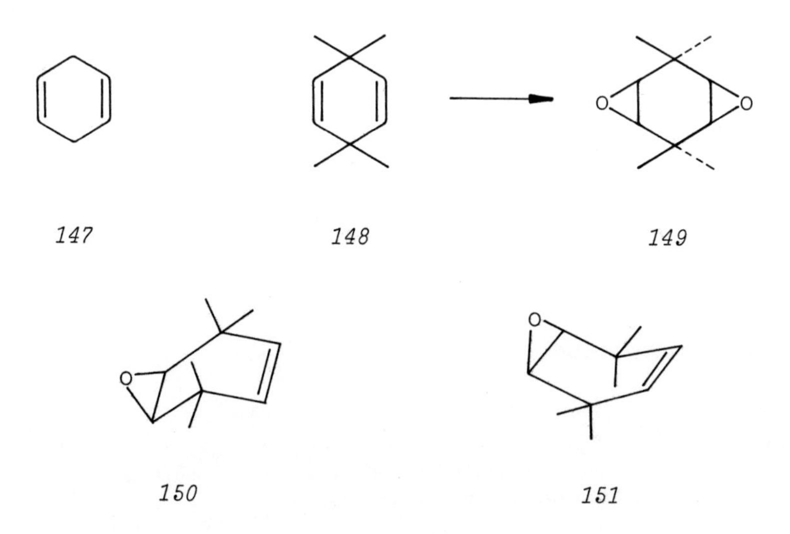

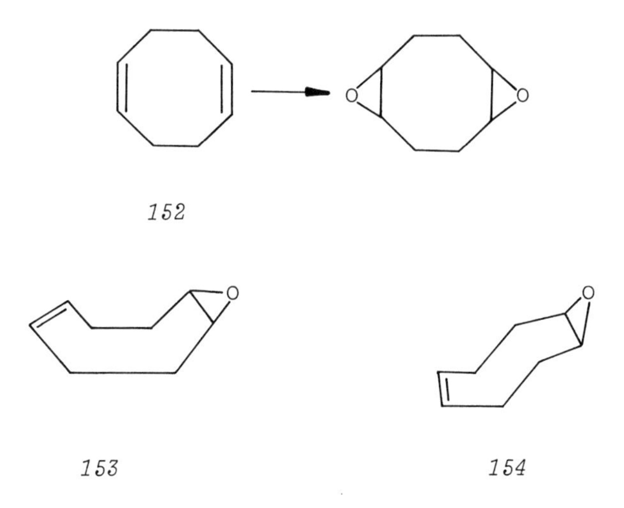

1,3,5-Cyclooctatriene (*155*) forms the 1,2-epoxide, containing only 5% of the 3,4 isomer (141). This preference could be exclusively of steric origin, since the hydrogen atoms in positions 2 and 5 make the central double bond more shielded than the lateral ones. It would be of interest to investigate the monoepoxidation of an acyclic conjugated triene, such as 2,4,6-octatriene, in order to assess the importance of conjugative effects in determining preferential epoxidation in systems of this type.

155 *156*

The exclusive formation of the *anti*-diepoxide *158* from the diene *157* has been attributed to unfavorable dipole-dipole interaction in the transition state leading from monoepoxide to *syn*-diepoxide, rather than to steric effects (142).

In the lanostadiene derivative *159*, CPBA attacks the 7,8 double bond exclusively from the β, and the 9,11 one from the α side (143); this selectivity can be accounted for on the basis

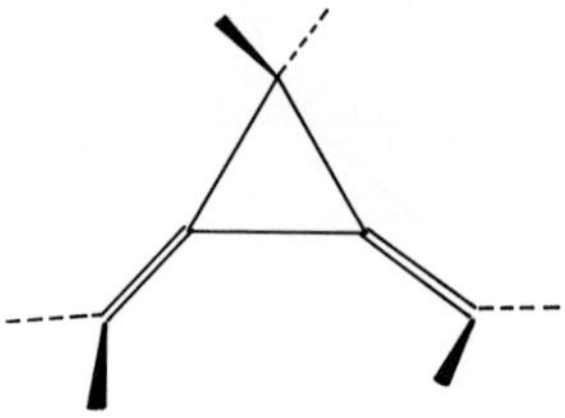

157 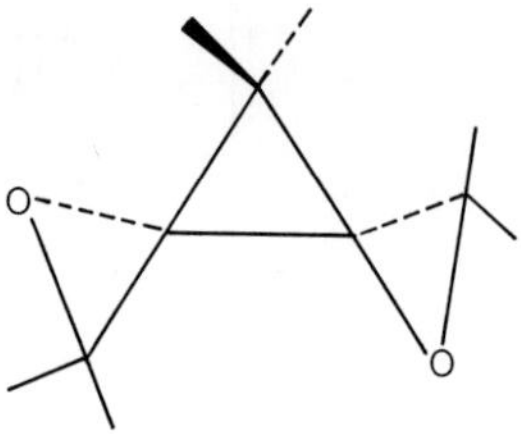 158

of the screening effects of the methyl groups.

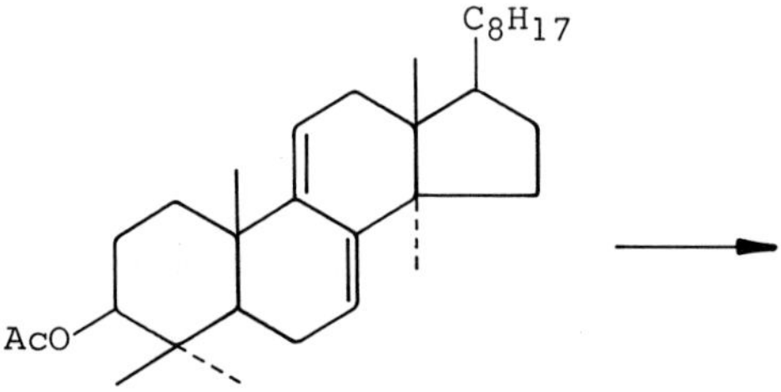

C_8H_{17}

AcO

159

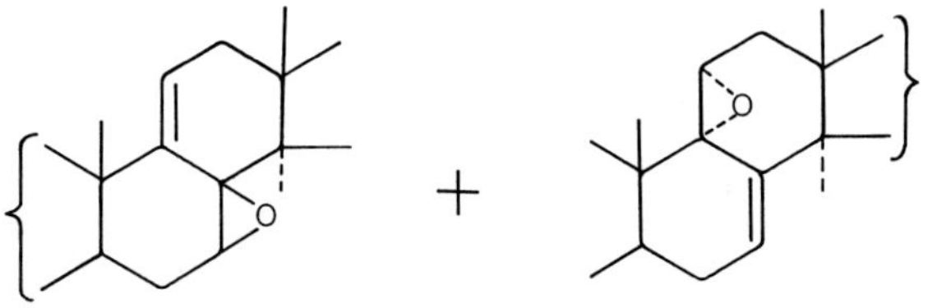

9. *Hydroxyalkenes*

In 1957 Henbest and Wilson (144) and Albrecht and Tamm (145) observed that cyclohexenes and unsaturated steroids having a free hydroxyl group in the allylic position exhibited a strong preference for epoxidation syn to the functional group. The corresponding ethers and esters, however, reacted at slower rates than the free alcohols and did not undergo preferential syn attack. These observations were reasonably explained in terms of a transition state involving hydrogen bonding between the unsaturated alcohol (H donor) and the peroxyacid (H acceptor)

(*160*). Such bonding determines the stereoselectivity of the

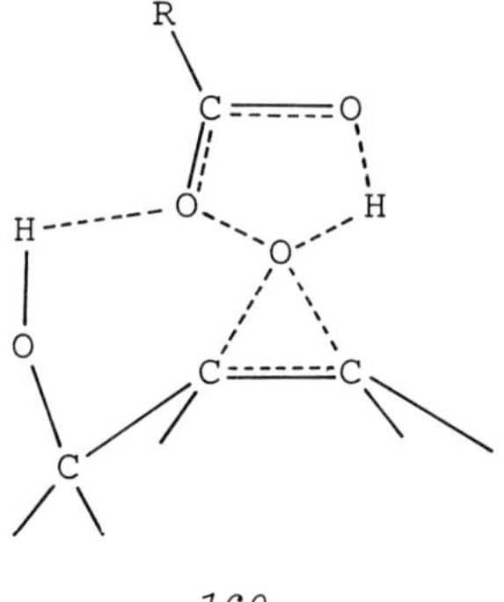

160

reaction on the basis of the geometry of the transition state and can account for the rate increase either by an intramolecular acid catalysis, or by influencing the entropy of activation. Although insufficient kinetic data are as yet available to assess the relative importance of these effects, the highly negative entropy of activation of the reaction of *161* with PBA ($\Delta S^{\ddagger}$ -41.0, as compared with -32.9 for cyclohexene, and -30.7 for *163*) (36) is in agreement with the mechanism outlined in *160*.

A considerable volume of experimental material has been published since 1957 on the stereochemistry of epoxidations of hydroxyalkenes, showing that they can provide a very useful route for stereoselective synthesis. 2-Cyclohexen-1-ol (*161*) is converted by PBA in benzene essentially into the *cis*-epoxide *164* (144); a recent accurate determination of the ratio of *164* to *167* has given 91:9, with very little dependence on solvent

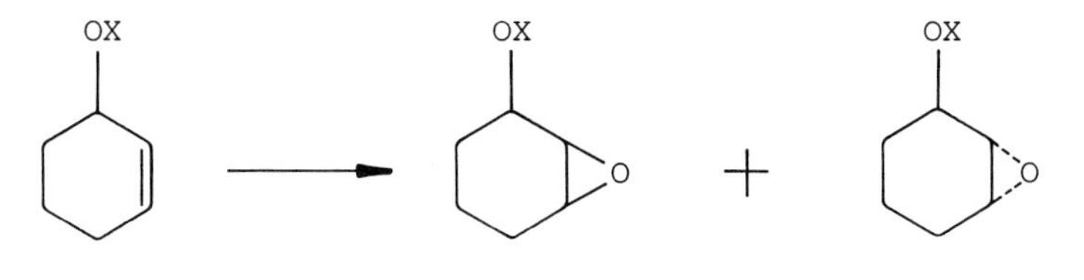

161: X = H
162: X = Ac
163: X = Me

164: X = H
165: X = Ac
166: X = Me

167: X = H
168: X = Ac
169: X = Me

(benzene to acetic acid) and peroxyacid (PBA, PAA, MPA) (146).

The epoxidation of the corresponding acetate *162* is much slower (144) (Table 4) and not very stereoselective (43:57 ratio of *165* to *168*) (146).

Table 4 Relative Rates of Epoxidation with PBA in Benzene

Compound	Relative Rate	Ref.
Cyclohexene	1.0	144
161	0.55	144
163	0.067	144
162	0.046	144
173	0.42	144
176	0.76	146
174	0.106	146
187	3.6	144
186	1.8	144
183	0.88	144

cis-Carveol (*170*) is epoxidized preferentially on the endocyclic double bond and forms the *cis*-epoxide *171*, whereas the acetate *172* is monoepoxidized selectively on the side chain (147). This shows that the inductive effect of the acetoxy group of *172* deactivates the trisubstituted double bond enough to make it less reactive than the disubstituted one, and confirms that the inductive effect of the hydroxyl group of *170*, which cannot be much smaller than that of the acetoxy group, is counterbalanced by the favorable effect of the hydrogen bonding. More evidence about this point is provided by the relative rate data shown in Table 4. The reaction rate of the free alcohol *161* is about

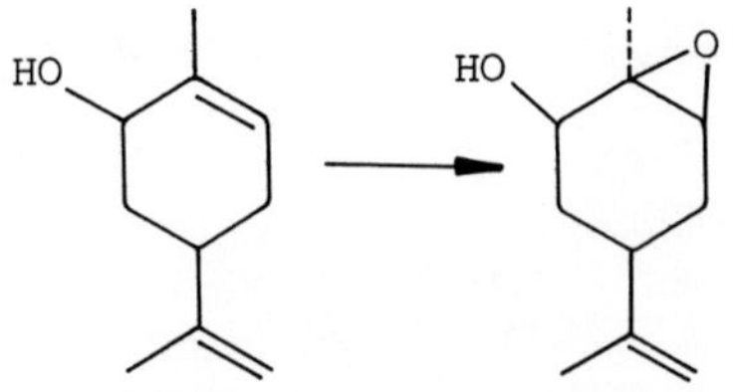

HO HO O

170 *171*

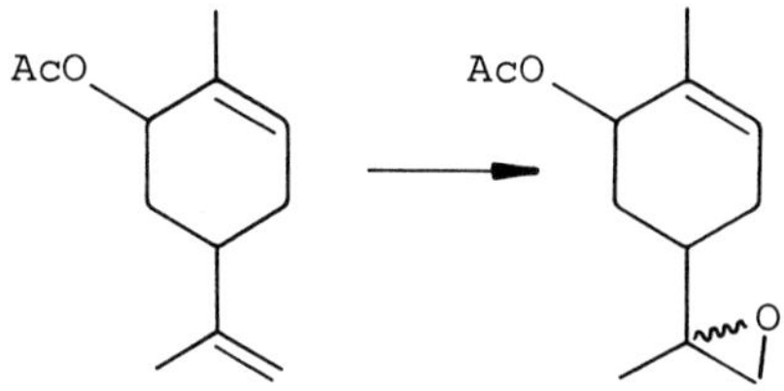

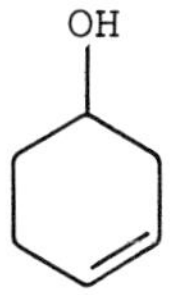

172 173

one half that of cyclohexene, but about ten times as large as those of the ether *163* and ester *162*, and even faster than that of the homoallylic alcohol *173*.

A significant insight into the importance of the axial or equatorial nature of the allylic hydroxyl group on the steric course of epoxidation is given by the *t*-butyl derivatives *174* and *176* (146). From a comparison of the rate data in Table 4 and the steric ones of Scheme 1, one can deduce that a pseudo-equatorial allylic hydroxyl group induces a higher degree of

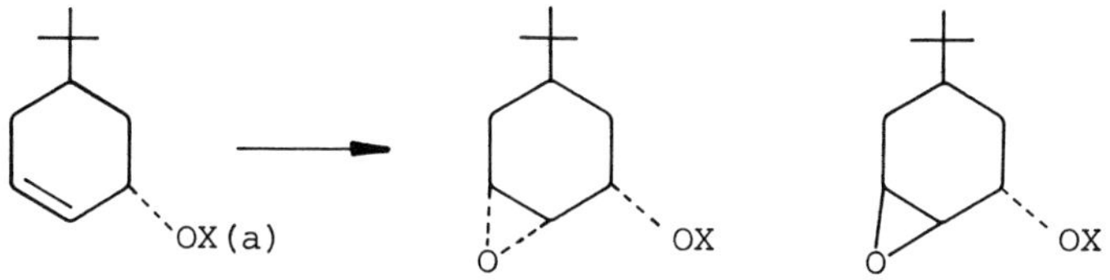

OX(a) OX OX

174: X = H	84%	16%
175: X = Ac	32%	68%

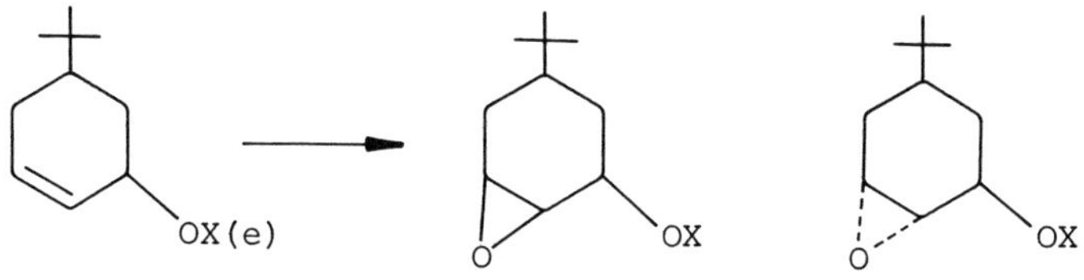

OX(e) OX OX

176: X = H	96%	4%
177: X = Ac	55%	45%

Scheme 1

stereoselectivity and enhances the rate more than a pseudoaxial one. Again no syn selectivity is observed in the epoxidation of the acetates *175* and *177*, and the course can be explained on simple steric terms: when the acetoxy group is axial (*175*) it directs epoxidation anti to itself; when equatorial it does not exert any influence, and only the slight preference for epoxidation syn to the *t*-butyl group, characteristic of the unsubstituted compound *20*, is observed.

Other cases of allylic cyclohexenols being epoxidized syn to the hydroxyl group include *178* (148), which in contrast to the corresponding acetate gives a 9:1 ratio of β- to α-epoxide; *179*, which forms exclusively the *cis*-epoxide (149); and in the steroid series the cholestane derivatives *180* (145), *181*, *182*, *183* (144), *184* (150), and *185* (151), for all of which a high or exclusive preference for β-epoxide formation is observed, in contrast with the usual rule of α attack. In these more complicated systems other steric effects can come into play; for

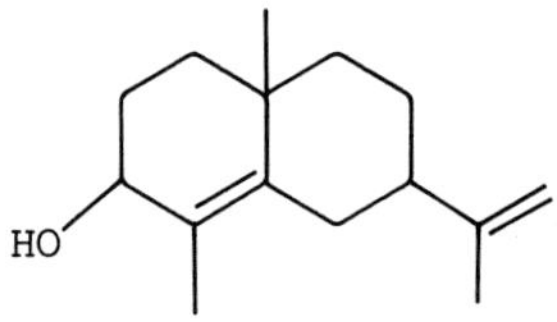

HO

178

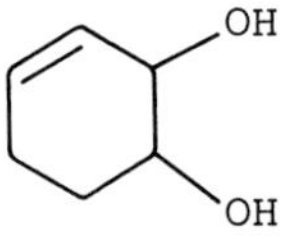

OH

OH

179

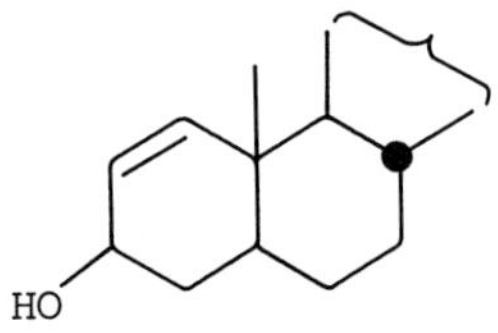

HO

180

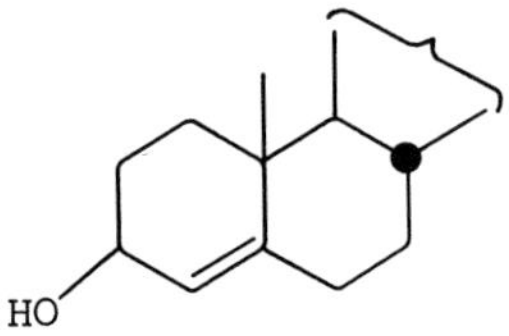

HO

181

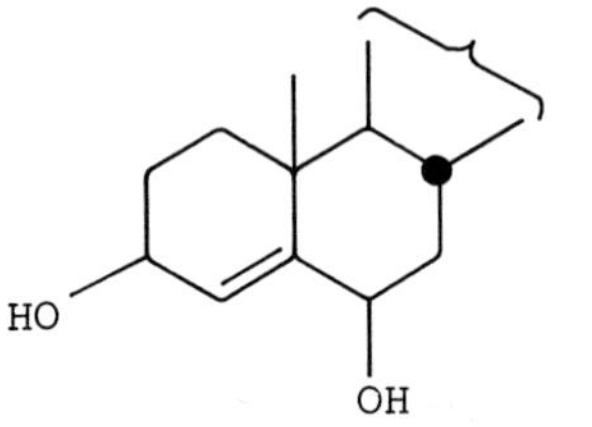

HO

OH

182

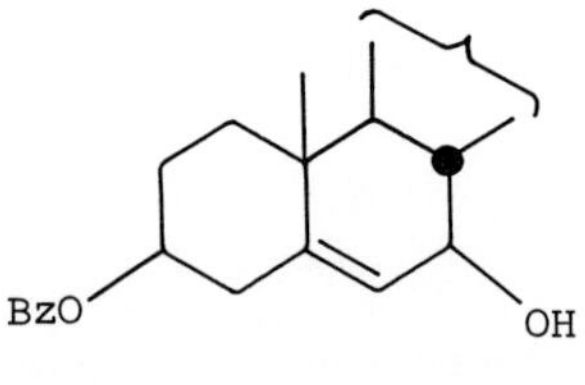

BzO

OH

183

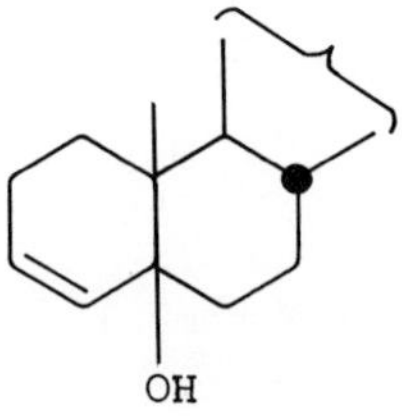

OH

184

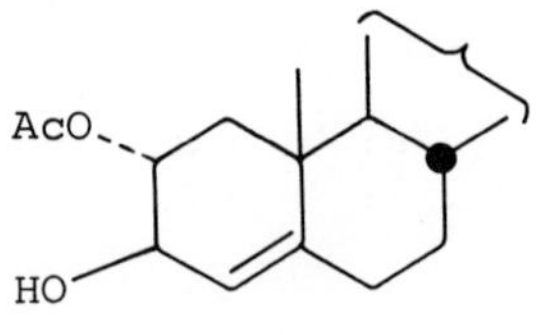

AcO

HO

185

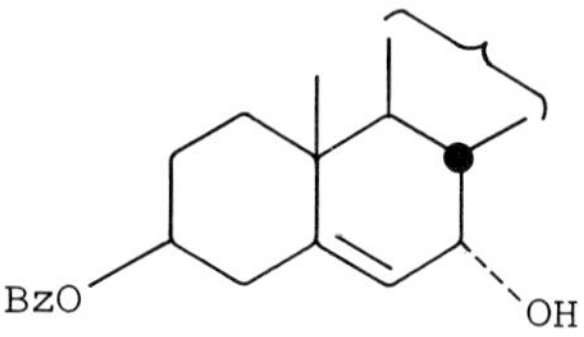

186 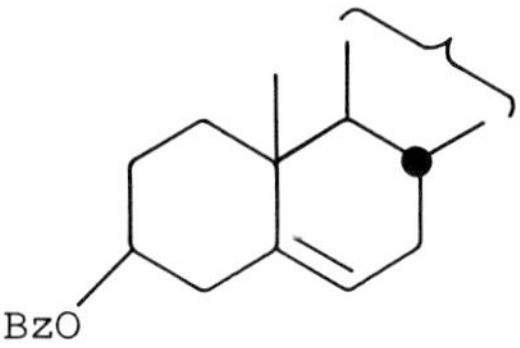 *187*

instance, the 7α analogue of *183* (*186*), which has a pseudoaxial hydroxyl group, forms the α-epoxide faster than *183* (pseudo-equatorial hydroxyl) produces the β-epoxide (Table 4). Evidently the steric shielding by the 19-methyl group influences the rate of the latter reaction, which, however, is still about twelve times faster than would be expected for β attack on a 5,6 double bond deactivated by the inductive effect of the allylic hydroxyl function (144).

Syn stereoselectivity is also found in the epoxidation of allylic cyclopentenols; *188* (152), *189*, and *190* (153) give a large excess of *cis*-epoxides, whereas *191* (154) and *192* (153) yield mainly the *trans*-epoxides.

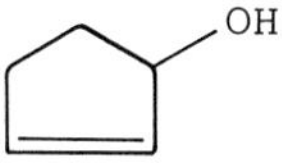

188 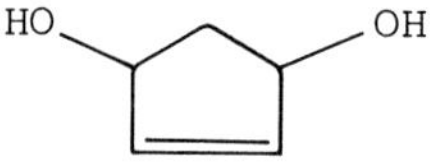 *189* 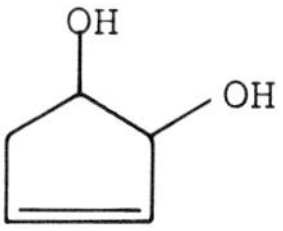 *190*

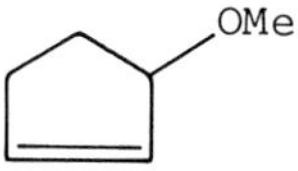

191 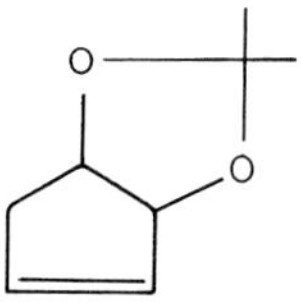 *192*

In the pinocarveols *193* (155) and *194* (156) epoxidation occurs anti to the bridge, independent of the configuration of the hydroxyl-bearing carbon, the shielding effect of the methyl groups being preeminent. On the other hand, the bornane derivatives *195* and *196* have been reported to be epoxidized exclusively syn the hydroxyl function (157), in accordance with the smaller shielding by the bridge methyl groups on the exocyclic double bond in this system.

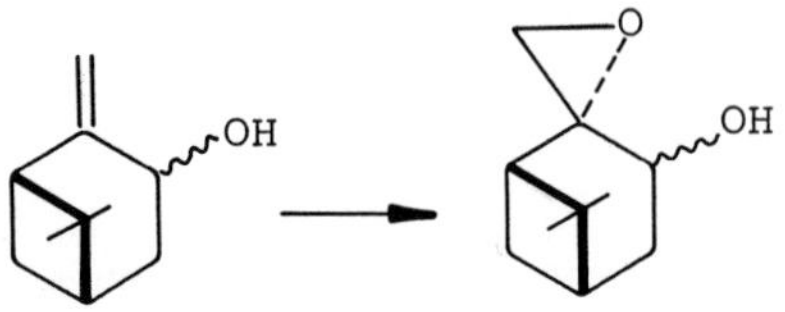

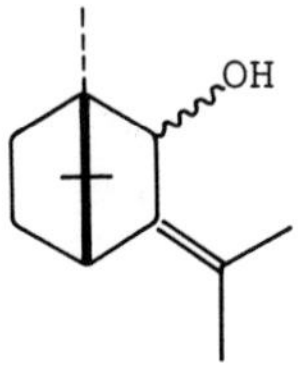

193: α-OH
194: β-OH

195: α-OH
196: β-OH

2-Cycloheptenol (*197*) forms with MPA or PAA a mixture of *cis*- and *trans*-epoxides in a ratio of about 2:1 (158), 2-cyclooctenol (*198*) only the *trans*-epoxide (159).

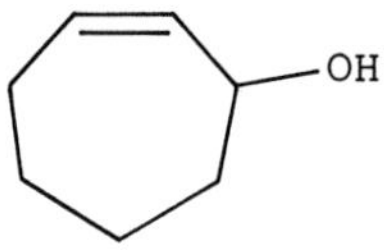

197

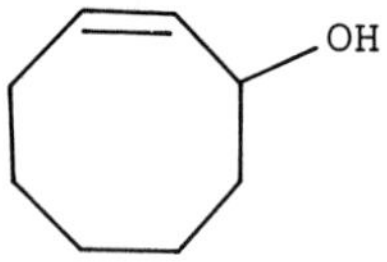

198

Some stereoselectivity can also be observed in the epoxidation of acyclic allylic alcohols. It has been mentioned in Section II-A-2 that very little or no asymmetric induction is present in the epoxidation of acyclic alkenes. The data in Table 5 show that a hydroxyl function may cause preferential formation of either the erythro (*199*) or the threo isomer (*200*), and in a few cases the stereoselectivity is quite high.

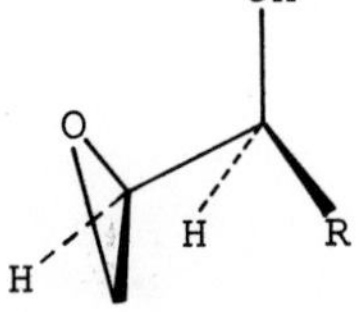

199

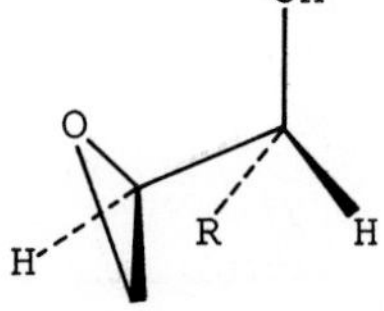

200

Table 5 Steric Course of Epoxidations of Acyclic Allylic Alcohols

Alkenes	% *erythro*-epoxide	% *threo*-epoxide	Ref.
CH_2=CH-CHOH-Me (*201*)	37	63[a]	160
	39	61[b]	31
CH_2=CH-CHOH-Et (*202*)	40	60[b]	31
CH_2=CH-CHOH-i-Pr (*203*)	32-38	68-62[b]	31
CH_2=CH-CHOH-*t*-Bu (*204*)	44	56[b]	31
	54	46[a]	160
CH_2=CH-CHOH-Ph (*205*)	34.1	65.9[b]	146
CH_2=CMe-CHOH-Me (*206*)	62	38[a]	160
trans-Me-CH=CH-CHOH-Me (*207*)	90	10[a]	160
cis-Me-CH=CH-CHOH-Me (*208*)	<2	>98[a]	160
Me_2C=CH-CHOH-Me (*209*)	4	96[a]	160
cis-Me-CH=CMe-CHOH-Me (*210*)	55	45[a]	160
Me_2C=CMe-CHOH-Me (*211*)	10	90[a]	160
HO CMe=CH_2	50	50[c]	161

[a]NPBA in ether.
[b]MPA in ether, except where stated otherwise.
[c]CPBA in $CHCl_3$.

Explanations based on conformational effects have been proposed (31,146,160), which account for many of the data. The most general one (146), which encompasses both the acyclic and the cyclic allylic alcohols, assumes that the preferred geometry of the alkenol at the epoxidation transition state is close to that depicted in *212*. When R" = H, the bulkier substituent will preferentially assume the position R' and attack by the peroxyacid on the face of the double bond nearer to the hydroxyl will produce the *threo*-epoxy alcohol. When R" is an alkyl group, interaction between R' and R" becomes more important and the α-alkyl group may prefer position R, with formation of an excess of the erythro product, as observed with *206* and *207*. A much higher degree of stereoselectivity is induced by an alkyl group (R"") cis to the hydroxy alkyl substituent (*208*, *209*, *211*), since this favors conformation *212* (R' > R), by increasing the nonbonding interactions. Although such a scheme is in satisfactory agreement with some of the experimental results, it

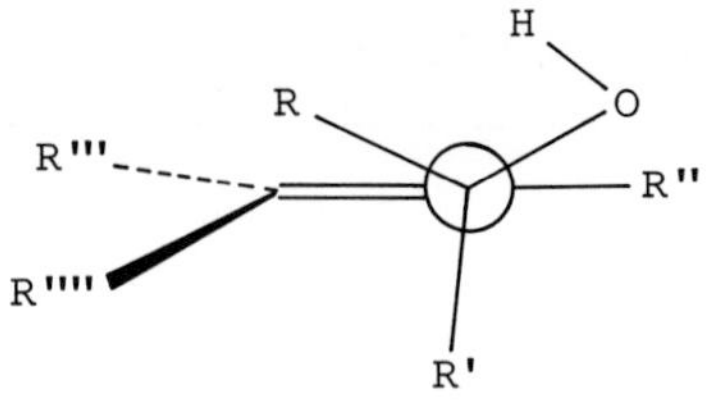

212

fails to provide an explanation for other ones, such as the decrease in stereoselectivity with increasing bulk of R' (*201* to *204*), and particularly the high preference for the *erythro*-epoxide in the reaction of *207*: no reason can be seen for the influence of the size of R''' on the steric course of the reaction, and an experimental checking of this result may be advisable.

A transition-state geometry of the alkenol resembling *212* is in particularly good agreement with the data on cyclic alcohols mentioned above. In the case of pseudoequatorial cyclohexenols, the conformation about $C_{(1)}$ and $C_{(2)}$ is very near to the optimal arrangement (R', R'' = H, R = ring CH_2) for hydrogen bonding with the peroxyacid and syn epoxidation, whereas in pseudoaxial ones the geometry implied in *212* cannot be reached without ring distortion. This explains why epoxidations are faster and more stereoselective in the former case. In a more flexible ring system, such as the one of cyclooctenol (*198*), the arrangement *212* can lead to the *trans*-epoxide. Such an arrangement is present in the chair conformation *213* of *198* and syn attack is also inhibited by the shielding effect of the α hydrogens at $C_{(5)}$ and $C_{(8)}$. The formation of *cis*-epoxide therefore appears to be characteristic only of cyclopentenols, cyclohexenols, and in a more limited way, cycloheptenols.

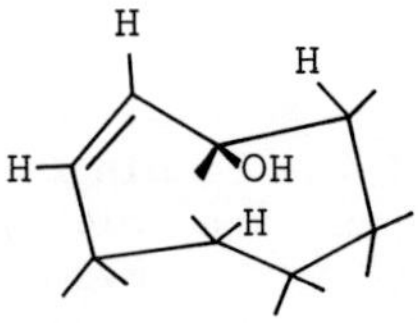

213

It is of interest in this connection to point out that in another reaction type, the Simmons-Smith cyclopropane synthesis (162) also involving the formation of three-membered rings from alkenes, a *syn*-directing influence of the hydroxyl group has been observed, with a trend that is very similar to that found in epoxidation (163). Thus 2-cyclohexenol gives exclusively the syn product *214*, and it has been shown that a pseudoequatorial hydroxyl group has better orienting properties than a pseudoaxial one (164). The parallelism is further evidenced by the fact that in the Simmons-Smith reaction, as in the epoxidation, 2-cyclooctenol (*198*) yields a 99.5:0.5 ratio of anti to syn adduct (165). These analogies between epoxidation and

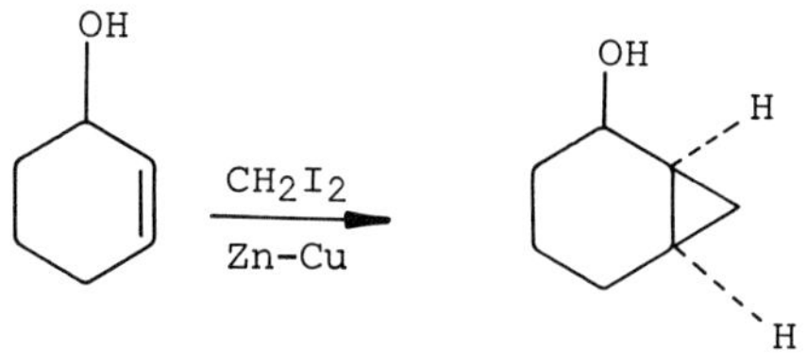

OH $\xrightarrow[\text{Zn-Cu}]{CH_2I_2}$ OH H H

214

I—Zn-----I CH_2

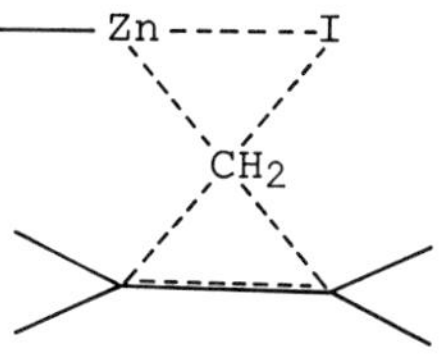

215

methylenation are not surprising, in view of the similarities in the end products and in the geometries of the proposed transition states: *215*, which has been suggested for the transition state of the Simmons-Smith reaction, is reminiscent of *7*, proposed for epoxidations (166). The directing effect in the former case can be explained by hydrogen bonding between the hydroxyl group and the zinc atom (164).

Allylic alcohols in which the hydroxyl and double bond are in different rings can deviate from the normal behavior. Whereas the cholestane derivatives *216* and *217* form exclusively α-epoxides, as expected for the additional steric shielding by the axial substituent on $C_{(6)}$, and *218* produces a mixture of β- and α-epoxides, in which the former predominates because of the free β-hydroxyl function, the formation of a substantial amount of β-epoxide from the 6α-hydroxy derivative *219* is unexpected (167). The explanation may be in the fact that in *219* the $C_{(6)}$-O bond is coplanar with the double bond and does not fit into the scheme *212*. Therefore the hydrogen-bonded complex with the peroxyacid can produce the epoxide from both the α and the β sides; this would be facilitated in conformation *220*, in which the 19-methyl group is pseudoequatorial to ring A, and β shielding is diminished.

The alcohols *221* and *222* are epoxidized exclusively from the α side (168), although they resemble *219* and *218*. However, in this case the unsaturation is in ring B, which is conformationally rigid because of its trans fusion with ring C, and the 19-methyl group, being pseudoaxial, provides more β shielding

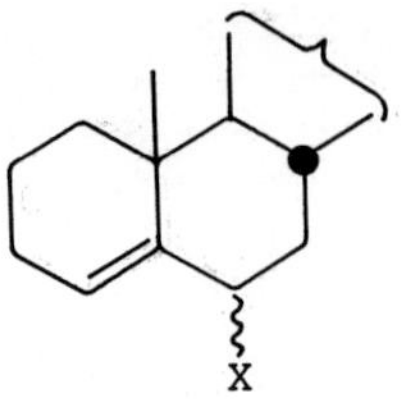

216: X = β-MeO
217: X = β-AcO
218: X = β-OH
219: X = α-OH

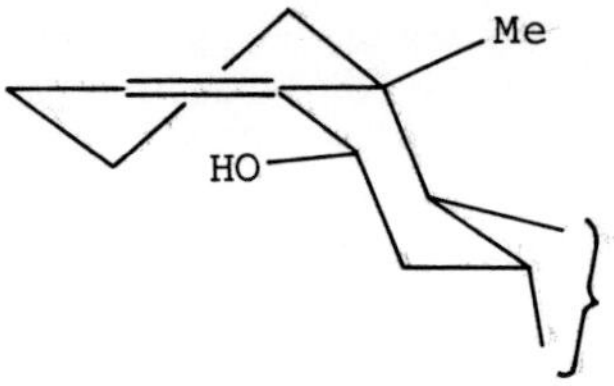

220

against attack on the double bond. Furthermore, in *222* intramolecular hydrogen bonding between hydroxyl and the methoxy group could compete with the intermolecular bonding to the peroxyacid.

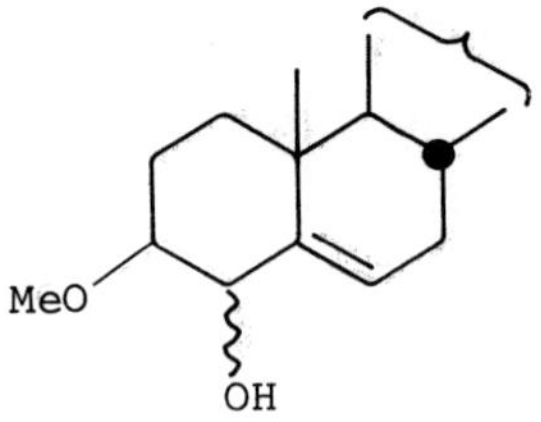

221: 4α-OH
222: 4β-OH

In the B-norcholestane series both the alcohol *223* and its acetate *224* are converted by PBA into the β-epoxides (169). This is expected for the former, but is less obvious for the latter. The five-membered ring B, imposing some limitations on the size of the dihedral angle at the A/B ring function, can account for this preference for β epoxidation, which takes place

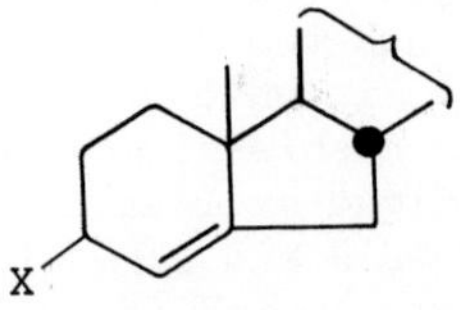

223: X = OH
224: X = OAc

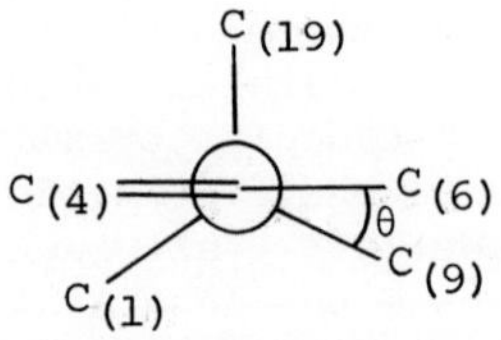

225

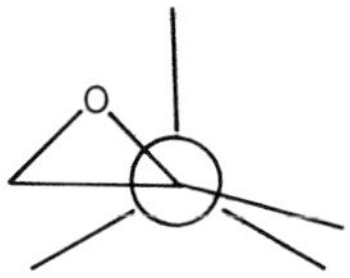

226 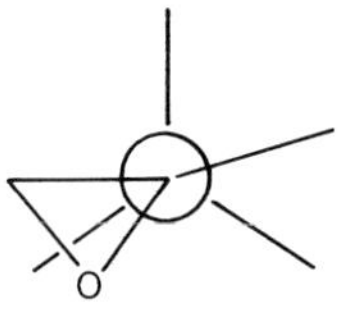 *227*

with little conformational change (*225* to *226*) and causes a decrease in θ. On the other hand, α epoxidation increases the same angle and the ring strain (*227*); even if this could be prevented by a conversion to a boat conformation, the advantage thus gained would be more than offset by the bad flagpole-bowsprit interaction between hydroxyl and methyl. Product development apparently plays a major role in this case.

When the double bond is in the five-membered ring no β preference is observed, because the cyclohexane ring is less sensitive than the cyclopentane one to deformations in its dihedral angles; this is seen in the almost exclusively α epoxidation of *105* (105), *228* (170), and *229* (171).

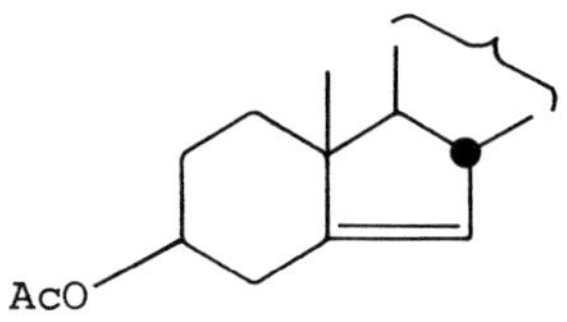

228 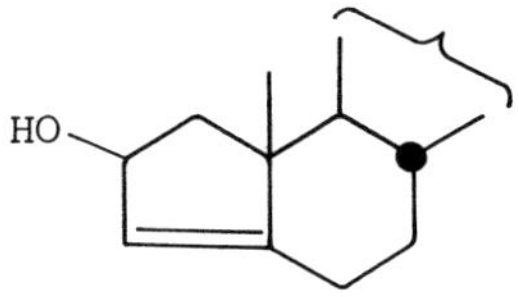 *229*

A hydroxyl group in the homoallylic position or farther away from the double bond can exert its *syn*-directing influence on epoxidation only if it can get sufficiently near for mutual interaction. No stereoselectivity was detected in the epoxidation of 3-cyclohexen-1-ol (*230*) with PBA in ether (172), but this epoxidizing system is not the most appropriate one for observing asymmetric induction, since in homoallylic alcohols, in contrast with allylic ones, the degree of stereoselectivity depends on the solvent used; basic solvents interfere, probably through competition with the peroxyacid as acceptors of the hydrogen bond from the hydroxyl group. Thus the unsaturated diol *231* forms only α-epoxide with CPBA in ether, but gives an about 50:50 mixture of the two diastereoisomeric epoxides when methylene

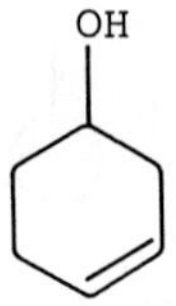

OH HO OH

230 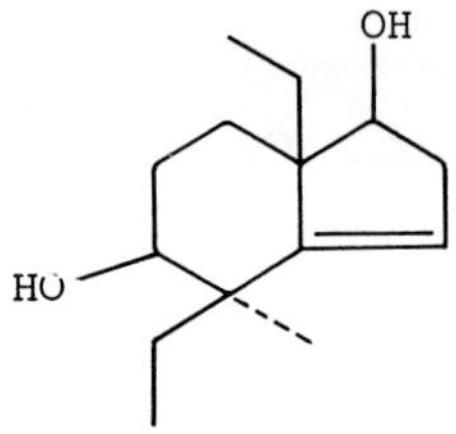 *231*

chloride is used as the solvent (173). Similarly, terpinen-4-ol (*232*), in which the conformation with axial hydroxyl should be preferred, produces an excess of the syn epoxidation product (*233*) with PAA in methylene chloride and with PLA in chloroform, of the anti one (*234*) with MPA in ether (174). The alcohol *235*, in which the hydroxyl group is axial to ring C, is converted by CPBA in chloroform entirely into the α-epoxide (175). The

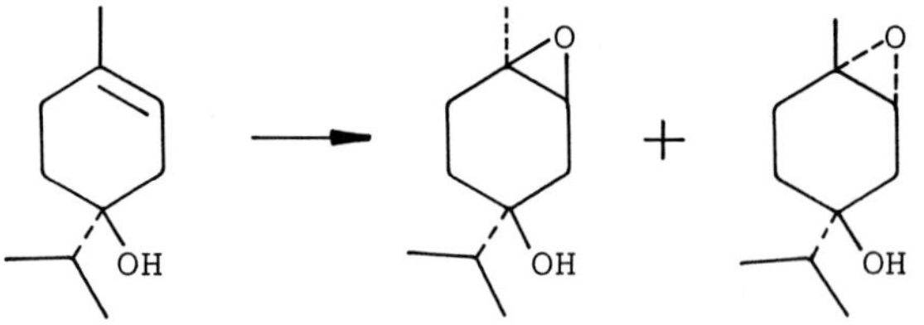

O OH O OH O OH

232 *233* *234*

OH CO_2Me

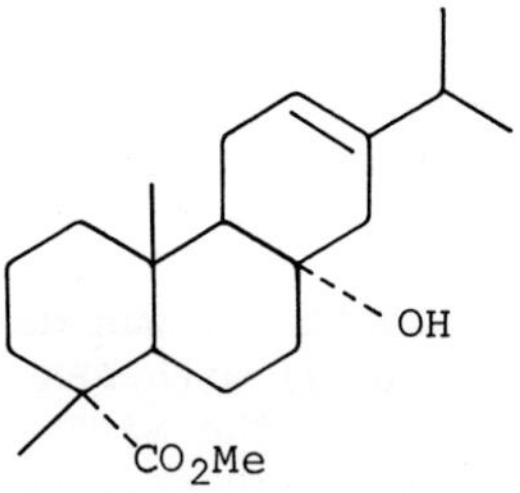

235

hexahydroindanols *236* (in which the conformational situation of the cyclohexene moiety is similar to that of *230*) and *237* are epoxidized by PBA in chloroform exclusively syn to the hydroxyl group (176). A strong preference for syn epoxidation is also exhibited by the octalin derivatives *238* (177), *239*, and *240* (178).

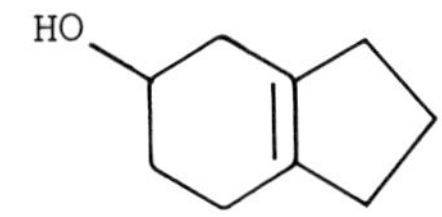

236 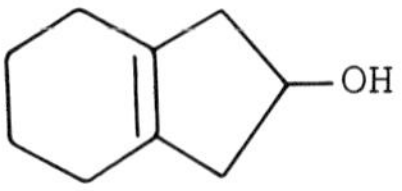 *237* 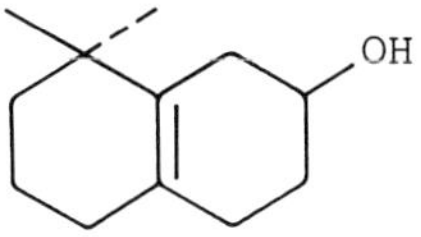 *238*

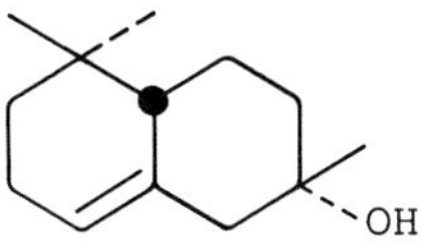

239 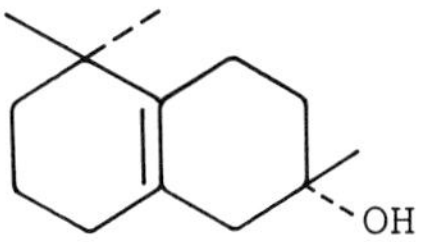 *240* 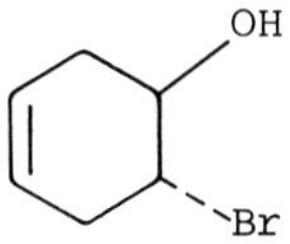 *241*

The formation of a 7:3 ratio of *cis*- and *trans*-epoxides (with respect to OH) from the bromohydrin *241* with MPA in ether (136), on the other hand, is less easy to explain, not the least since intramolecular hydrogen bonding would be expected to favor the diequatorial conformation.

A very definite, even if somewhat erratic, solvent effect was also observed for 3-cyclopentenol (*242*) and its methyl ether (*243*) (Table 6). In the least polar solvent, cyclopentane, the reaction is highly stereoselective, both with the free alcohol (syn attack, hydrogen bonding to peroxyacid) and with the ether (anti attack, polar effect of the substituent, Sect. II-A-11). In the polar protic solvent (methanol) stereoselectivity is practically absent and the two substrates behave identically (36,179).

The greater efficiency of an allylic with respect to a homoallylic hydroxyl group in directing syn epoxidation is indicated by the nearly quantitative conversions *244* → *245*, *246* → *247* (153,180), and *248* → *249* (181). The diacetate of *244*, on the other hand, yields a mixture of the two diastereoisomeric epoxides (182). The triol *250* is converted into a 2:1 mixture of *251* and *252* (183).

Cholesterol (*253*) forms the α- and β-epoxides with MPA in ether in a ratio of 60:40 (184), which is rather similar to that obtained from 5-cholestene (*255*) with PBA in chloroform (70:30) (81), in accordance with the lack of a directing influence by an equatorial homoallylic hydroxy group. However, the almost

Table 6 % syn Epoxidation with PLA

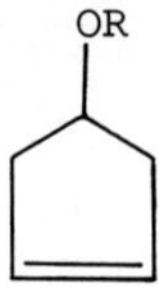

Solvent	*242*: R = H	*243*: R = Me
$(CH_2)_5$	90	8
MeCN	79	33
Et_2O	37	9
MeOH	57	57

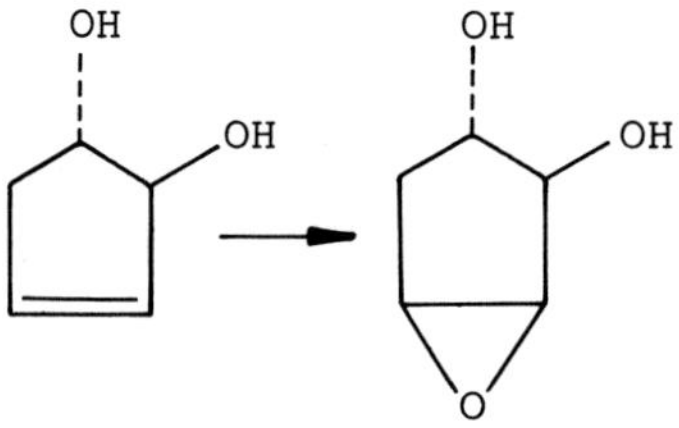

244 *245* 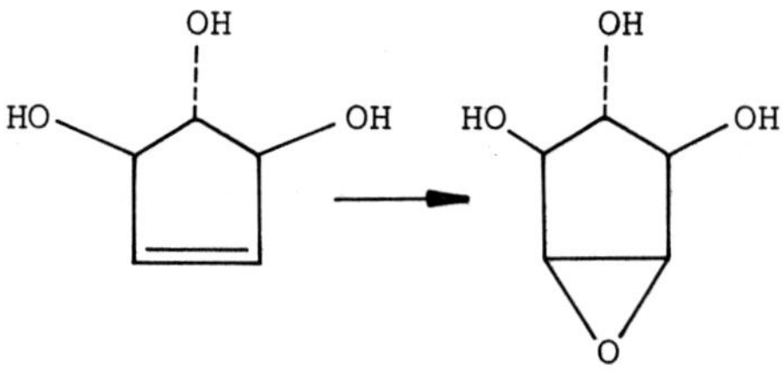 *246* *247*

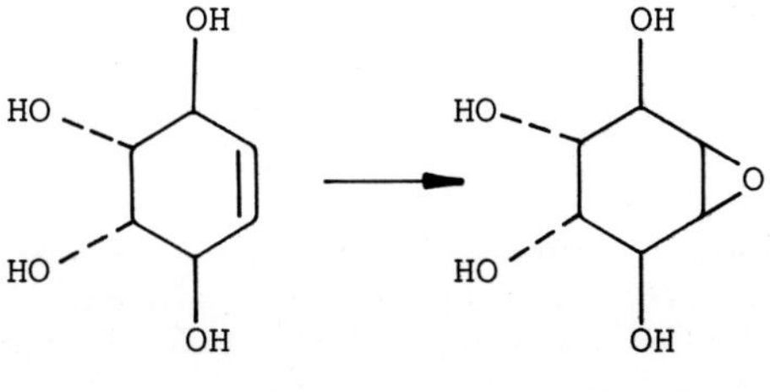

248 *249*

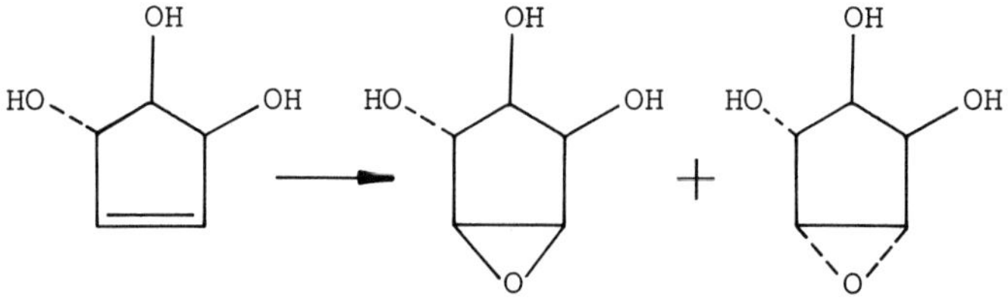

250 *251* *252*

exclusive formation of the α-epoxide from epicholesterol (*254*) (184,185), compared with the 4:1 β/α ratio obtained in the epoxidation of the corresponding methyl ether *256* (81), together

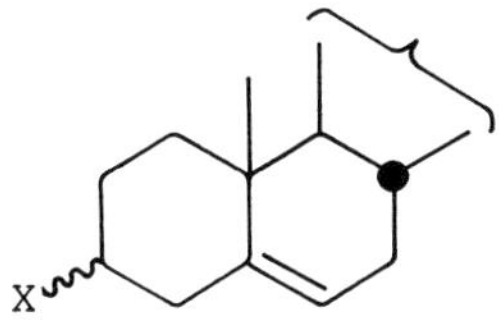

253: X = β-OH
254: X = α-OH
255: X = H
256: X = α-OMe

with the fact that the epoxidation of *254* is 3.8 times faster than that of *256*, demonstrate the favorable effect of an axial homoallylic hydroxyl group on syn epoxidation. A similar situation is found in lumisterol (*257*), which gives the 5,6β-monoepoxide (*258*) with PBA in benzene, the rate being about four times faster than for the corresponding acetate ester (186,187).

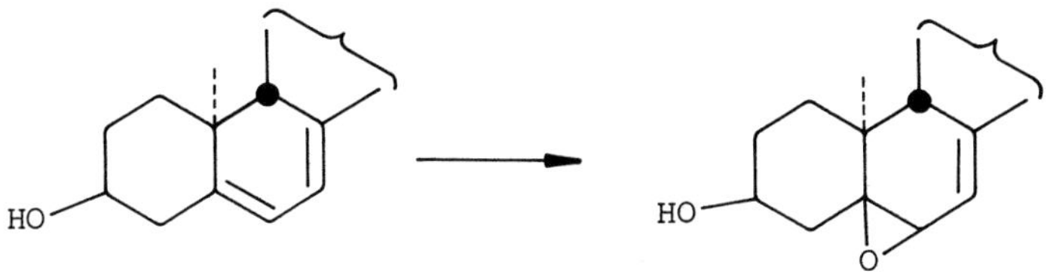

257 *258*

Intramolecular catalytic effects of the same type are

observed in the bicyclic derivative *259*, which gives a 7:3 ratio of β- to α-epoxide, as compared with 4:6 for the acetate (177), and in the steroids *260*, *261* (188), and *262* (189), which yield more than 80% of the β-epoxides. The rate of epoxidation of *260* is again much higher than that of the corresponding 19-acetate, which produces an excess of the α-epoxide. These results prove that a good *syn*-directing effect is exerted by a hydroxyl group, when the OH-π distance is of the order of 2-2.5 Å. This effect is completely absent in drimenol (*263*), which gives only the α-epoxide because the hydroxymethylene group is equatorial (190).

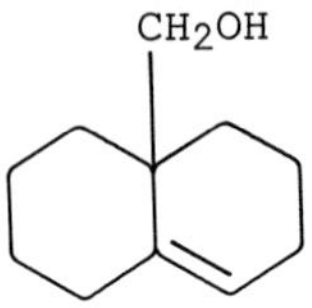

259 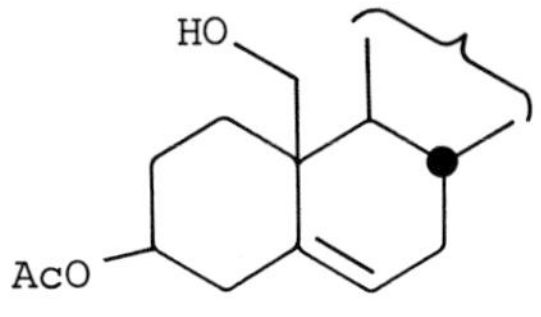 *260* 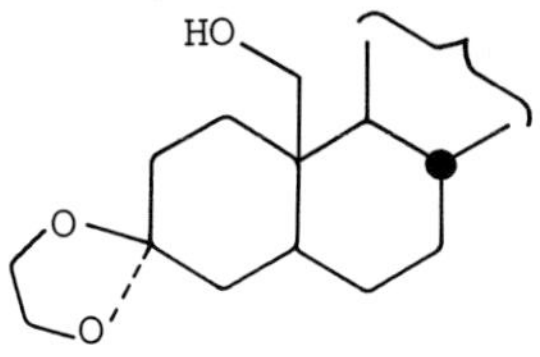 *261*

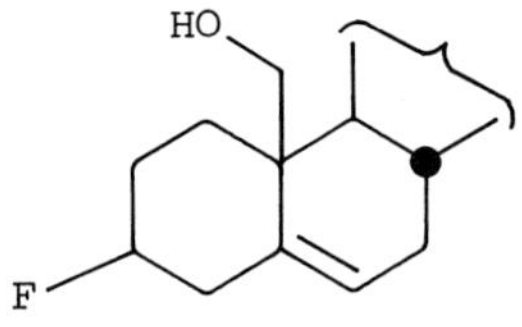

262 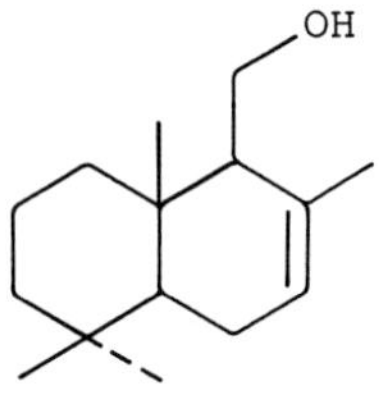 *263*

Particular steric effects can interfere with the *syn*-directing influence of the hydroxyl group, even if it is favorably situated with respect to the double bond. Thus the norbornene derivatives *264* and *265* are attacked by peroxyacid entirely from the exo side (191) because of the highly unfavorable nature of endo attack (Sect. II-A-7), and 11β-hydroxy-Δ^7-steroids (*266*) form only about 25% of the β-epoxide (192), since the axial methyl groups hinder the formation of the hydrogen-bonded transition state.

The tricyclic epimers *267* and *268*, in which ring A is conformationally mobile and the equatorial preference of the methyl group favors the axial disposition of the hydroxyl, are both

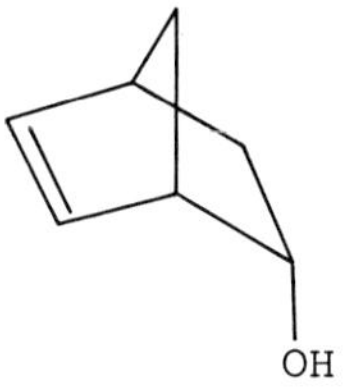

264 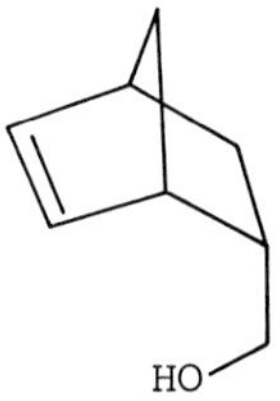 *265* 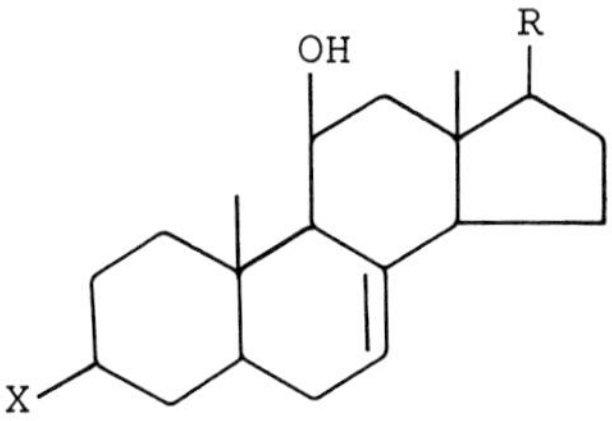 *266*

epoxidized exclusively syn to hydroxyl (193). However, in analogous steroid derivatives the situation is less straightforward: the epimers *269* and *270* produce only β-epoxide, even though the rate of epoxidation is about four times faster for the β- than for the α-alcohol (193); the same high preference for β epoxidation is also found for *271* (194) and *272* (195), which

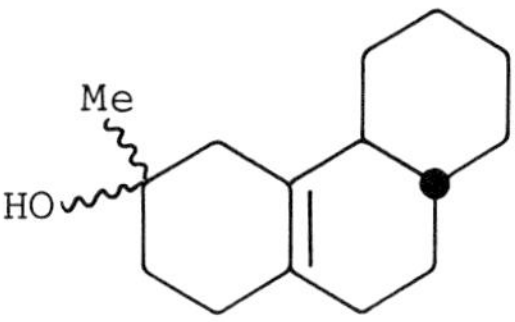

267: α-Me, β-OH
268: β-Me, α-OH

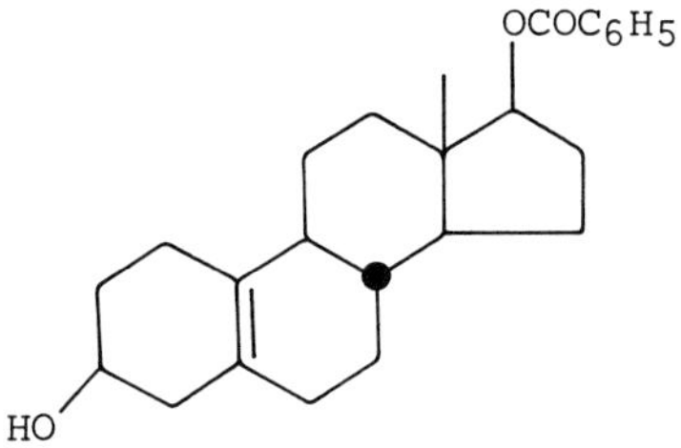

269: α-OH
270: β-OH

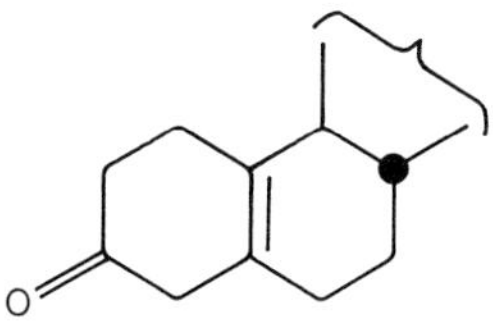

271 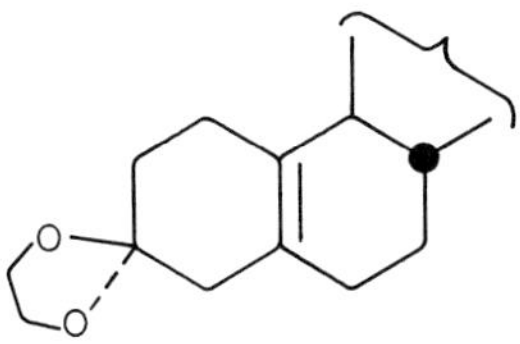 *272*

have no free hydroxyls. An explanation for this preference, based on the tendency to form the more stable 9,10-anti disposition (194), does not appear entirely satisfactory. Also in reactions not involving the ring junction, such as the LAH reduction of *271*, attack takes place exclusively from the β side (196).

271 is an unhindered ketone and should therefore be subject to axial attack by hydride; this implies a preference for conformation *274* over *273* in $\Delta^{5(10)}$-steroids, which can be justified by the very short distance between the hydrogens in positions 1β and 11α (ca. 1.8 Å) in *273* (196). In conformation *274* only a 3β (axial) hydroxyl group can facilitate syn epoxidation, and this explains the higher rate of epoxidation of *270*. An accurate

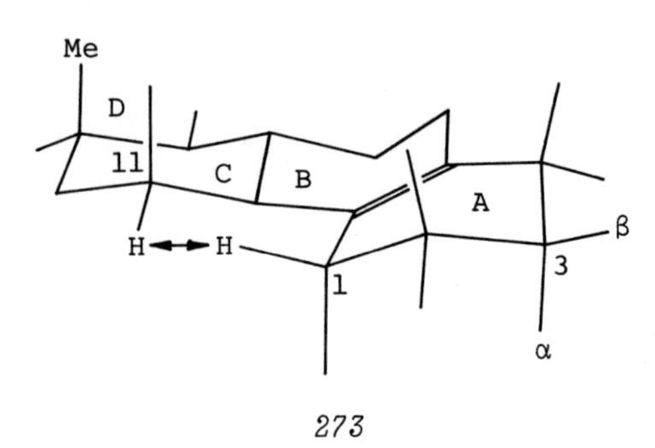

273

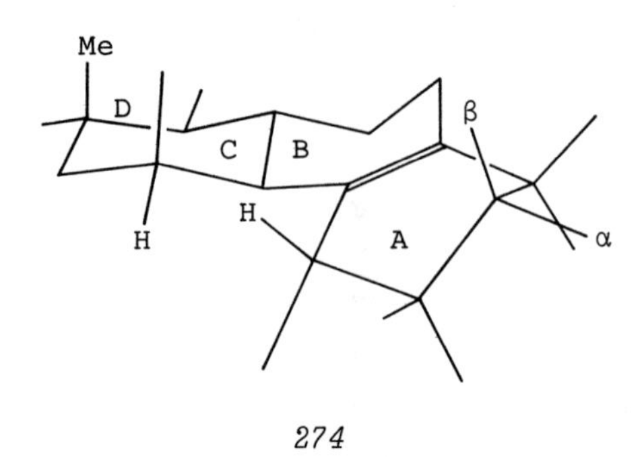

274

conformational analysis (193,197) also reveals that the distances between the hydrogen atoms at $C_{(1)}$ and $C_{(11)}$ decrease in passing from the olefin to the α-epoxide, but increase in going to the β-epoxide; this may justify at least in part the general preference for β epoxidation. Introduction of a methyl group in position 3 does not change the situation for *276* (equatorial Me, axial OH, only β-epoxide formed), but allows for the formation of 60% α-epoxide in the case of *275*, since conformation *273* (equatorial Me) can now compete with *274* (axial Me). The syn stereoselectivity is not, however, as high for *275* as for the tricyclic analogue *268*, and this has been attributed to a long-range effect of the trans C/D ring junction in *275*. This imposes a decrease in the $C_{(17)}$,$C_{(13)}$,$C_{(14)}$,$C_{(15)}$ dihedral angle, which is conformationally transmitted through all the ring system, causing an increase in the nonbonding interactions of the hydrogens at $C_{(1)}$ and $C_{(11)}$ (193).

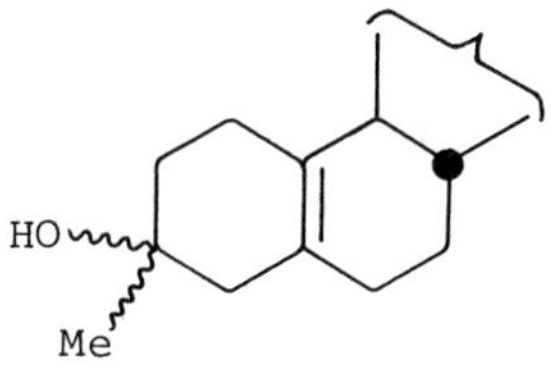

275: β-Me, α-OH
276: α-Me, β-OH

Interestingly enough, in the 5(10),9(11)-dienes *277*-*279* α approach to the 5(10) double bond is the preferred course (Table 7) (197). In this case the trigonal structure at $C_{(11)}$ decreases the difference in energy between the two half-chair conformations corresponding to *273* and *274*, but calculations on the changes in nonbonding interactions in passing from the dienes

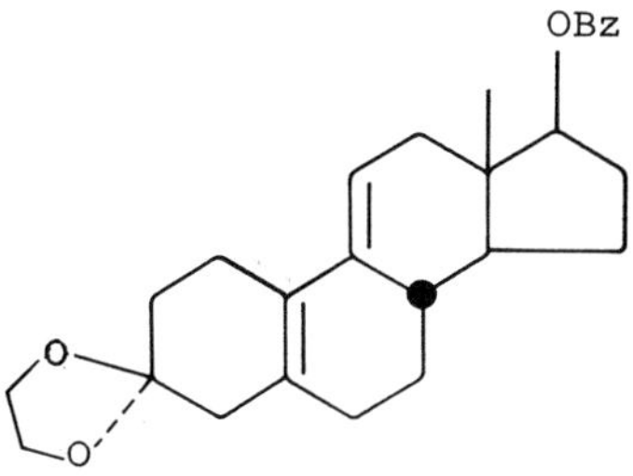

Table 7 Epoxidation of 5(10),9(11)-dienes with CPBA

Substrate	% 5,10α-epoxide	% 5,10β-epoxide	% 9,11α-epoxide
277	50-55	10-15	25-30
278	80	5-10	10-15
279	43	43	—

277

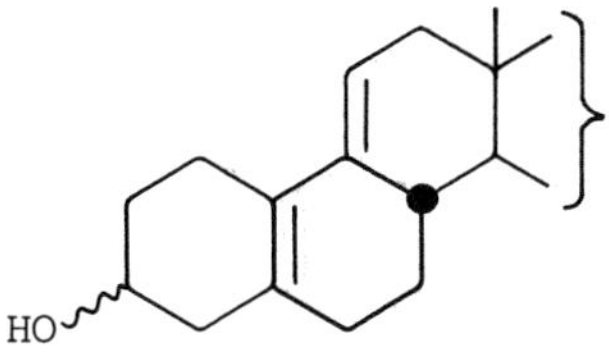

278: α-OH
279: β-OH

to the 5,10-monoepoxides indicate that, no matter what the initial conformation, the transition states leading to the α-epoxides are favored over those leading to the β-epoxides. Therefore, even if a *syn*-directing effect of the hydroxyl is observed both in *278* and in *279*, it is more pronounced in the α-alcohol. It must be pointed out, however, that four axial or pseudoaxial hydrogens (or substituents) around the double bond or the β side and only three on the α side, could also account in part for the preference for α attack.

In the rearranged steroids *280*-*282* the 6α-hydroxy derivative *280* produces with MPA in benzene-ether the α-epoxide exclusively, whereas the 6β-hydroxy (*281*) and 6β-acetoxy (*282*) derivatives react more slowly to give about equimolar amounts of α- and β-epoxides (198). Although ring B can be conformationally mobile, a calculation of the nonbonding interactions indicates that conformation *283* with an axial 6β substituent is

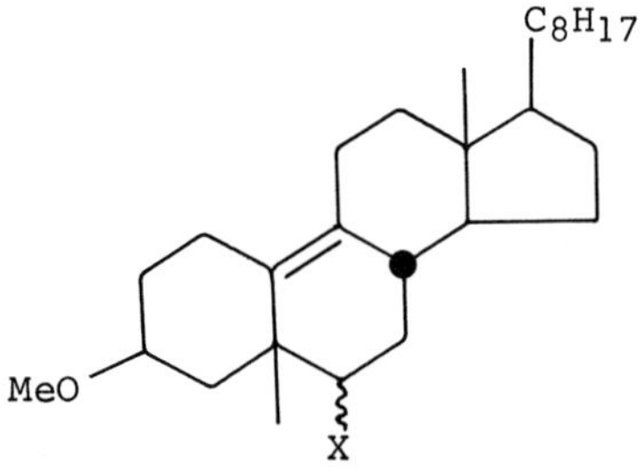

C_8H_{17}

MeO

X

280: X = α-OH
281: X = β-OH
282: X = β-OAc

very unfavorable, and that the most stable conformation for both alcohols should have ring B in a slightly distorted boat form (*284*); in it an α-, but not a β-hydroxyl group would be suitably disposed for syn epoxidation.

Hydroxyl groups that are separated from the double bond by four or more bonds could get near enough to it to exert their *syn*-directive influence only in very special cases. Compounds *285* (199) and *286* (200) give the same ratios of α- to β-epoxide (60:40). Compound *287*, which would have the right OH•••π distance in the boat conformation, is epoxidized to the extent of 70% anti to the hydroxyl (201), in accordance with the expected preferred axial attack on the exocyclic double bond in the conformation with equatorial methyl group (Sect. II-A-6).

α-Terpineol (*288*) was reported to give with PAA a mixture

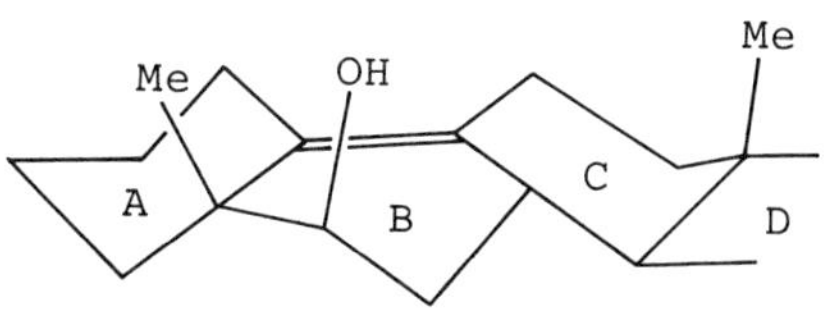

283

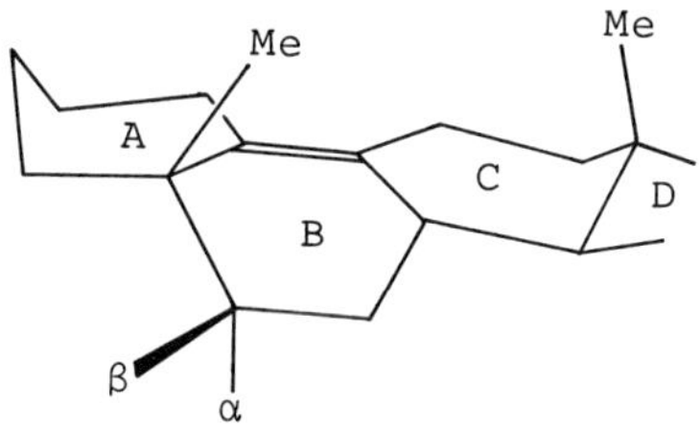

284

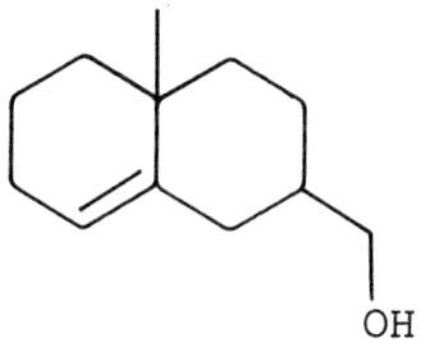

285

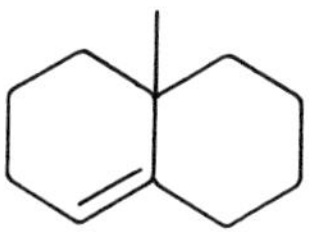

286

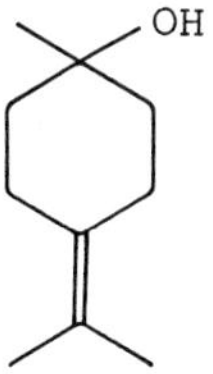

287

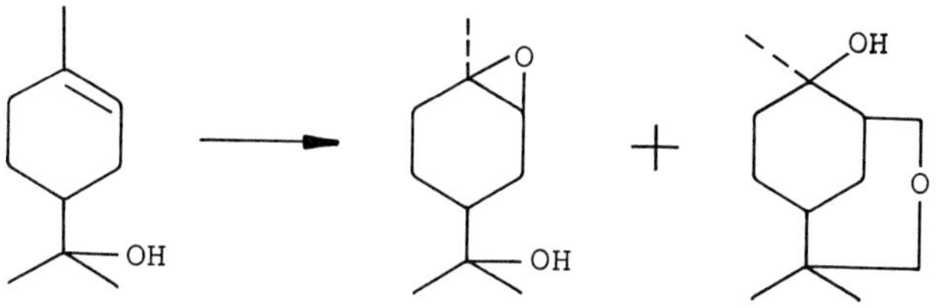

288

289

290

of *289* and *290* (202); this would correspond to exclusive syn attack (which would require an axial position for the hydroxy-isopropyl side chain in the transition state), but the unusual stereochemistry of *290* involving intramolecular cis opening of the epoxide ring of *289* requires a more convincing confirmation. The second product could be the cineol derivative *292*, rather than *290*, and be formed in the reaction medium by an acid-catalyzed cyclization of the α-epoxide *291*. This would be in accordance with a recent reformulation of supposed dihydropinols as cineols (203).

Some evidence of asymmetric induction involving a remote hydroxyl in a noncyclic molecule was provided by the fact that the epoxide *294*, prepared by the epoxidation of active linaloöl

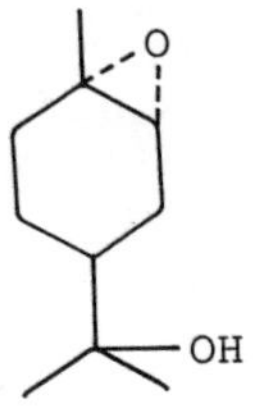

291

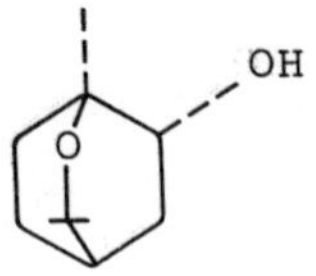

292

(*293*), followed by removal of the original chiral center, exhibited a weak optical rotation; preferential attack on one side of the double bond by the hydrogen-bonded peroxyacid could explain this result (204).

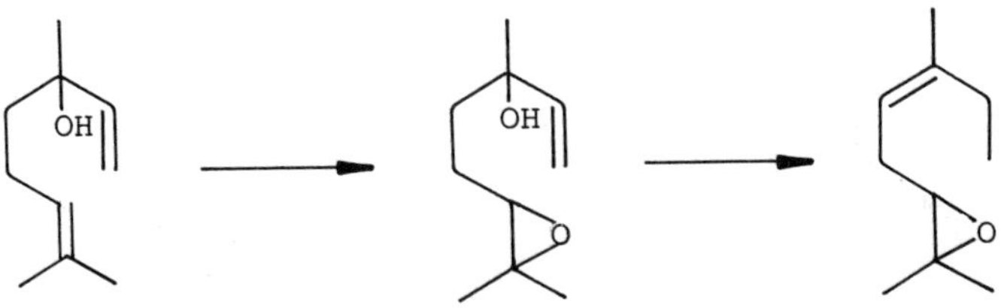

293

294

10. *Unsaturated Amides*

Acylamino groups can exert a *syn*-directing influence on epoxidations, which is very similar to that discussed for alkenols. 3-Benzamidocyclohexene (*295*) is converted by PAA exclusively into the *cis*-epoxide (205), and compound *296* into the oxide *297* (206), showing that the stereoselectivity associated with an allylic acetamido group outweighs those of an allylic plus a homoallylic hydroxyl group, probably because the higher acidity of the amide hydrogen atom provides stronger binding to the peroxyacid molecule. In the steroid field *298*, *299* (207, 208), *300*, and *301* (209) are also epoxidized exclusively syn to the amidic function. This holds also for the ureido derivatives *302* and *304*, whereas the acetyl derivatives *303* and *305* produce mixtures of diastereoisomeric epoxides (210); this could be due to preferential hydrogen bonding of the peroxyacid with the more acidic hydrogen α to the acetyl group, which would keep the peroxyacid too far from the double bond for stereoselective epoxidation.

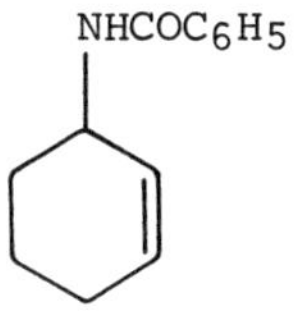

295

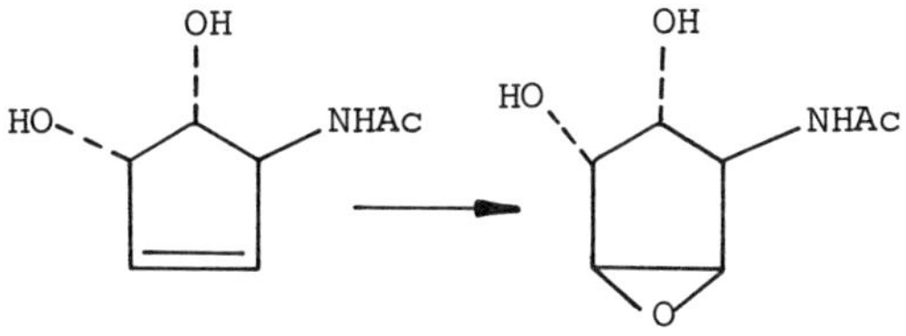

296

297

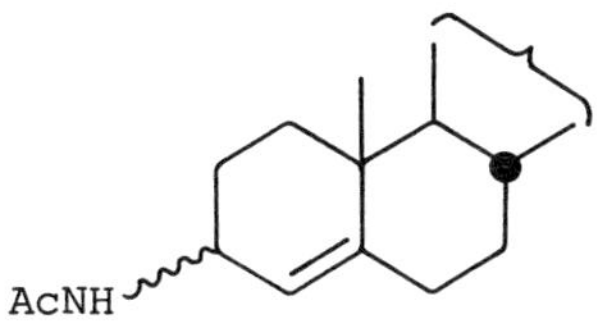

298: α-AcNH
299: β-AcNH

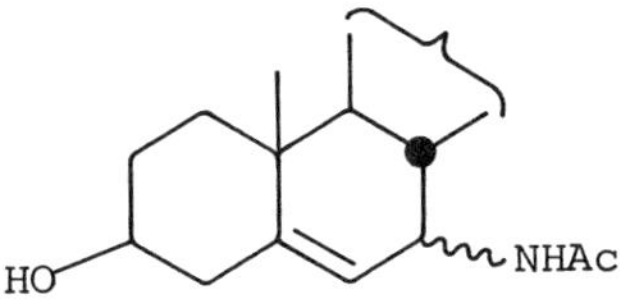

300: α-AcNH
301: β-AcNH

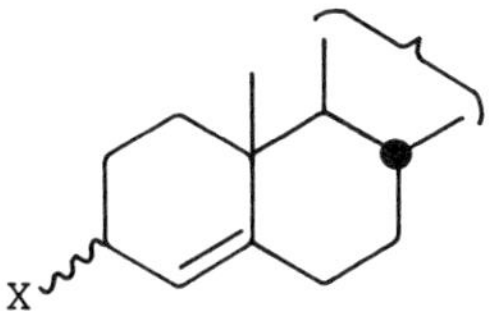

302: X = α-$NHCONH_2$
303: X = α-NHCONHAc
304: X = β-$NHCONH_2$
305: X = β-NHCONHAc

11. *Alkenes Substituted with Other Types of Electronegative Groups*

When an unsaturated molecule carries an electronegative substituent having no hydrogen atom capable of bonding with the peroxyacid, stereoselectivity effects can still be observed, which may be of polar or steric origin. Thus the nitriles *306* and *307* give with NPBA in chloroform a 96:4 ratio of *trans*- to *cis*-epoxide, and only *trans*-epoxide (211), respectively. Henbest (36) has used cyclopentene derivatives as models to investigate the relative importance of steric and polar effects in these systems. 4-Cyanocyclopentene (*308*) is epoxidized anti to the substituent to a greater extent than the 4-methyl derivative (*29*) when cyclopentane is used as the solvent, although the size of CN is smaller than that of the methyl group; this difference in steric course disappears in the more polar solvent, acetonitrile (Table 8). Furthermore, a certain preference for anti

Table 8 Epoxidation of Cyclopentene Derivatives with PLA

Substrate	Solvent	% *trans*-epoxide
29	$(CH_2)_5$	73
	MeCN	74
308	$(CH_2)_5$	95
	MeCN	76
309	$(CH_2)_5$	87
	MeCN	52
310	$(CH_2)_5$	80
	MeCN	26
311	$(CN_2)_5$	69
	MeCN	54
312	$(CH_2)_5$	72
313	$(CH_2)_5$	57
	MeCN	46

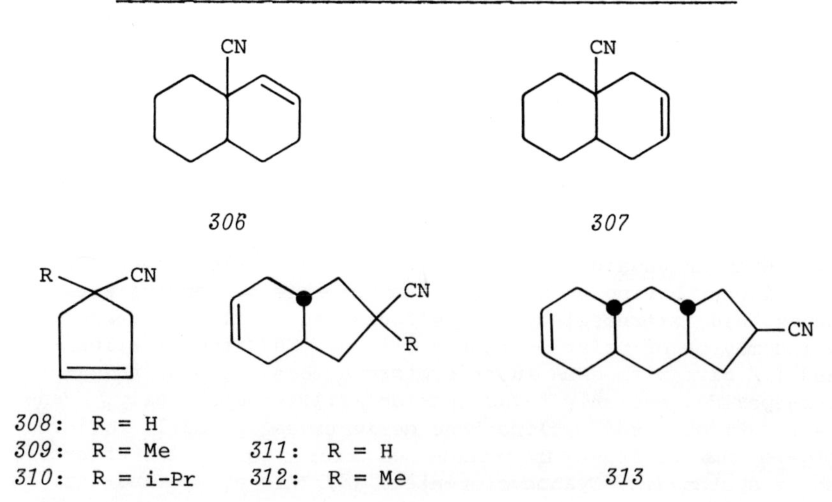

epoxidation is also found in the bi- and tricyclic derivatives *311* and *313*, in which the cyano group is too far from the double bond to exert any direct steric influence. Polar effects therefore play a major role in the epoxidations conducted in the nonpolar solvent and favor anti attack, even when an alkyl group shielding the anti side is present, such as in compounds *309* and *310*. This can be interpreted on the basis of a transition state *314* in which the dipoles of the substituent and of the attacking peroxyacid molecule are directed into opposite directions (212). This hypothesis is supported by the fact the percentage of anti attack decreases with distance, with increasing solvent polarity,

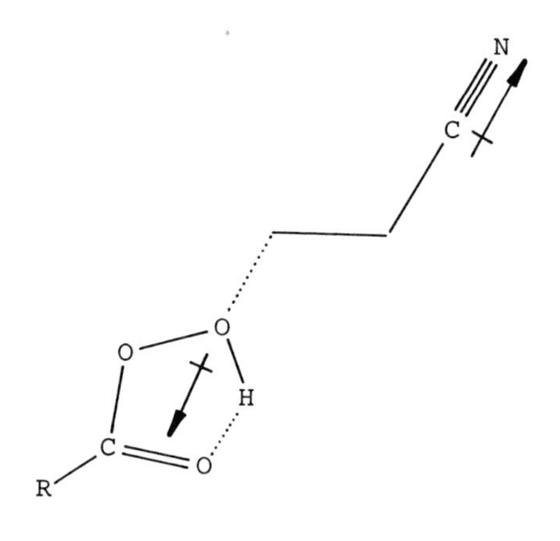

314

and with a decrease in the dipole moment of the substrate (36).

A preference for anti epoxidation was also observed for 4-cyanocyclohexene (*315*), its solvent dependence being in accordance with that expected for a polar effect (82% *trans*-epoxide in cyclopentane, 69% in acetonitrile and nitromethane) (36). However, a report on the exclusive anti epoxidation of compounds *316*, *317*, and *318* (191) is not correct, at least not for *316*, which gives a 66:34 ratio of *trans*- to *cis*-epoxide (213).

The exclusive formation of the *trans*-epoxide from the *cis* form of *319*, and the nonstereoselectivity in the epoxidation of the corresponding *trans* isomer (214), can be explained on similar terms, since in the former but not in the latter, the conformation with an axial cyano group (*320*) is favored and leads to the *trans*-epoxide because of the polar and of the steric effect. The same can be said for the formation of a 9:1 ratio of *322* to *323* from *321*, whereas the reason for a 2:1 preference for *325* over *326* in the epoxidation of *324* is not at all clear (215).

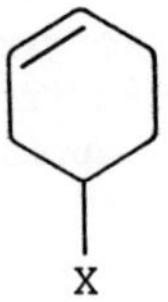

X

315: X = CN
316: X = CO_2Me
317: X = CH_2OH
318: X = CH_2OAc

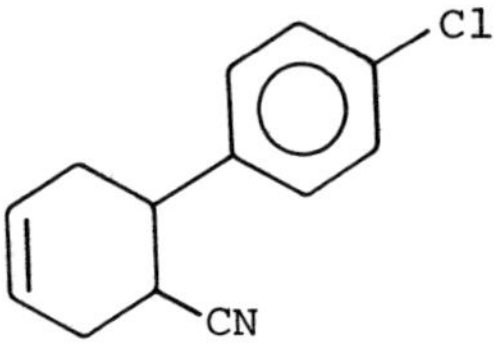

Cl CN CN Ar

319 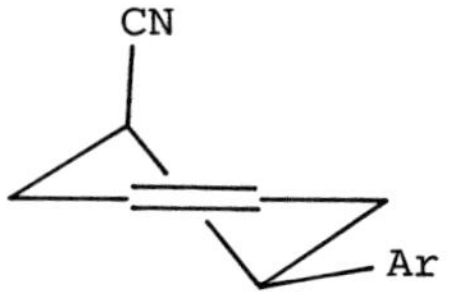 *320*

The 99% epoxidation of *327* syn to the acetoxy group can be explained entirely on the basis of steric effects in the most stable conformation (216).

The preference for the formation of the *cis*-epoxide from the lactone *328*, a key step in one of the syntheses of prostaglandins, and the erratic solvent dependence of this reaction (Table 9) (217) resists any interpretation on either steric or polar grounds.

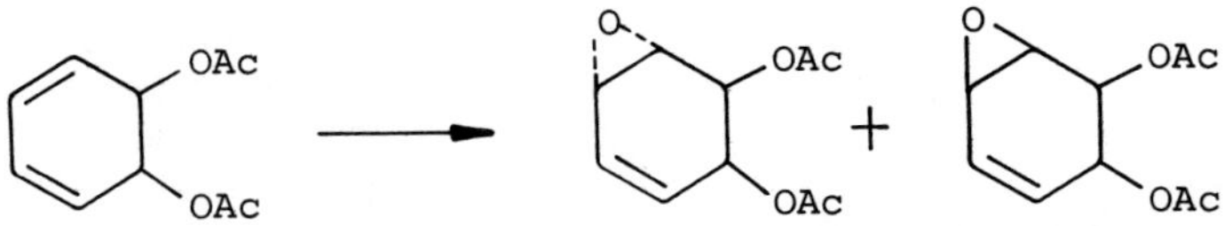

OAc OAc → O OAc OAc + O OAc OAc

321 *322* *323*

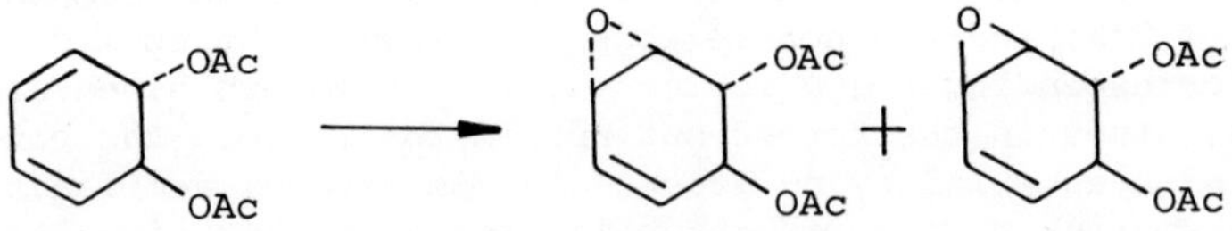

OAc OAc → O OAc OAc + O OAc OAc

324 *325* *326*

327 *328*

Table 9 Solvent Dependence in the Epoxidation of *328*[a]

Solvent	cis/trans ratio
AcOH	8
n-hexane	4
cyclohexane	2.7
CH_2Cl_2	2.3
$CHCl_3$	1.3
t-BuOH	1.1
AcOEt	1.0
C_6H_6	0.89
CCl_4	0.67

[a]CPBA was used in all solvents, except for acetic acid, in which the epoxidation was carried out with PAA.

Whereas the tetrahydrophthalic anhydride *329* is epoxidized exclusively to the cis isomer in several different solvents, the corresponding diester *330* forms an excess of the *trans*-epoxide (80% in cyclohexane, 73% in MeCN) (36,218). This difference can be explained in conformational terms, since *329* can only exist in two boat conformations, of which *331* should be the more stable and would be attacked preferentially from the exo side (in analogy with norbornene). In the half-chair conformation of the diester *332*, epoxidation will occur more easily from the anti side, in analogy with the case of *320*. The stereoselectivity is less pronounced in the epoxidation of the imide *333*, which gives the *cis*- and *trans*-epoxides in a ratio of 2:1, possibly because the greater bulk of the phenethylamino group hinders approach to the syn side in the conformation analogous to *331* (218).

In the bicycloheptane series the *endo*- and *exo*-anhydrides

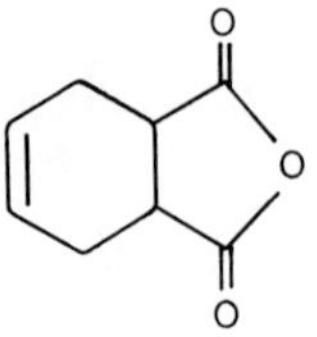

329

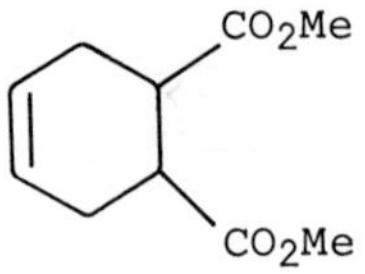

330

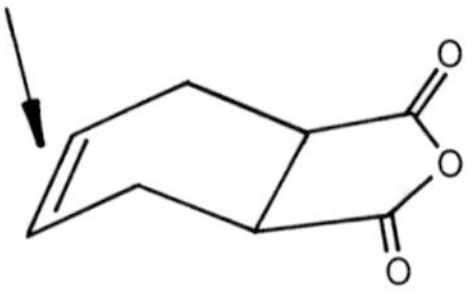

331

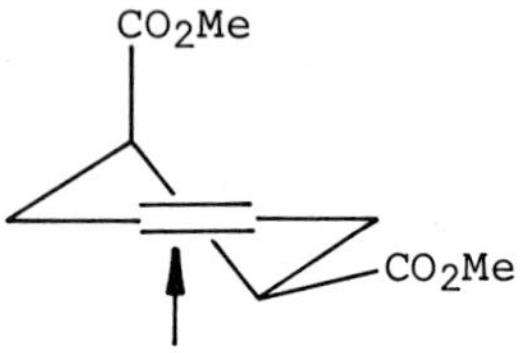

332

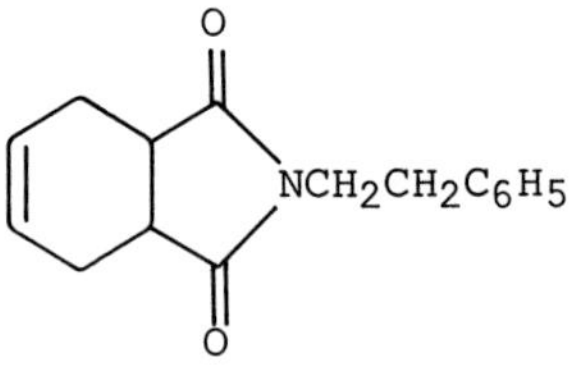

333

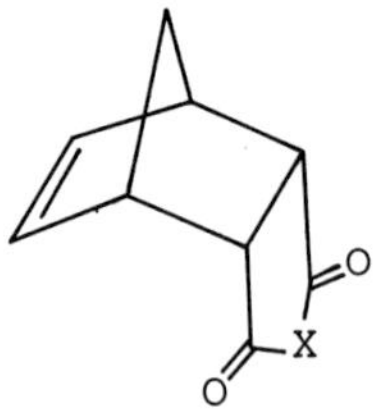

334: X = O
336: X = $NCH_2CH_2C_6H_5$

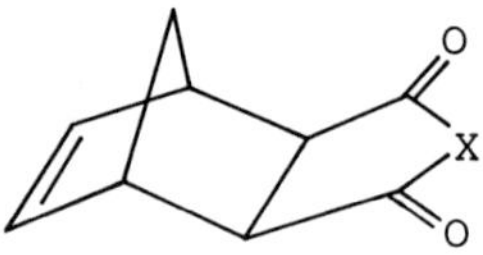

335: X = O
337: X = $NCH_2CH_2C_6H_5$

and imides *334-337* are all epoxidized only from the exo side (219,220). The kinetic data shown in Table 10 (220) indicate that the norbornene derivatives react more slowly than *329*, as expected for the steric effect of the methylene bridge, and exo isomers (*335*, *337*) faster than endo ones (*334*, *336*), even if the shielding to exo attack should be about the same for both types of isomers. This was explained with the assumption of a field effect of the carbonyl groups providing an electron-deficient center acting across space to deactivate the double bond toward electrophilic attack; the endo anhydride or imide groupings, being nearer to the π cloud, would have a stronger deactivating effect than the exo ones (220). A field effect of the imide function acting either on the endocyclic or on the exocyclic double bond was also postulated to explain the formation of *339* from *338*, and of *341* from *340* (221). Although the fact

Table 10 Relative Rates of Epoxidation with CPBA in Chloroform

Substrate	Relative rate
329	80
334	1.0
335	3.2
336	7.5
337	18.5

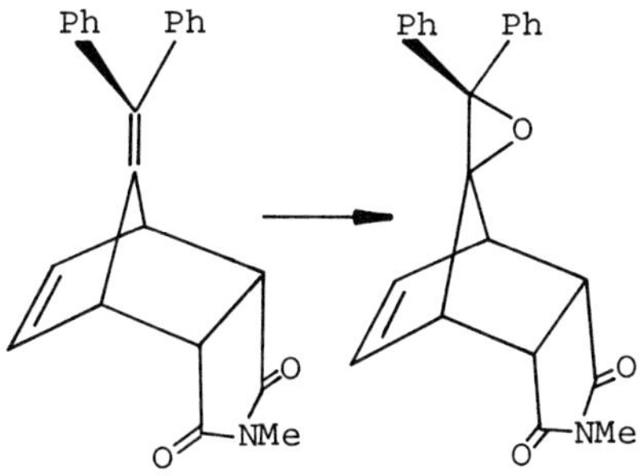

338 *339*

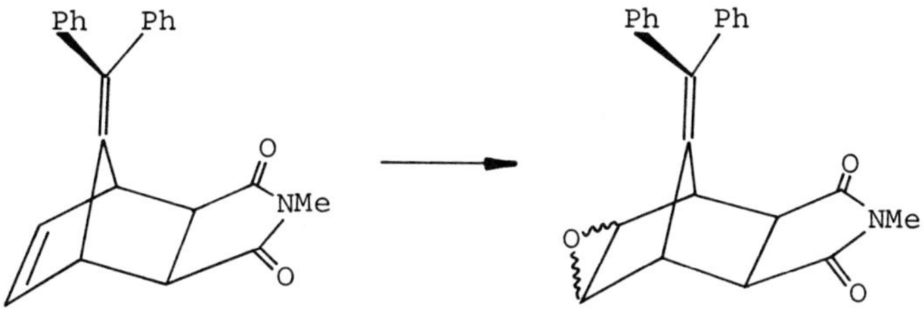

340 *341*

that *342* gives an unresolved mixture of epoxides deriving from attack on either double bond appears to support this interpretation, alternative explanations based on steric and torsional effects cannot be completely ruled out.

The remarkable difference in behavior toward PBA of the reserpine synthesis intermediates *343* and *344* was also explained by an interaction through space between double bond and carbonyl

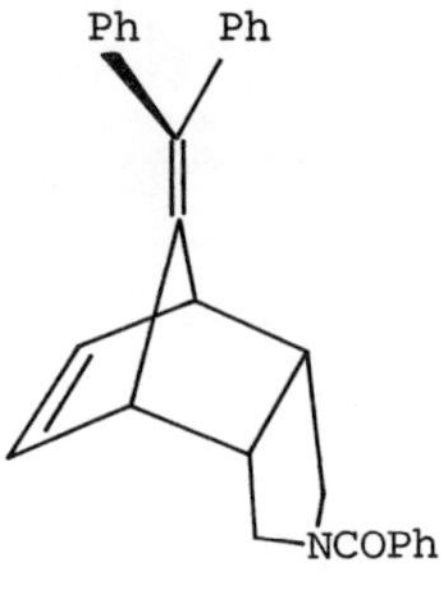

342

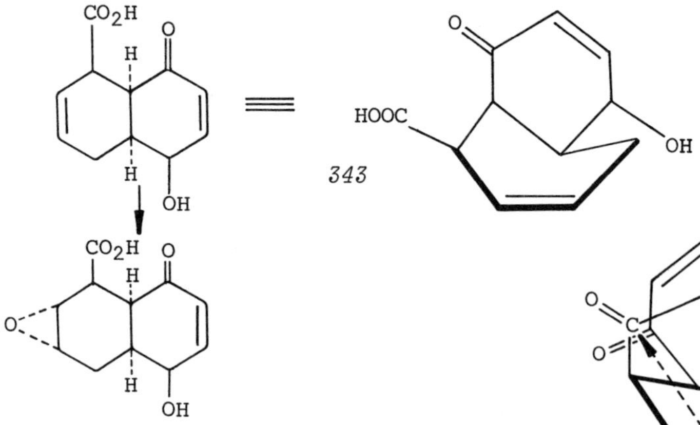

343

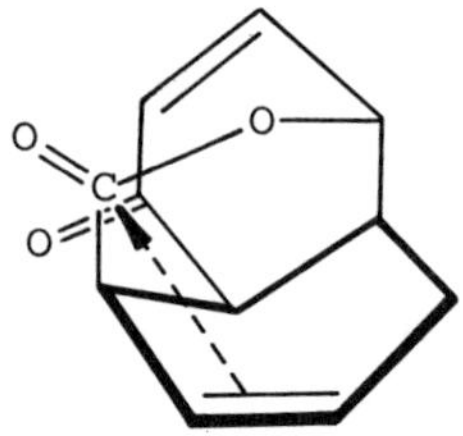

344

345

group (222). Whereas *343* easily gives the expected epoxide *345*, the corresponding lactone *344* reacts much more slowly and does not yield definite products. It was assumed that in the rigid cage structure of *344* the geometry favors electron release from the isolated double bond to the electron-deficient pseudoaxial carbonyl carbon more than in the less rigid hydroxy acid *343*.

In the bicyclo[2.2.2]octene series the anhydride *346* is epoxidized entirely anti to the functional group, but additional steric effects can reverse this preference, as shown by *347* and *348*, which are attacked by peroxyacids syn to the anhydride grouping (223), because of the presence of the additional ring, and by *349* in which the 10-methyl group exhibits a stronger shielding effect than the anhydride grouping (224).

The introduction of a second epoxy function with NPBA in

346 347 348

CO_2Me

349

OAc

HO HO HO

350 351 352

HO HO HO

353 354 355

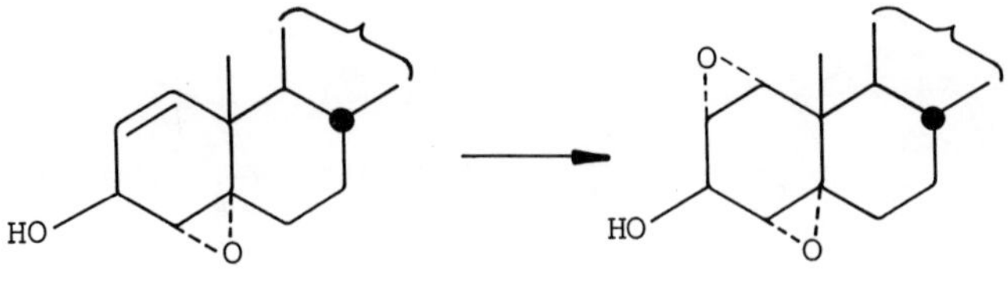

HO O HO O O

356 *357*

the epoxyalkenes *350*, *352*, *353*, *355*, and *356*, which takes place exclusively syn to the first one, can be attributed to the conformational situation, rather than to a specific *syn*-directing effect of the epoxide oxygen, as proposed (225), or of the hydroxyl group (which would not explain the conversion *356* → *357*). An α-epoxy group imposes conformation *358*, a β one conformation *359*, to ring A, in each of which the side syn to the existing epoxide group is less hindered.

Another case in which steric effects alone do not account for the observed stereochemistry is that of the epoxidation of the Δ^4-octalin derivatives. Although *360*, *361*, and *362* are epoxidized by PAA exclusively anti to the functional group, *363* gives an about equimolar mixture of *cis*- and *trans*-epoxides (226,227).

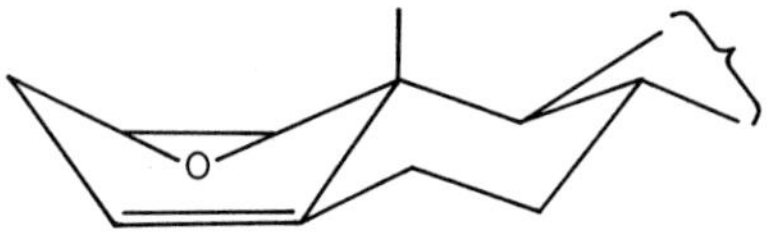

358 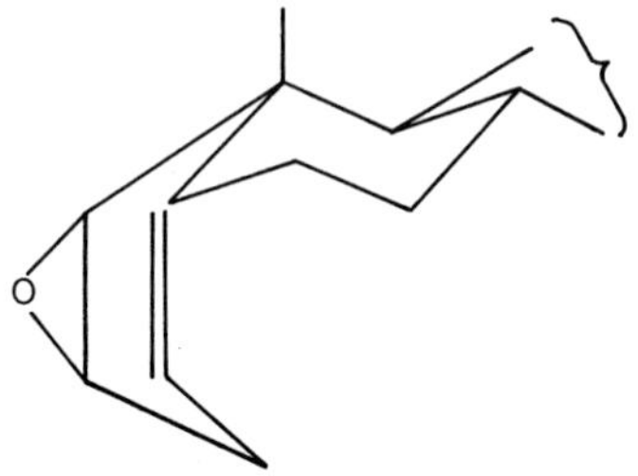 *359*

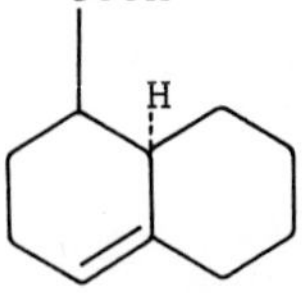

COOH H COOH H O O H O O O H O O

360 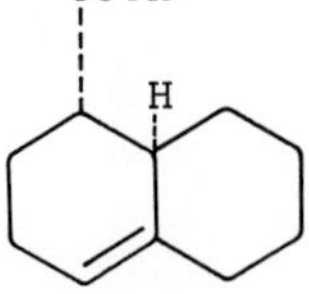 *361* 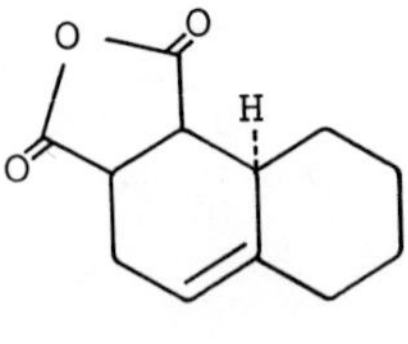 *362* 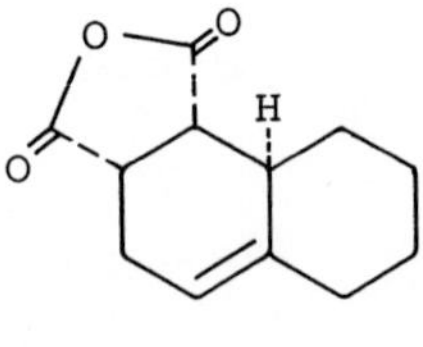 *363*

12. *Asymmetric Epoxidation*

When the epoxidation of prochiral alkenes is carried out with chiral peroxyacids, optically active epoxides can be obtained, but the optical yields are usually low (228). For instance, styrene gives with (+)-peroxycamphoric acid the (-)(S)-epoxide, and the optical yields range from 2.0 in ether to 4.4 in chloroform (51). Under the same conditions compound *364* forms the corresponding epoxide (an intermediate in the asymmetric synthesis of the alkaloid orixine) of 2.4% optical purity (229).

A correlation between the structures of the alkene and of the peroxyacid and the absolute configuration of the predominant epoxide enantiomer could be established on the basis of a transition-state model *365*, in which it was assumed that the bulkier group bonded to the alkene faces the least hindered region of the peroxyacid, between S and M, near S. The highest optical yield (7.5%) was achieved in the reaction between *trans*-stilbene and (+)-peroxycamphoric acid (230).

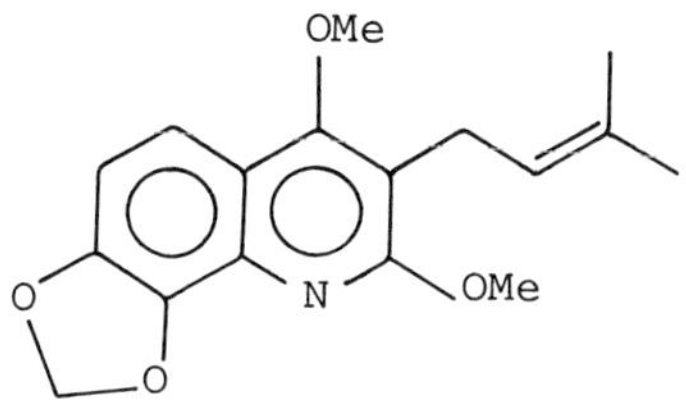

OMe N OMe O O

364

R" L M H R' C_6H_5 S

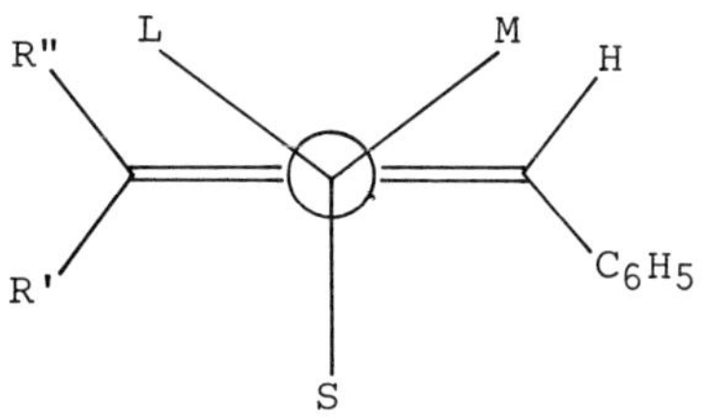

365

The asymmetric epoxidation of 1,2-dihydronaphthalene with (+)-peroxycamphoric acid provides a route for the preparation of the interesting naphthalene-1,2-oxide (*366*) in 10% optical yield (231).

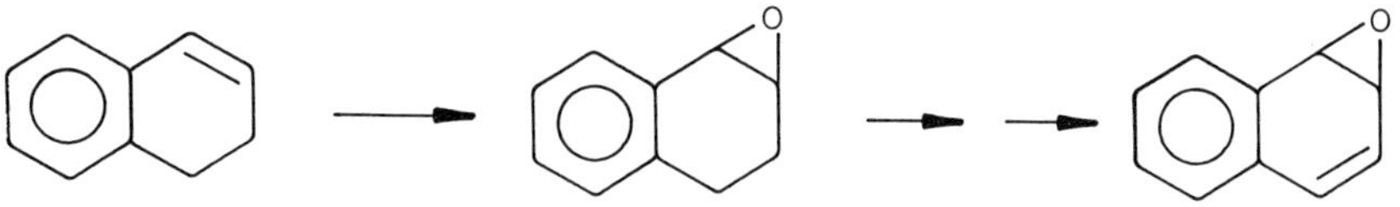

O O

366

B. Oxidation of Alkenes with "Peroxycarboximidic Acids"

Payne (232) found that epoxides are formed when alkenes

are treated with hydrogen peroxide in a medium consisting of a nitrile buffered at pH 8, preferably with sodium hydrogen carbonate (233) or phosphate (234). The reaction probably involves the formation of peroxycarboximidic acids *367*, which have, however, never been isolated. In the absence of the olefin these acids, which can be considered as imino analogues of peroxyacids, are reduced by a second molecule of hydrogen peroxide to give the amides by the well-known Radziszewski reaction; if an

$$RCN + H_2O_2 \longrightarrow R\text{-}\underset{\underset{NH}{\|}}{C}\text{-}OOH \xrightarrow{H_2O_2} RCONH_2 + O_2 + H_2O$$

367

$+ \; RCONH_2$

olefin is present in the medium, the acids act as epoxidizing reagents, presumably through a mechanism of the same type as that proposed for the peroxyacid oxidations. This reaction is particularly useful when the substrate or the product is not stable under acidic conditions. For instance, *368* produces a good yield of the epoxide with benzonitrile/hydrogen peroxide, but only the Baeyer-Villiger product *369* with PAA (233).

368 *369*

There are several other differences between the two methods of epoxidation. For instance, 1-butene and 2-methyl-2-butene are epoxidized at about the same rate with benzonitrile/hydrogen peroxide, but at greatly different rates with PAA, and acrolein diethyl acetal is epoxidized much faster by the former than by the latter. Apparently, the formation of *367* is the slow step (232), and peroxycarboximidic acids have greater steric requirements than peroxyacids. Limonene, which is monoepoxidized exclusively on the endocyclic double bond by peroxyacids, gives with acetonitrile/hydrogen peroxide an about equimolar mixture of *370* and *371*, and β-pinene (exocyclic double bonds) forms the epoxide faster than α-pinene (endocyclic double bond) (235). The greater steric requirements of peroxybenzimidic acid are also evident if one compares the epoxidations of rigid methylenecyclohexanes with this reagent and with peroxyacids. In contrast with the latter, the former always lead to an excess of equatorial attack, as shown for compounds *108*, *109*, *110*, and *114* (106) and for the cholestane derivatives *120*, *122*, and *123* (119); these data can be compared with those relative to peroxyacid epoxidations in Table 3, Sect. II-A-6. The bicyclo[3.3.1]-nonane *372* also gives only the epoxide *373* with benzonitrile/hydrogen peroxide, by equatorial attack (236).

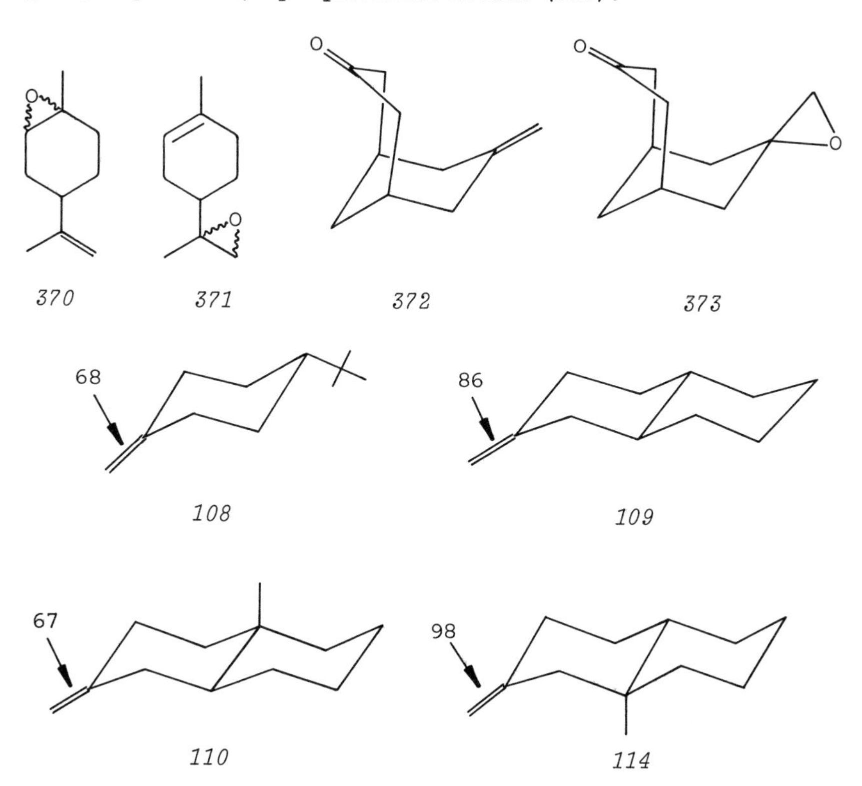

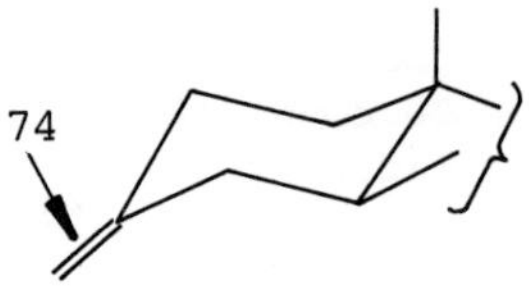

74 83 65

120 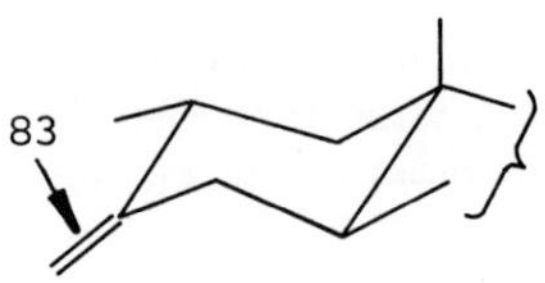 *122* 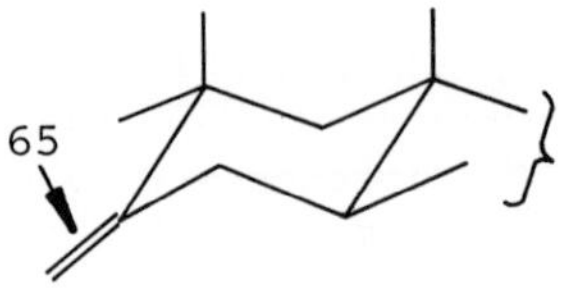 *123*

Although only few data are available, it appears that the *syn*-directing effect of a hydroxyl group is also present in this type of epoxidation. Thus 2-cyclohexenol (*161*) gives 82% *cis*-epoxide with benzonitrile/hydrogen peroxide (146), and *374* produces with acetonitrile/hydrogen peroxide the epoxides *376* and *377* in a 73:27 ratio (preferential epoxidation syn to the allylic hydroxyl), as compared with a 39:61 ratio in the analogous reaction of the diacetyl derivative *375* (237).

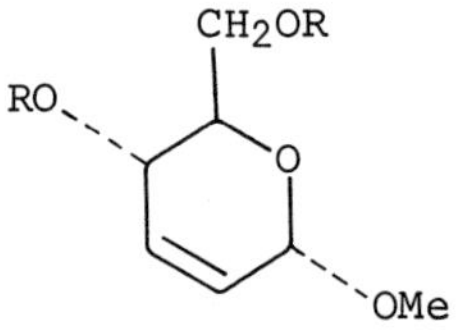

CH_2OR RO O OMe

374: R = H
375: R = Ac

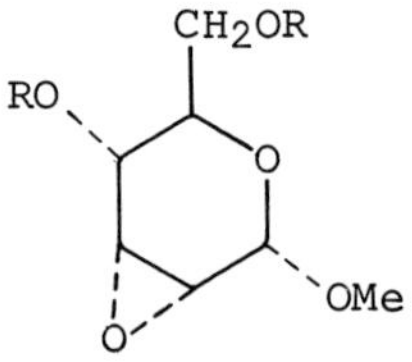

376 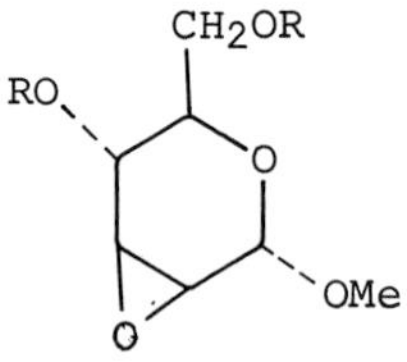 *377*

C. Oxidation of Electrophilic Alkenes with Alkaline Hydrogen Peroxide (Weitz-Scheffer Reaction)

Compounds in which a double bond is conjugated to electron-withdrawing groups, such as enones and unsaturated acids and nitriles, undergo only very slowly, or not at all, electrophilic attack by peroxyacids; other reactions, e.g., the Baeyer-Villiger oxidation, can interfere with the formation of epoxides. In these cases the oxidation with hydrogen peroxide under basic conditions is one of the methods of choice for the conversion into epoxides. This reaction, which was first reported by Weitz and Scheffer (238), has been applied to many α,β-unsaturated aldehydes and ketones, and to a more limited range of α,β-unsaturated nitriles, dicarboxylic esters, sulfones, etc. (239, 240). Yields are usually rather good if the alkalinity and reaction time are adequately controlled; otherwise side reactions leading to dicarboxylic acids and other products can become important (241).

Bunton and Minkoff (242) found that the reaction is first order both in unsaturated ketones and O_2H^-, and that, in contrast with the peroxyacid oxidation, its rate is depressed by the introduction of a β-methyl group, in accordance with a mechanism involving a Michael-type addition of the hydroperoxide anion to the conjugated system, followed by ring closure of the intermediate enolate anion *378* with expulsion of OH^-. The reaction may be stereoselective, but is usually not stereospecific. This is easily understood if the enolate ion *378* has a sufficiently long lifetime for a rotation about the α,β bond to take place before cyclization. Furthermore, if there is a hydrogen atom on the α carbon, cis-trans equilibration can occur, both

$$H_2O_2 + OH^- \rightleftharpoons HO_2^- + H_2O$$

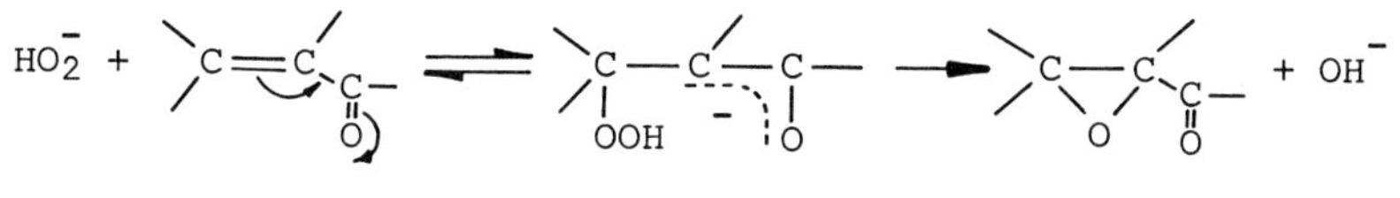

378

on the starting enone and on the product epoxide. House and Ro (243) established that the diastereoisomeric epoxides *379* and *381* are equilibrated by sodium ethoxide in deuterated ethanol,

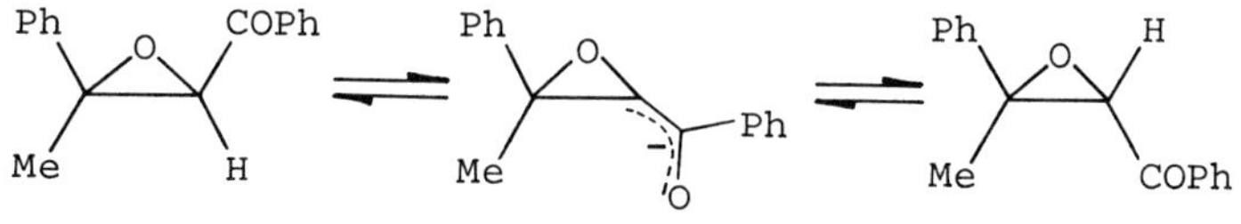

379 *380* *381*

with incorporation of one atom of deuterium, as expected, if *380* is the intermediate in the equilibration. It may appear surprising that the treatment of the trans isomer of *382* with sodium methoxide converts it to a large extent into the cis isomer; this obviously nonequilibration conversion is probably due to the lower solubility of the latter isomer, which precipitates from the reaction medium (244). That enolization of the starting or final compounds is not the sole cause for the lack in stereospecificity is shown by the conversion of both isomers of *383*, which have no α-hydrogen, into the single epoxide *384* (243). Although (*Z*)-*383* is very stable under the action of alkali, it is converted into the (*E*) isomer at an appreciable rate if hydrogen peroxide is also present, as shown by recovering

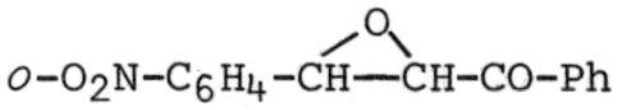

o-O_2N-C_6H_4-CH—CH-CO-Ph (epoxide, O bridging the two CH)

382

MeCH=CMe-CO-Me ⟶ (epoxide: Me, H on one carbon; Me, CO-Me on the other)

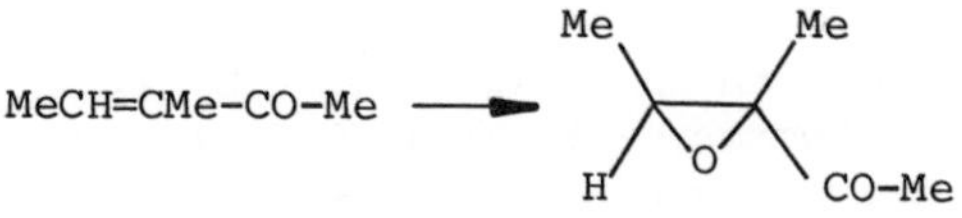

383 *384*

the unreacted enone from an incomplete reaction. These findings can be explained in terms of the reversibility of the first step in the Weitz-Scheffer reaction, a fact that has been independently confirmed (241,245).

On the basis of the data reported above one might conclude that the stereochemistry of the Weitz-Scheffer epoxidation of an enone system which is not included in a ring is determined exclusively by the relative energies of the transition states leading to the diastereoisomeric epoxides, and that the most stable isomer should always be formed in excess. This is very often the case, but the exclusive formation of *384* from (Z)- and (E)-*383* cannot be accounted for exclusively on the basis of the steric interactions between the substituents during the formation of the oxirane ring. The differences in the effective sizes could not justify such a high preference for the (E) isomer, since COMe would be expected to be, if not smaller, certainly not much bigger than Me. Similarly, the formation of the (E)-epoxides *387* from (Z)- and (E)-*385* and *386* indicates that the thermodyanimcally less stable diastereoisomers can be the main products. It rather appears that the diastereoisomeric epoxide which has the less hindered carbonyl group is the favored product. This has been interpreted (246) in terms of overlap control in the transition state leading from the intermediate *378* to the epoxide. The negative charge on the anionic α-carbon

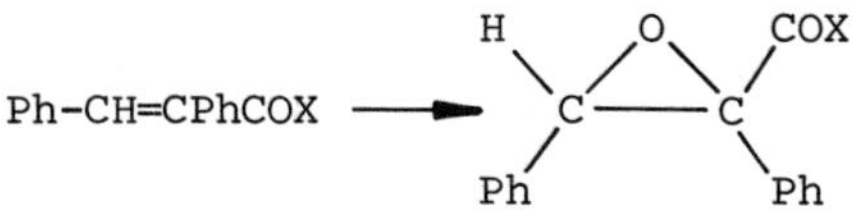

Ph-CH=CPhCOX ⟶ (epoxide: H, Ph on one C; COX, Ph on the other C)

385: X = Me
386: X = Ph

387

requires a delocalization, which imposes a conformational restriction about the acyl to α-carbon bond such that there is coplanarity between carbonyl and the adjacent carbon atoms, allowing for maximum overlap. Of the two conformers in which this demand is satisfied *388* should have the lower energy, since the ordinary steric interaction between the two phenyl groups can be minimized by assumption of the best conformations about the C-phenyl bonds. In *389* the R/phenyl interaction would tend to

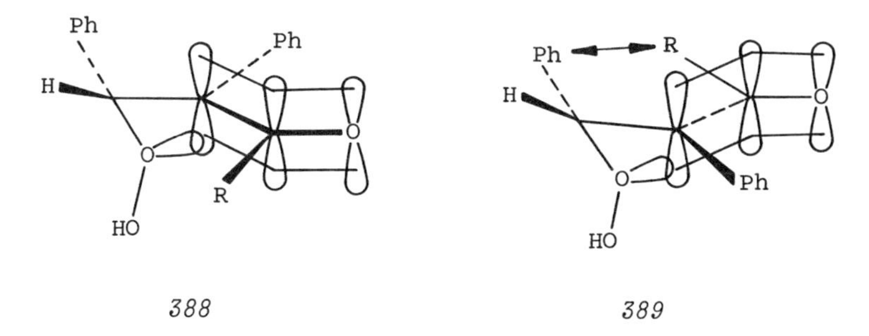

twist the carbonyl out of coplanarity, thus increasing considerably the transition-state energy. With groups of lesser steric demand than acetyl, such as CHO (*390*) and CN (*391*), the (*Z*)-epoxides are the only products. The (*E*)-epoxyaldehyde *393* must therefore be prepared through a route first suggested by Wasserman and Aubrey (247), involving peroxyacid epoxidation of the (*E*)-alcohol *392*, followed by mild oxidation; dicyclohexylcarbodiimide in DMSO is particularly good for the latter step (248).

Ph Ph H CHO

390

Ph CN H Ph

391

Ph Ph H CH_2OH —AcO_3H→ Ph O Ph H CH_2OH —DCC, DMSO→ Ph O Ph H CHO

392 *393*

In β,β-disubstituted chalcones the isomer with the carbonyl function cis to the smaller β substituent reacts with alkaline hydrogen peroxide much faster than its diastereoisomer; for instance, whereas *394* gives *395*, *396* is not converted into the epoxide. This may be due to the fact that in *396* the interference between phenyl and COR prevents the coplanarity that is necessary for the electron delocalization favoring the nucleophilic attack by O_2H^- (249).

The Weitz-Scheffer epoxidation of α-ionone (*397*) is reported to be remarkably stereoselective and to produce only one (*398*) of the possible four diastereoisomers (250).

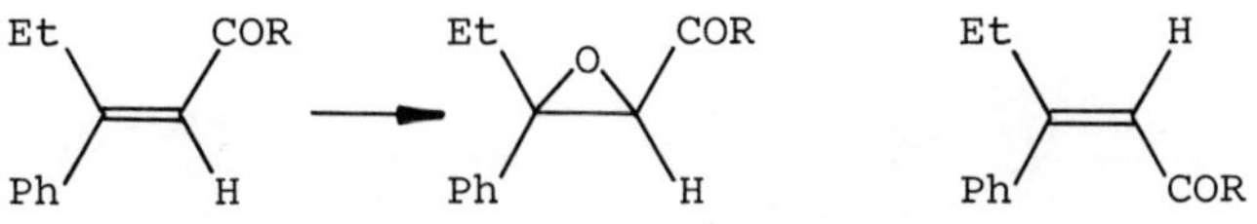

394 *395* *396*

397 *398*

Besides the carbonyl group, other strongly electron-withdrawing groups can activate the double bond for attack by the hydroperoxy anion. One carbalkoxy substituent is not sufficient for such an activation, but alkylidenemalonates *399* undergo epoxidation (251). Also acrylonitriles (*400*) can be epoxidized, but hydrolysis of the nitrile function often takes place concurrently to give the epoxyamides *401* by the Radziszewski reaction. It appears that epoxidation precedes the hydrolysis of the nitrile group, since amides of the type *402* are not converted into epoxides under Weitz-Scheffer conditions (252). It has been proposed that the reaction involves attack by hydrogen peroxide on the nitrile group to give an intermediate peroxycarboximidic acid (*403*), which undergoes an intramolecular epoxidation (232,253). Such a mechanism does not appear as very likely, because it would involve an electrophilic attack on an electron-

$RR'C{=}C(CO_2Et)_2$ (*399*) $RCH{=}CR'CN$ (*400*) $RCH{-}CR'CONH_2$ (epoxide, *401*) $RCH{=}CR'CONH_2$ (*402*)

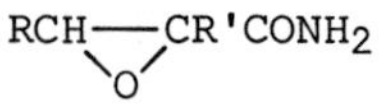

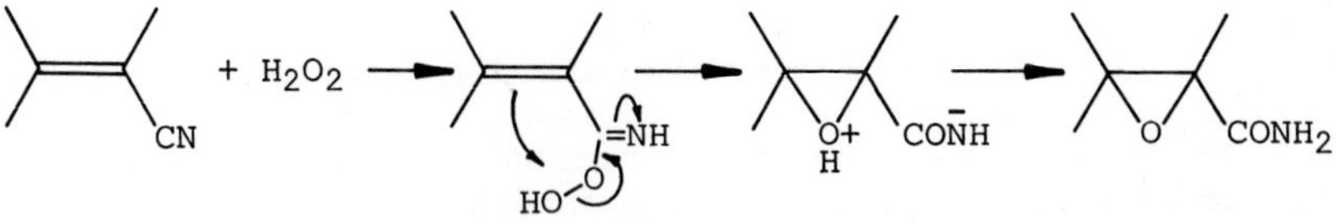

403

poor double bond and should lead to stereospecificity, which is not observed. It has actually been disproved for aryl-substituted cyanocarbethoxy derivatives of the type *404*, which with hydrogen peroxide and sodium bicarbonate form mixtures of the diastereoisomeric cyanocarbethoxy epoxides, accompanied by minor amounts of carbamoylcarbethoxy epoxides; when Ar = phenyl and R = H, the reaction produces exclusively the epoxide *405* (254). The conversion of the cyano into the amido group can also be completely suppressed if the oxidation is conducted in ethanol in the absence of base: *406* (R = H) gives only *407*, whereas when R is an alkyl group, mixtures of diastereoisomeric epoxides are formed (255).

$ArRC{=}C(CN)CO_2Et$

404

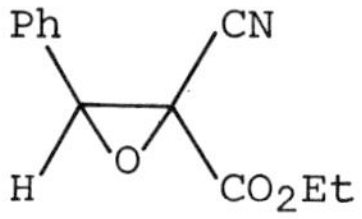

Ph, CN, H, O, CO_2Et

405

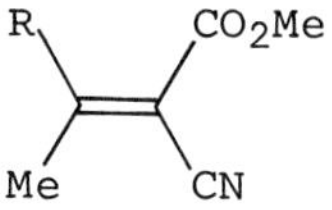

R, CO_2Me, Me, CN

406

H, CO_2Me, Me, O, CN

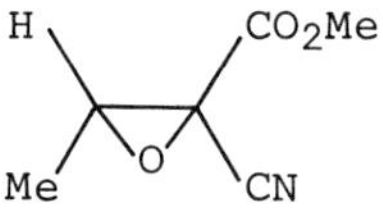

407

The nitroolefins *408* and *409* are converted by hydrogen peroxide and sodium hydroxide into diastereoisomerically pure epoxides, the configurations of which have not been determined. The isomer *410* does not react under these conditions, probably because of the noncoplanarity between the nitro group and the double bond, as seen for *396* (256). Unsaturated sulfones, such as *cis*- and *trans-411*, give under Weitz-Scheffer conditions 100% stereoselectivity the more stable *trans*-epoxide (257).

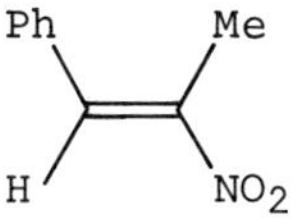

Ph, Me, H, NO_2 — *408*

Ph, Ph, H, NO_2 — *409*

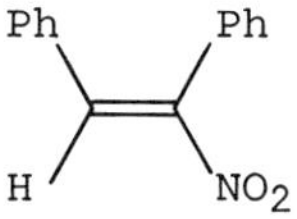

Ph, NO_2, H, Ph — *410*

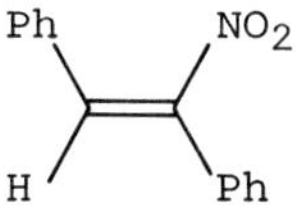

$PhCH{=}CHSO_2Ph$ — *411*

A partial asymmetric synthesis of α-cyanoglycidic acids can be achieved by Weitz-Scheffer epoxidation of alkylidenecyanoacrylic acid (-)-menthyl esters (*412*). Mild hydrolysis of the products gives optically active acids (258).

RCH=C(CN)CO_2Menthyl(-)

412

The epoxidation of an enone with an endocyclic double bond can be entirely stereoselective. Thus carvone (*413*) gives only the epoxide *414* (259), and 4-menthen-3-one (*415*) only *416* (260). This is in accordance with the fact that the hydroperoxy group must be as nearly axial as possible near the transition state for the cyclization step. Of the two axial conformations of the anions derived from *413*, *417* (leading to *414*) will be definitely more stable than *418* (261).

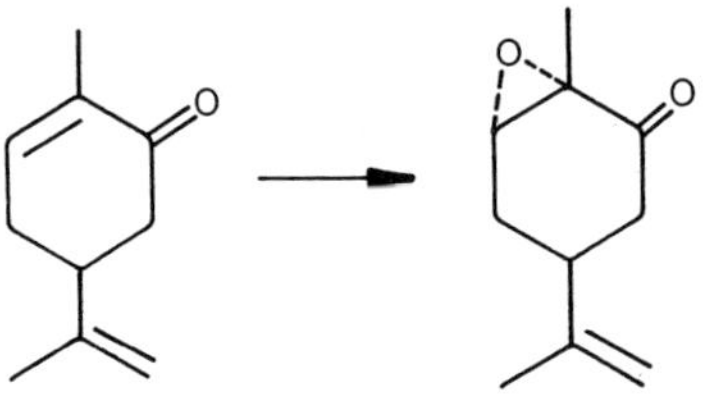

413 *414* 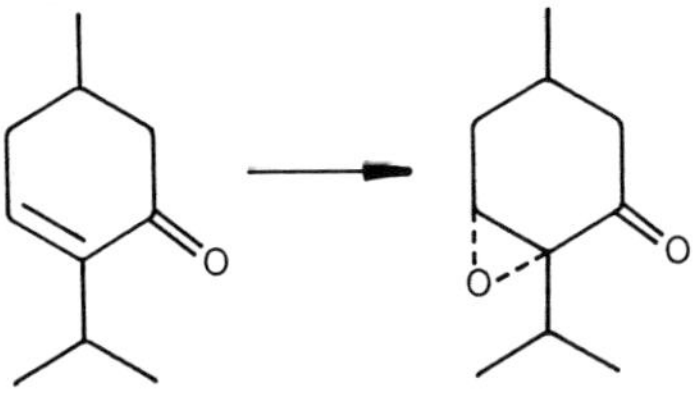 *415* *416*

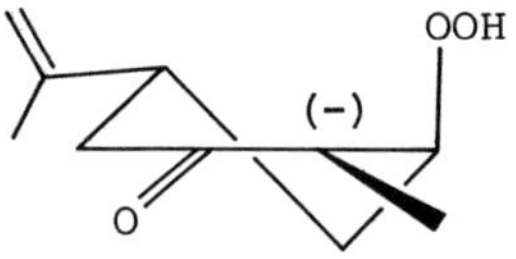

417 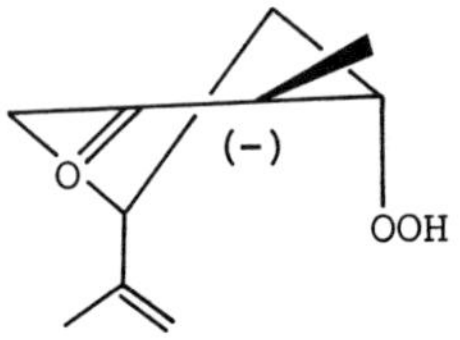 *418*

The formation of an 8:2 mixture of *420* and *421* from piperitone (*419*) could be rationalized on the basis of a preference for conformation *422* of the hydroperoxy anion (261), but the presence of a chiral center α to the carbonyl allows for facile epimerization by bases; it is actually found that *420* and *421* equilibrate under the reaction conditions to an 8:2 mixture (262). Epimerization also involves the starting enone, since optically pure (-)-*419* gives partially racemized *420* (259). On the other hand, the exclusive epoxidation from the sides indicated by the arrows of the nonepimerizable compounds peryllaldehyde (*423*), verbenone (*424*), myrthenal (*425*) (259), and umbellulone (*426*) (263) can be explained by the theory of optimal overlap control, as for *413* and *415*.

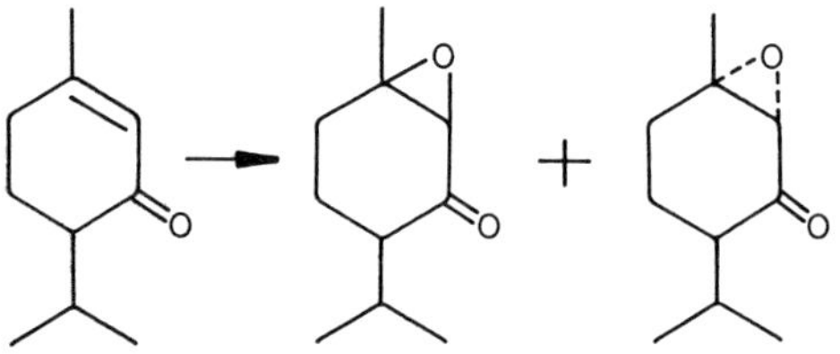

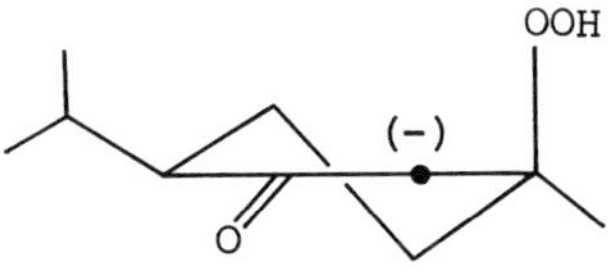

419 *420* *421* *422*

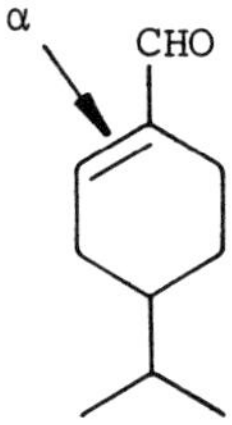

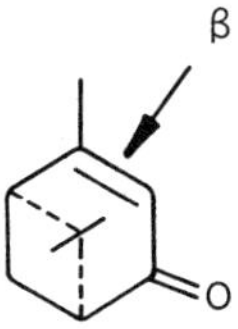

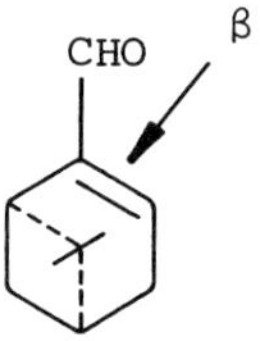

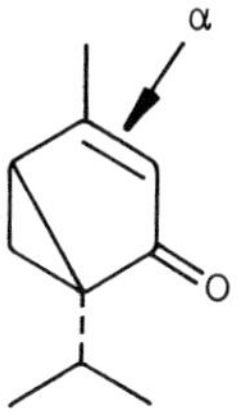

423 *424* *425* *426*

Griseofulvin (*427*) forms only one epoxide in the Weitz-Scheffer reaction, of probable structure *428* (264).

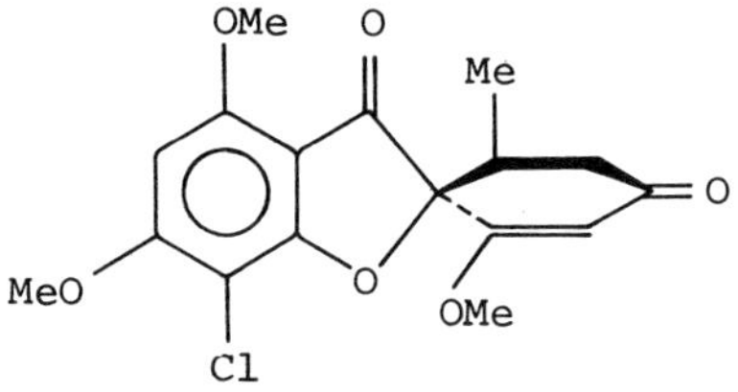

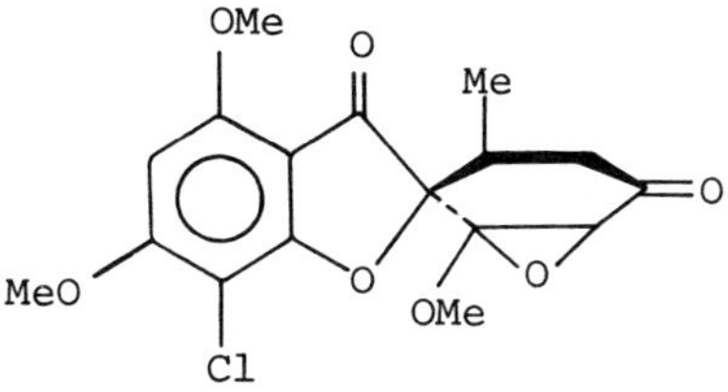

427 *428*

The Weitz-Scheffer reaction is less stereoselective for an enone with an exocyclic double bond because of the possibility of rotation of the hydroperoxyalkyl side chain in the intermediate carbanion. An excess of the more stable epoxide is usually formed, as exemplified by the epoxidations of pulegone (*429*) (261), pinocarvone (*430*) (265), and isopropylidenecamphor (*431*) (157).

The exclusive formation of *433* from *cis*- and *trans*-*432* (266) and of mixtures of *435* and *436* from *cis*- and *trans*-*434* (122) is again in accordance with the rule that the keto-epoxide with the

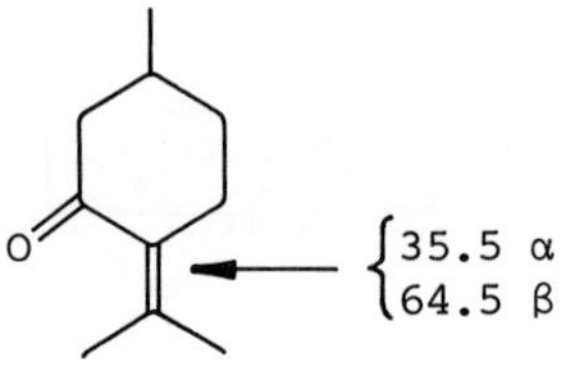

$\begin{cases}35.5\ \alpha \\ 64.5\ \beta\end{cases}$ $\begin{cases}35.5\ \alpha \\ 64.5\ \beta\end{cases}$ $\begin{cases}33\ \alpha \\ 67\ \beta\end{cases}$

429 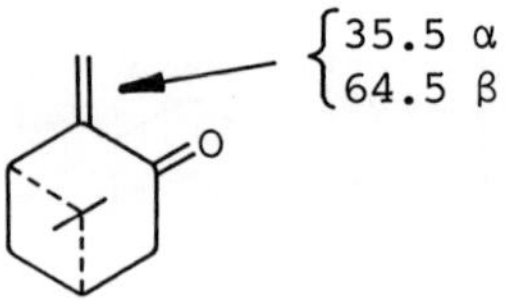 *430* 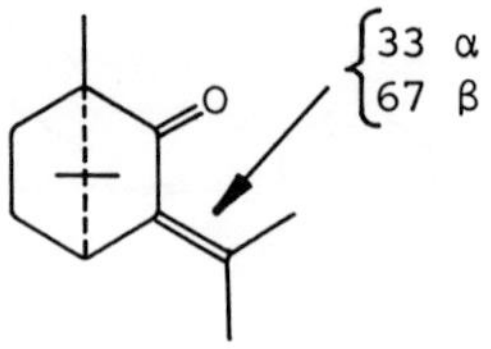 *431*

least hindered carbonyl group is preferentially obtained in the Weitz-Scheffer reaction. When the interference between the side-chain phenyl and the substituents on $C_{(2)}$ gets too large, such as in *437*, the compound does not epoxidize.

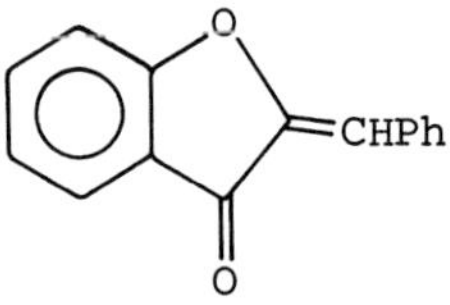

CHPh Ph H Ph H CHPh

432 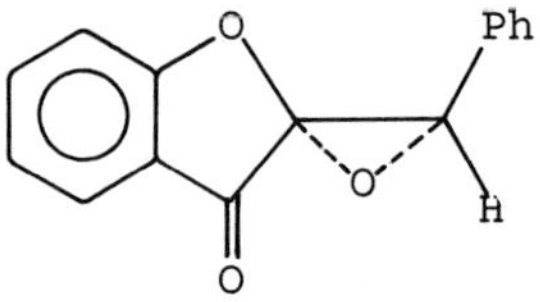 *433* 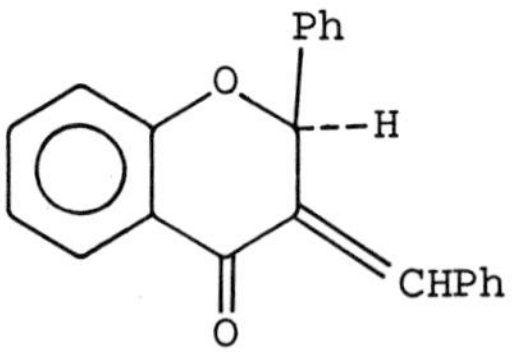 *434*

Ph H Ph H Ph H Ph H Ph Ph CHPh

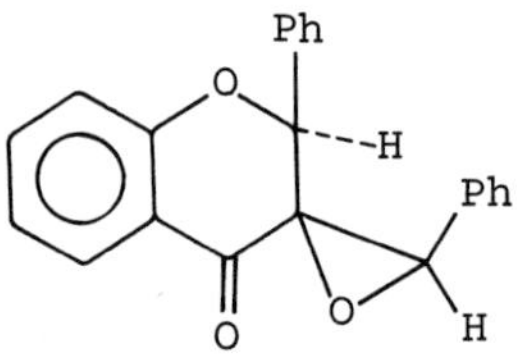

435 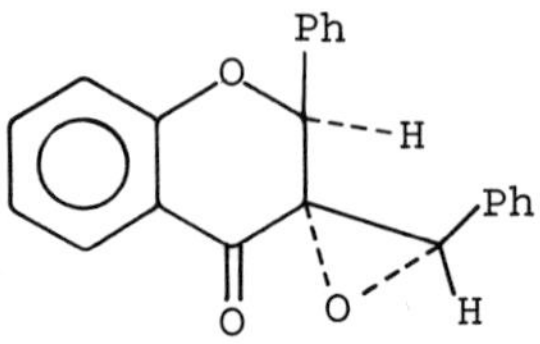 *436* 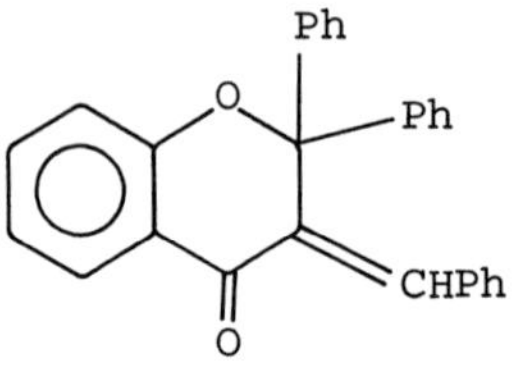 *437*

In the steroid series the tendency to α attack is much less important in the Weitz-Scheffer, than in the peroxyacid, oxidation. Thus, whereas Δ^1-3-oxo derivatives such as *438* (267), *439* (268), and Δ^{16}-20-oxo derivatives such as *440* (269) form exclusively the less hindered α-epoxides with hydrogen peroxide and alkali, Δ^5-4-oxo (*441*) and Δ^4-6-oxo (*442*) steroids give only the β-epoxides (270). It must be assumed that the hydroperoxy group has a relatively small steric demand and that its interference with the 19-methyl group is less important than the stereoelectronic requirements in the cyclization step, which, as pointed out above for *417*, imply an axial disposition of the hydroperoxy group. Whereas β attack on $C_{(6)}$ of *441* (or $C_{(4)}$ of

442) gives the right disposition of this group (*443*), α attack would produce an equatorial hydroperoxy group (*444*), and the transition state leading from it to the α-epoxide would involve a rather unfavorable twisting of ring B.

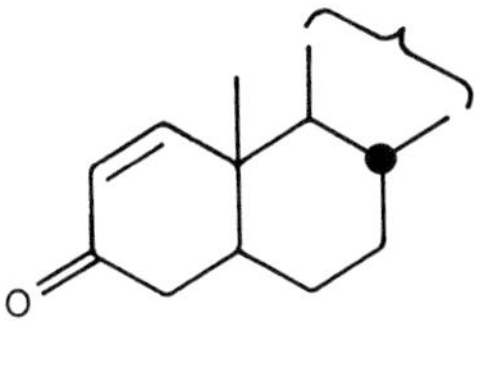

438 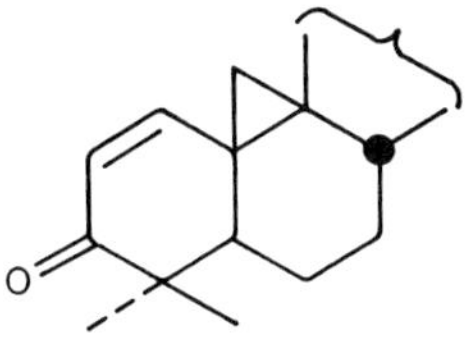 *439* 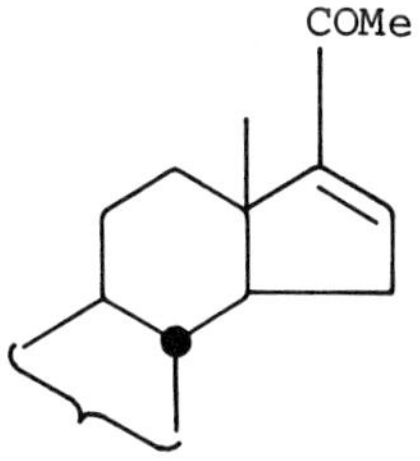 *440*

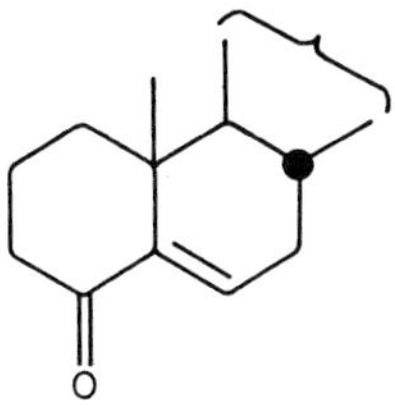

441 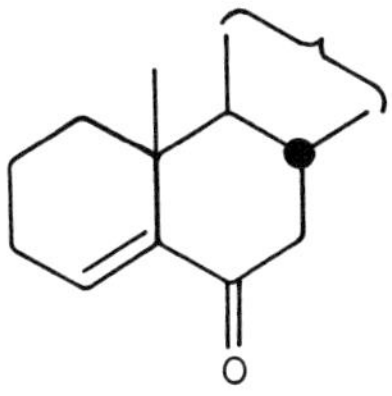 *442*

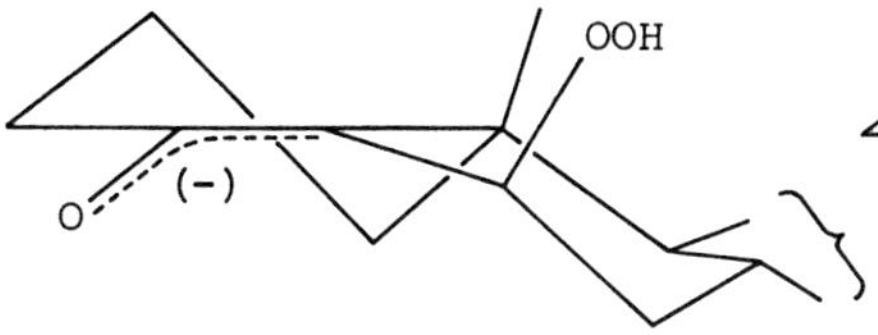

443 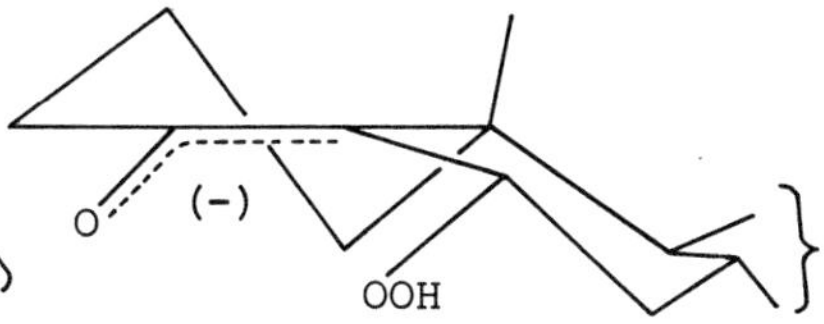 *444*

It is less obvious why the β-epoxides are the main or only products in the Weitz-Scheffer oxidation of Δ^4-3-oxo steroids (*445*) (271-273), since the hydroperoxy group can assume the axial disposition with respect to ring A both in the 4α- (*446*) and 4β-hydroperoxides (*447*). Two effects could be responsible for this stereoselectivity (274). Stabilization of the carbanion should be greater in *447* than in *446*, since 5β-3-oxosteroids enolize preferentially toward $C_{(4)}$, and 5α ones toward $C_{(2)}$. Furthermore, in the transition state for ring closure, which can be considered as an intramolecular nucleophilic displacement on oxygen, the departing OH^- must be anti to the enolate π bond. In such a geometry the reaction will be more hindered with *446* (O_2H axial to ring B) than in *447* (O_2H equatorial to ring B). In accordance with the latter explanation is the fact that when an additional double bond is present at 9,11, such as in *448* (171) and *449* (275), the 4,5α-epoxide is by far the main product, probably because in these compounds the 9α-hydrogen is absent and there is only one *syn*-axial interference between the leaving OH^- and hydrogen in the transition state analogous to *446*.

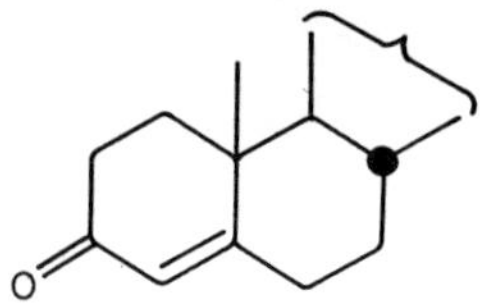

O (-) O OH H 1 H

445 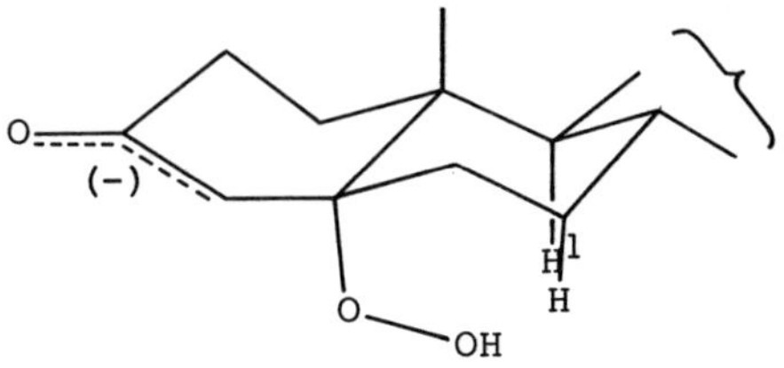 *446*

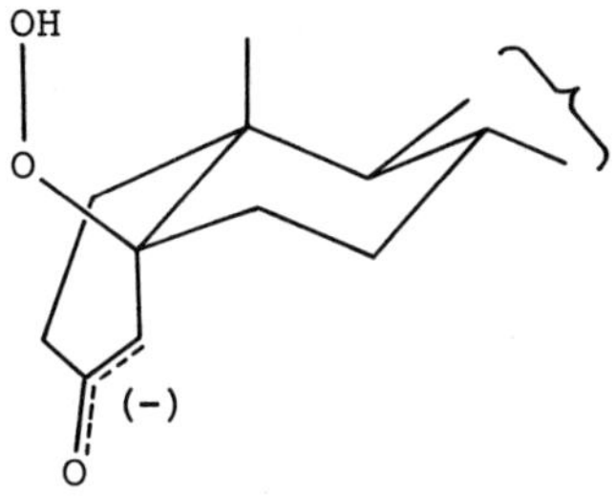

OH O (-) O OAc O

447 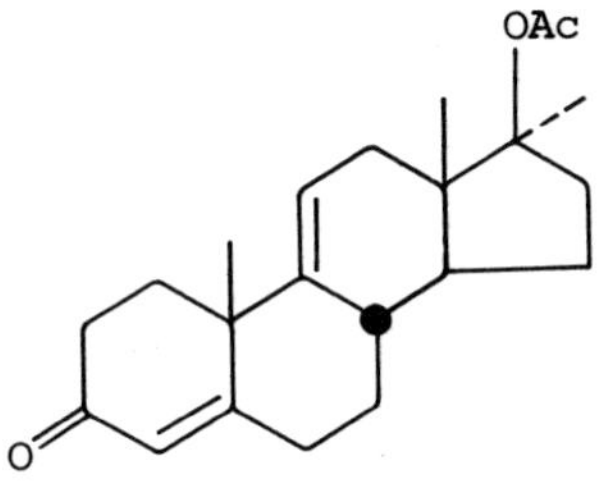 *448*

OH O R' R" O

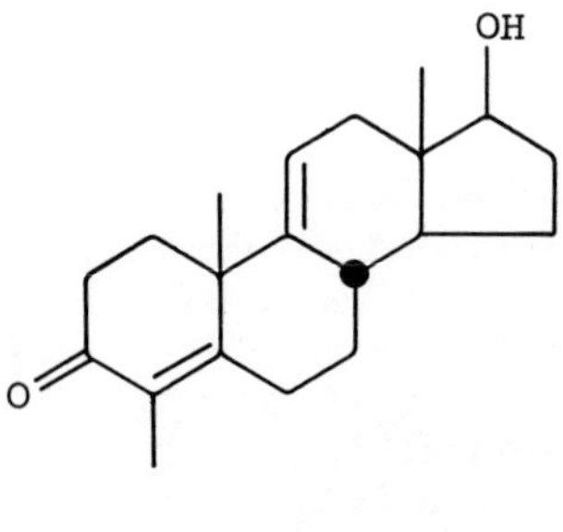

449 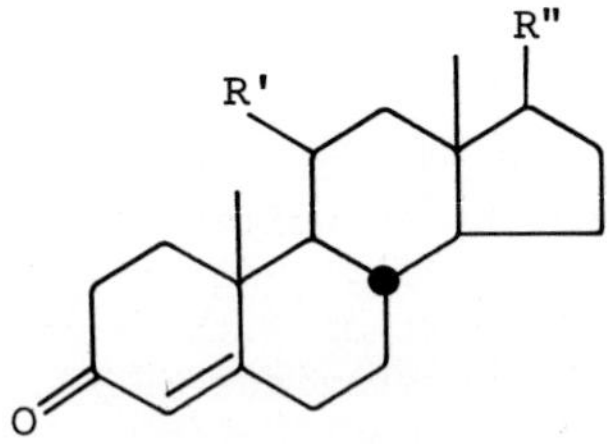 *450*

A very peculiar polar effect of remote electronegative substituents on the steric course of the Weitz-Scheffer epoxidation of Δ^4-3-oxosteroids (*450*), which was observed by Henbest and Jackson (245), increases the percentage of α-epoxide, and in some cases makes it the main product, as shown in Table 11. In

Table 11 Weitz-Scheffer Epoxidation of Δ^4-3-oxosteroids (*450*)

Substrate		% α-epoxide	Substrate		% α-epoxide
R'	R"		R'	R"	
H	H	0	H	oxo	30
H	β-C_8H_{17}	0	H	β-OH	31
H	β-$CHMe(CH_2)_3OH$	11	H	β-CO_2^-	60-65
H	α-Br	26	α-OH	β-OH	0
H	β-COMe	26	α-OH	α-Me, β-OH	0
H	α-OH	27	α-OH	β-COMe	25
H	β-CN	28	β-OH	oxo	51
H	β-CO_2Me	30	oxo	oxo	86

an attempt at an explanation it was proposed that the directive effect of the remote substituent was due to differences in the electrostatic interactions between this substituent and the two alternative reactions occurring in ring A. In the transition state of each of the two irreversible steps leading to the diastereoisomeric epoxy ketones the negative charge on the changing enolate group will interact with the fractional positive charge on the carbon carrying the electronegative substituent. This interaction will be stronger in the transition state leading to the α-epoxide since the enolate group is equatorial to ring B and a line drawn from it to $C_{(17)}$ passes through the steroid molecule, a region of low dielectric strength, whereas in the transition state leading to the β-epoxy ketone this line passes through the higher dielectric of the solvent. A rough calculation leads to plausible values of the energy differences for some of the cases in Table 11, even if the particularly high preference for α epoxidation of the 17β-carboxy and of the 11,17-dioxo derivatives, and other features cannot be justified on the basis of this hypothesis alone. Also, a 4-methyl substituent has a favorable effect on the formation of α-epoxide, as indicated by an increase from 0 to 37% α-epoxide in going from cholestenone to 4-methylcholestenone, and from 30 to 50% in going from testosterone to 4-methyltestosterone (276). This effect can be attributed to an unfavorable interaction between the methyl group and the 6α- and 7α-hydrogens in the transition state leading to the β-epoxide.

The epoxy enone *451* is epoxidized exclusively to the β,β-diepoxide *452* (225). This is understandable on the basis of the boat conformation of ring A, which must be similar to that of *359*. Epoxidation of *453* gives the diepoxide *454* (277). The Weitz-Scheffer epoxidation of tetraphenylcyclopentadienone produces only the *trans*-diepoxide *455* (278).

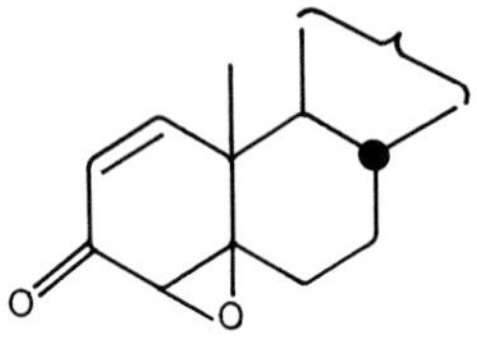

451 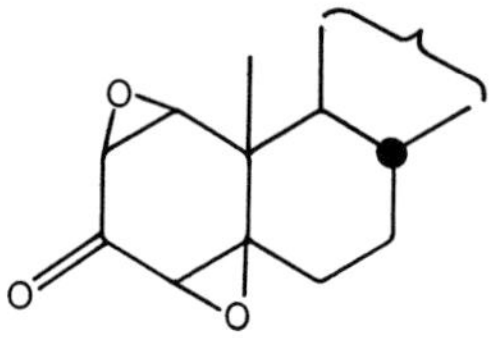 *452* 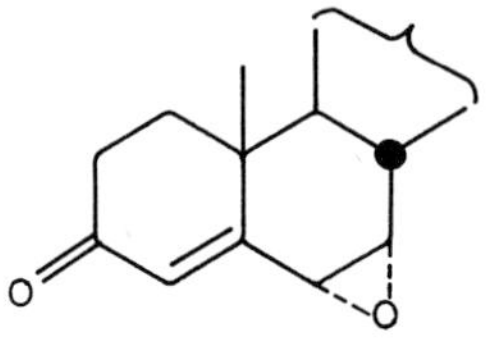 *453*

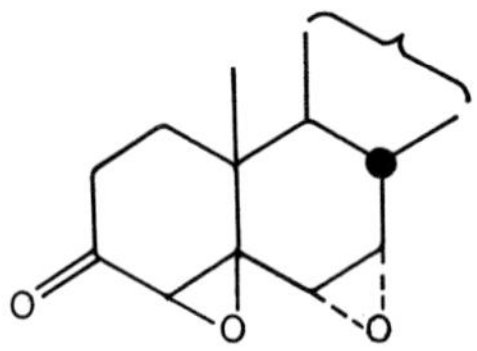

Ph Ph O O Ph Ph O

454 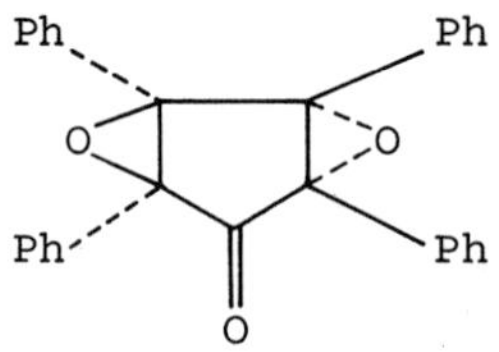 *455*

D. Other Oxidative Methods

1. *Oxidation with Hydrogen Peroxide and Catalyst*

Hydrogen peroxide can epoxidize electron-poor double bonds in the presence of certain transition-metal salts. Maleic and fumaric acid are converted stereospecifically by hydrogen peroxide/sodium tungstate into *cis*- and *trans*-epoxysuccinic acid, respectively (279). The reaction does not require the presence of such strongly electron-withdrawing groups as are necessary for the Weitz-Scheffer oxidation, since crotonic acid is easily converted into 2,3-epoxybutanoic acid, and even 2-heptene is slowly epoxidized by the above system. Sodium tungstate is a better catalyst than other transition-metal derivatives and pH 5 is the optimum for the reaction, which apparently involves the peroxytungstate monoanion formed by the equilibrium (280):

$$HWO_4^- + H_2O_2 \rightleftharpoons HWO_5^- + H_2O$$

Nucleophilic attack by the peroxytungstate anion can give the intermediate *456*, which cyclizes to the epoxide (281). The stereospecificity in the epoxidation of maleic and fumaric acid requires a very rapid cyclization step, or some kind of stabilization of the intermediate against rotation about the C-C bond, possibly through hydrogen bonding. The fact that olefins without negative substituents are epoxidized too implies that peroxytungstate acid can also behave as an electrophile, as do organic peroxyacids.

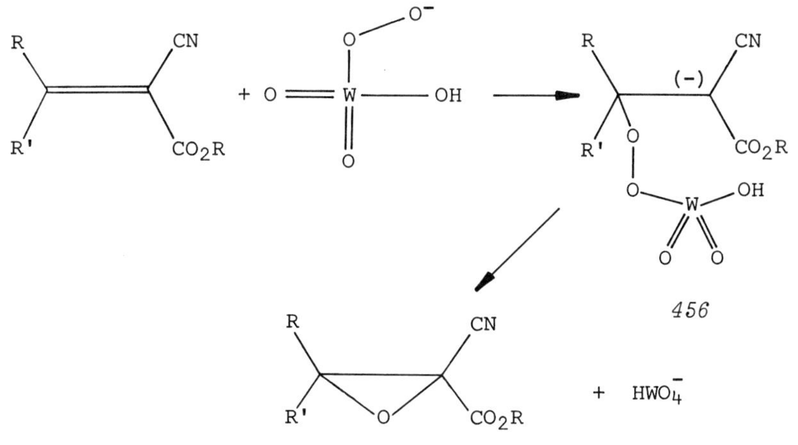

456

The hydrogen peroxide/tungstate method has also been employed for the stereospecific conversion of the unsaturated phosphonic acid *457* into the antibiotic phosphonomycin (*458*) (282). When the salt of *457* with (+)-α-phenethylamine is employed in the epoxidation, the corresponding salt of (-)-*458* of 92% optical purity crystallizes from the reaction mixture, and can be converted into (-)-*458*, the natural enantiomer of the antibiotic (283).

457 458

2. *Oxidation with* t-*Butyl Hydroperoxide*

t-Butyl hydroperoxide can be used in place of hydrogen peroxide in the epoxidation of enones, and allows one to conduct the reaction in homogeneous nonpolar media, in the presence of Triton B as the basic catalyst (284). Mesityl oxide gives 65% of the epoxide under these conditions, but the reaction is more sensitive to steric effects than the ordinary Weitz-Scheffer oxidation, because of the greater bulk of the reagent; it cannot, for instance, be used for the epoxidation of 4-cholesten-3-one. The same reagent system converts quinones into mono- and diepoxides (285); *459* gives a mixture of *cis*- and *trans*-diepoxides in which one of the isomers (presumably cis) predominates. Other substrates that have been epoxidized in this way include *460* (286) and cinnamaldehyde (287). Almost nothing is known about the mechanism of these reactions, which should, however, be analogous to that of the Weitz-Scheffer oxidation.

t-Butyl hydroperoxide can also epoxidize double bonds that are conjugated with aromatic rings, under neutral or basic conditions; both *cis*- and *trans*-stilbene are thus converted into *trans*-stilbene oxide (288), and *461* into the corresponding epoxide when the reaction is carried out in hexamethylphosphoric triamide (289). The latter reactions are most probably of radical type.

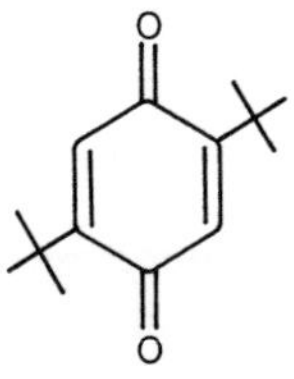

$CH_2{=}CH{-}PO(OEt)_2$

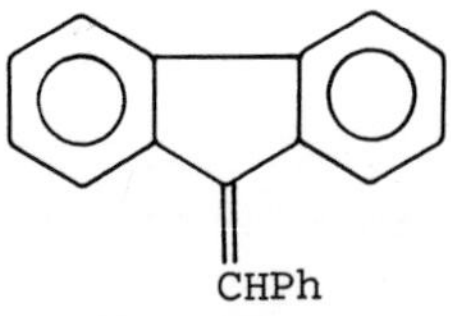

459 460 461

3. *Oxidation with Ozone*

Epoxides are sometimes obtained in the treatment of olefins with ozone (290), and may be the main products when double bonds are very hindered. For instance, *462* forms with ozone in pentan 71% of the corresponding epoxide (291). In the case of *463* the yield in epoxide is only 4%, but the reaction is entirely stereoselective since the epoxide has the trans configuration (292); it was deduced from the indeed rather limited evidence of this single observation that the reaction involves the formation of an olefin-ozone adduct in which the configuration is preserved, possibly a π- (*464*) or σ-complex (*465*). A high degree of hindrance about the C-C bond can prevent the conversion of the intermediate into the molozonide, and lead rather to its collapse into epoxide and molecular oxygen.

Recent cases of exclusive formation of epoxides instead of ozonides in natural product chemistry are those of compounds *466* (293), *467* (104), and *54* (58), which are all attacked from the less hindered side, as in peroxyacid epoxidations.

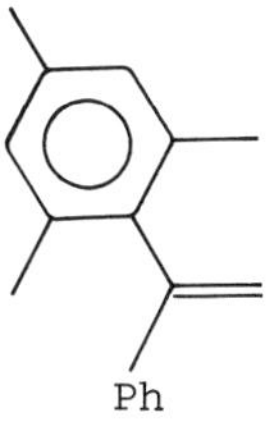

Ph

462

463

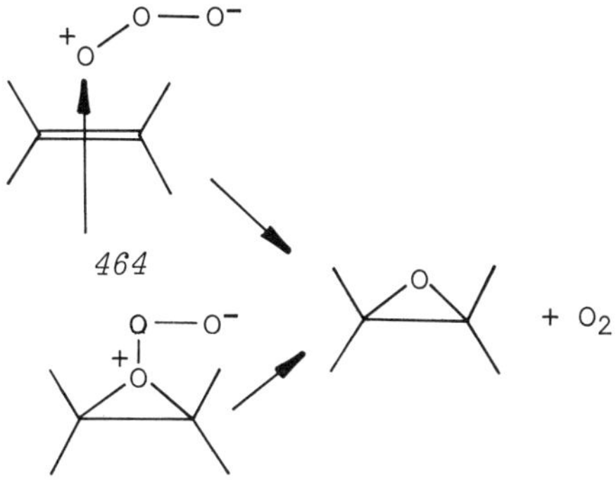

464

+ O_2

465

MeO_2C

AcO

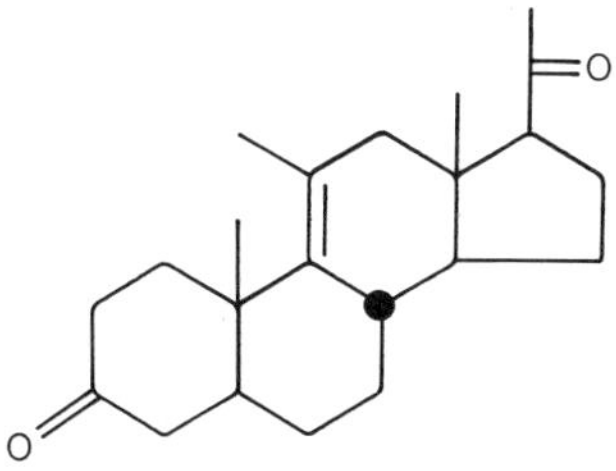

466

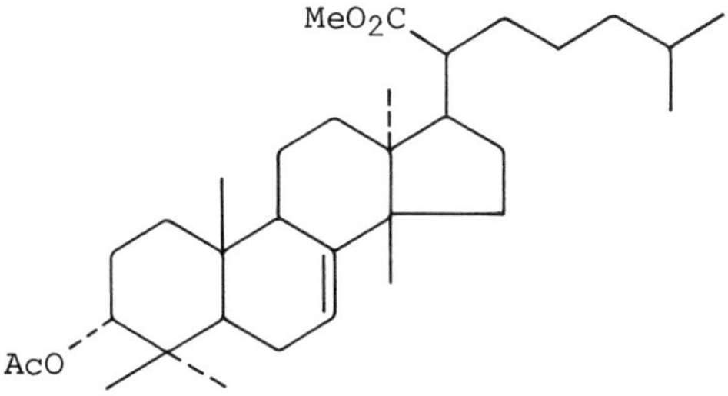

467

4. *Oxidation with Chromic Acid*

Epoxides are often the first products of the chromic acid oxidation of olefins (294,295). This reaction is usually not very important for preparative purposes, since further oxidation and rearrangement in the medium normally reduces, often to nothing, the yield in epoxide. Only in the case of tri- and tetrasubstituted (particularly aryl-substituted) olefins, such as *468* (296), *469*, *470* (297), and *471* (298), is it possible to isolate the epoxides in satisfactory yields. The intermediacy of epoxides in the oxidation of simpler olefins has also been demonstrated by trapping them with acetic acid; the chromic oxidation of cyclohexene in acetic acid, in the presence of perchloric acid, gives among other products a mixture of *cis*- and *trans*-2-acetoxycyclohexanol in the same ratio as that obtained in the acetolysis of epoxycyclohexane (299). It has been proposed, on the basis of comparisons between rates of oxidation of differently strained cycloalkenes, that the transition state of the chromic acid oxidation in acetic acid is three- rather than five-membered, and that it resembles the cyclic mechanism of the peroxyacid oxidation, as shown, for instance, in *472* (300). Although this mechanism could be valid for the reaction of aliphatic alkenes, it does not appear to be so for aryl-

$Me_2C{=}CHCHMe_2$ *468* $Me_2C{=}CHPh$ *469* $Me_2C{=}CPhMe$ *470* $Ph_2C{=}CPh_2$ *471*

O, O=Cr, O, H, O, C, O, Me

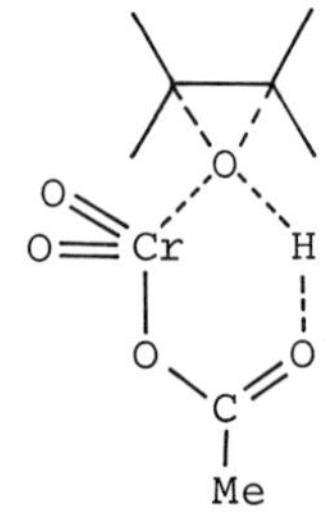

472

substituted ones, since the chromic acid oxidation of *cis*- and *trans*-1,2-diphenyl-1,2-di(*p*-chlorophenyl)ethylene is not stereospecific and seems rather to involve an intermediate with carbonium ion character (301,302). Determination of the steric course of the chromic oxidation of appropriate aliphatic olefins would be useful for testing mechanism *472*.

Epoxidation of sterically hindered cyclic olefins can be preceded by α oxidation and double-bond migration. For instance, compound *473* is converted into a mixture of *474* and *475* by chromi[c] acid in acetic acid (303).

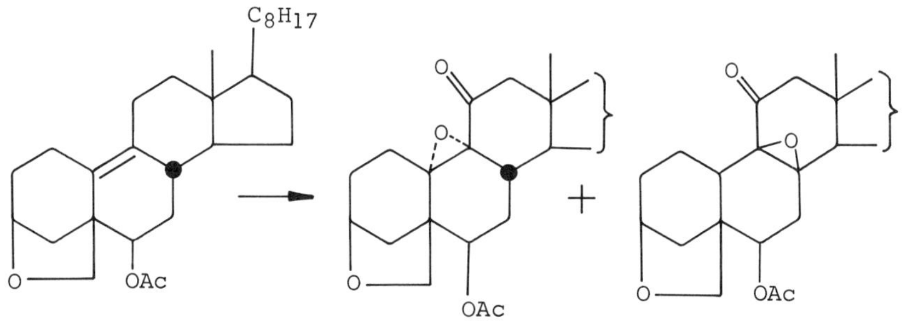

473 *474* *475*

The conversion of allylic alcohols into epoxy ketones with Jones reagent presents some interesting stereochemical features (304). The reaction occurs only when the hydroxy group is free and axial. Compounds *476* and *477* are converted in 50-60% yields into the epoxide *478* and into a mixture of *479* and *478*, respectively, the latter reaction involving an allylic shift. The corresponding α-hydroxy compounds (equatorial OH) are oxidized only to the stage of enones. The fact that epoxidation takes place only syn to the hydroxy function and that enones are not further transformed into epoxy ketones under the reaction conditions strongly suggests the formation of a chromate ester, followed by the transfer of an oxygen atom from the ester to the double bond.

The epoxidation of an isolated and of a conjugated double bond was observed in the chromic oxidation of zerumbol (*480*), which produced the diepoxy ketone *481* (305).

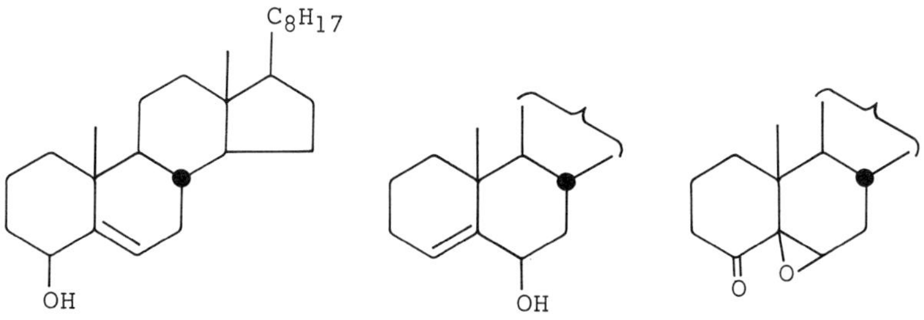

476 *477* *478*

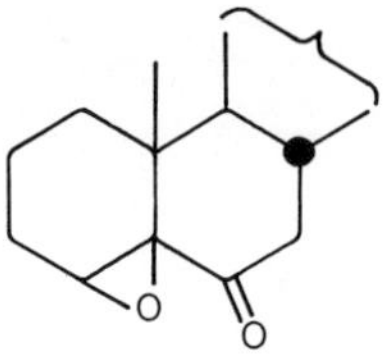

479 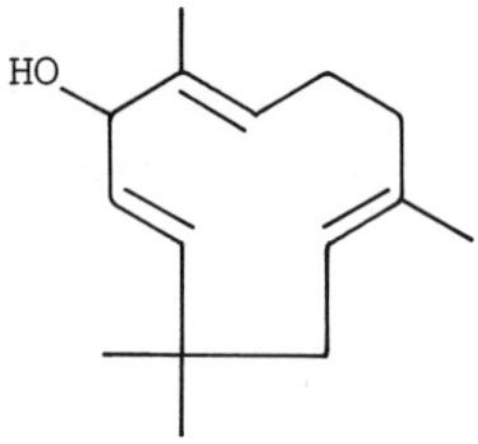 *480* 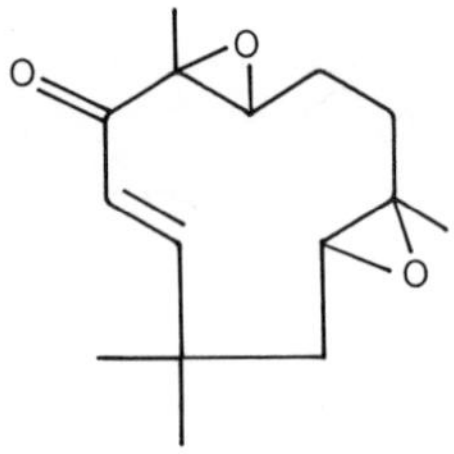 *481*

5. *Oxidations with Permanganate*

Only very few cases of formation of epoxides in the permanganic oxidation of olefins have so far been reported. $\Delta^{9,11}$-unsaturated steroids of the 5α and 5β series, such as *482* and *483*, are converted in good yields by potassium permanganate in acetic acid into the corresponding α-epoxides (306); tetraphenylethylene reacts similarly (301). The all-cis epoxydiols *484* and *485* are isolated in fair yields in the oxidation of cyclopentadiene and 1,3-cyclohexadiene with neutral or alkaline permanganate (149). Nothing is known about the mechanism of these reactions.

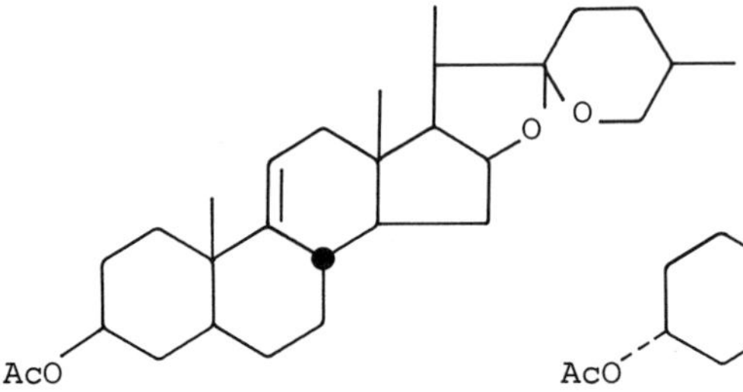

482 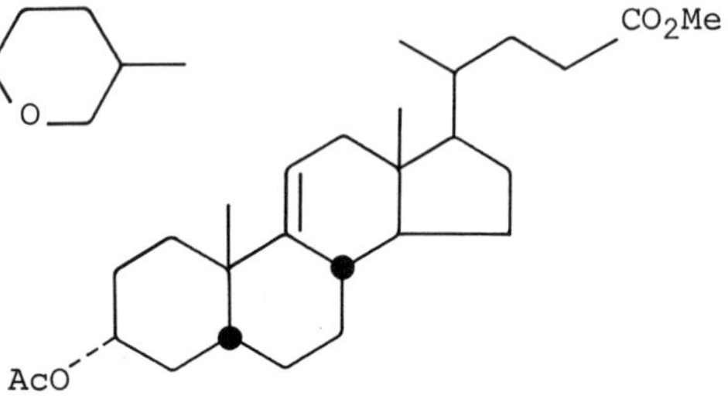 *483*

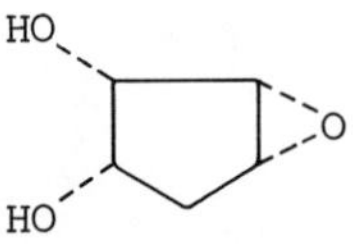

484 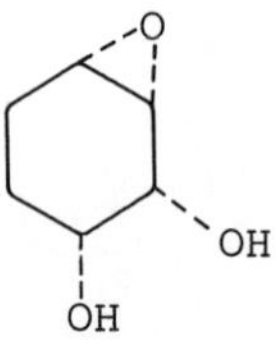 *485*

6. *Oxidation with Hypochlorite Ion*

Although hypochlorous acid is an electrophilic agent, the corresponding anion can behave as a good nucleophile, and has been occasionally used for the epoxidation of electron-poor double bonds. Benzalacetophenone is converted into the epoxide by sodium hypochlorite in aqueous pyridine (307), and *486* undergoes the same reaction in aqueous acetonitrile (308). This type of reaction has been investigated from the kinetic and stereochemical point of view by Robert and Foucaud (309), who found that in a series of α-cyanoacrylates (*487*) reaction with hypochlorite in aqueous ethanol at pH 7 produces the epoxides *488* in an entirely stereospecific way, a definite advantage over the nonstereospecific Weitz-Scheffer reaction. The observed stereochemistry would be in accordance with the intermediate formation of chlorohydrins by anti addition, followed by elimination of hydrogen chloride, but compounds of this type react only very slowly, or not at all with electrophiles, or with hypochlorous acid at lower pH; therefore the reaction must be initiated by a nucleophilic attack. Rosenblatt and Broome (308) proposed a mechanism similar to that accepted for the Weitz-Scheffer reaction, involving a slow attack by ClO^-, followed by a rapid cyclization of the intermediate anion *489*. On the other hand, the stereospecificity of the reaction was explained (309) by assuming a concerted attack by OH^- and HClO (*490*) to give the chlorohydrin.

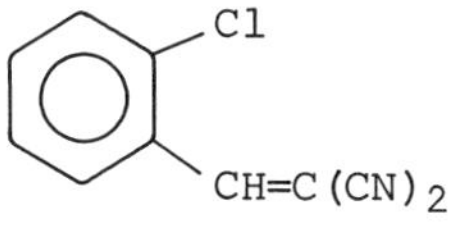

Cl, $CH{=}C(CN)_2$ — *486*

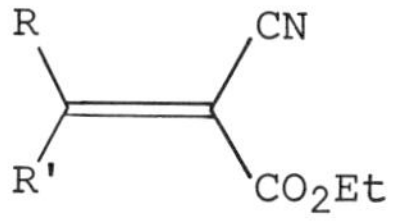

R, R', CN, CO_2Et — *487*

R, R', CN, O, CO_2Et — *488*

OCl^- → (-), OCl → O + Cl^-

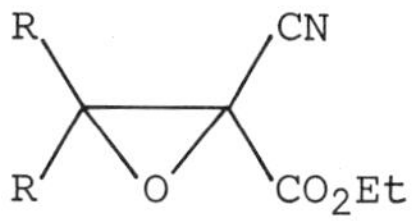

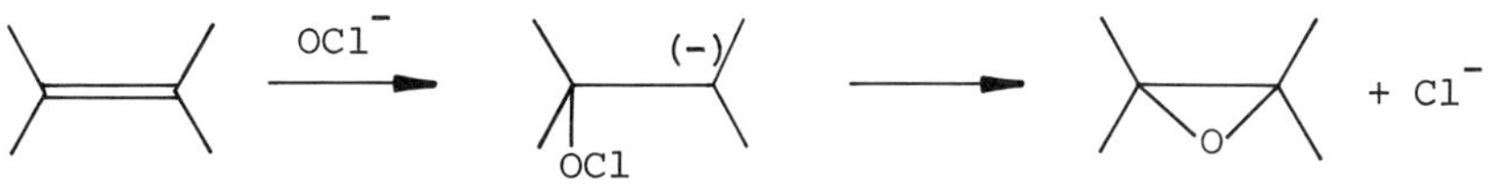

489

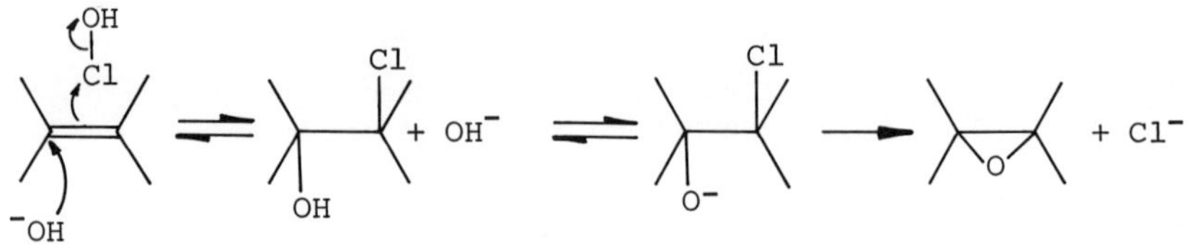

490

7. *Photochemical Oxidation of Allylic Alcohols*

Steroidal allylic alcohols can be converted into epoxy ketones by hematoporphirin-sensitized photooxygenation in pyridine. The reaction is stereospecific, the configuration of the epoxy function being opposite to that of the hydroxyl in the starting alcohol (310); cholest-4-en-3β-ol (*491*) gives the α-epoxide *492*, the α-epimer *493* the β-epoxide *494*. Although this type of reaction has not been investigated in great detail, it may involve a cyclic abstraction of the allylic hydrogen by oxygen (*495*), with formation of the enol hydroperoxide *496*, which can give the epoxy ketone in a manner similar to that postulated for the Weitz-Scheffer reaction.

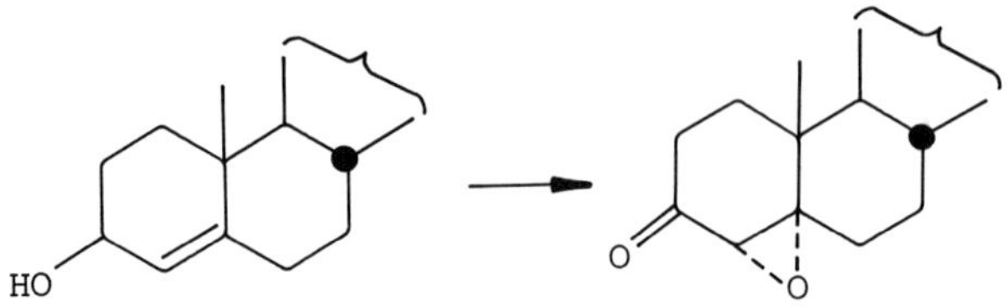

491 *492*

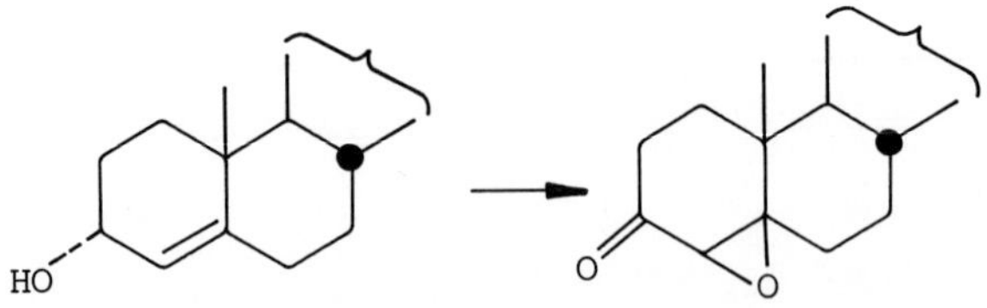

493 *494*

HO H O O 495 → HO HOO 496 → 492

E. 1,3-Elimination Reactions

The oldest and still one of the most important methods for the formation of the oxirane ring is based on the base-promoted dehydrohalogenation of a vicinal halohydrin. This type of reaction can be depicted as an intramolecular nucleophilic displacement, as shown in *495a* → *496a*. The leaving group X is not necessarily halogen, but can be RSO_3, RCO_2, OH, NR_3^+, N_2^+, etc. The literature concerning these reactions is extremely extensive, and only some recent work with particular stereochemical implications will be mentioned here; reference is made to detailed reviews (311,312) for the older literature.

OH X ⇌ (B⁻) O⁻ X → O + X⁻

495a *496a*

1. *Cyclization of Open-Chain 1,2-Difunctional Compounds*

The formation of epoxide rings in open-chain compounds by the elimination method is quite generally feasible. The *anti*-coplanar disposition *495a* that is necessary to reach the transition state for the conversion into epoxide is usually possible in bifunctional compounds of this type. The starting halohydrins can be obtained by the addition of hypohalous acids to alkenes. Since there can be some difficulty in handling these acids because of their limited stability and of solvent incompatibilities, they are usually generated *in situ*, a wide range of reagents being available for this purpose (312), such as calcium hypochlorite, several *N*-chloroamides, *N*-chlorourea, and *t*-butyl hypochlorite for chlorohydrins, whereas the reagents of choice for the preparation of bromohydrins are *N*-bromosuccinimide (NBS) and *N*-bromoacetamide (NBA), which usually give excellent yields when treated with olefins in aqueous dioxan containing catalytic amounts of perchloric acid (313). Although some of the compounds mentioned above can act as actual sources of hypohalous acid when treated with mineral acids, it appears that protonated NBS (*497*) behaves rather as a direct source of electrophilic bromine (314).

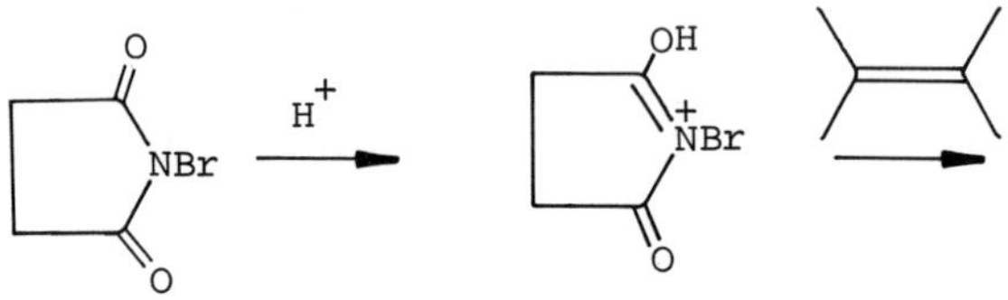

497

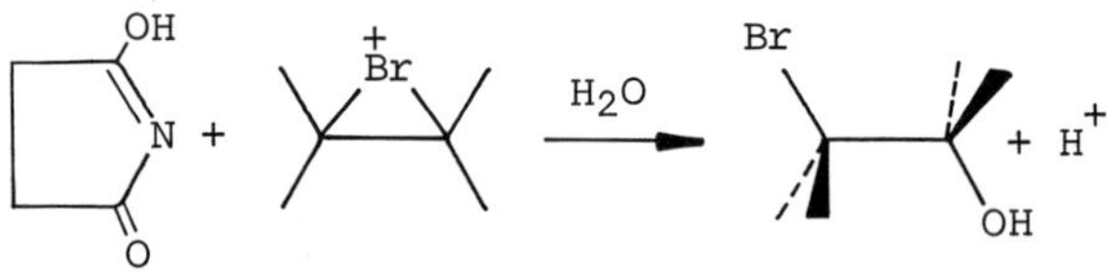

The formation of iodohydrins from olefins and iodine in water is not satisfactory because of the reversibility of the reaction, but it can be a useful route to epoxides if the iodination is carried out in the presence of an oxidizing agent, such as iodic acid, or oxygen and nitrite, which prevents the accumulation of the iodide ion. Treatment of the iodohydrins thus formed with calcium hydroxide gives high yields of epoxides (315).

No matter which method is employed for the preparation of the halohydrin, halogen and hydroxyl add in an anti fashion to the double bond, and the halohydrin is then converted into the epoxide with retention of configuration at the hydroxyl-bearing carbon and inversion at the halogen-carrying one. The overall result is that the epoxide (*499*) has the same configuration as the starting olefin (*498*), the steric course being the same as that of the peroxyacid epoxidation. Stereospecificity is usually complete in both steps. For instance, (*S*)-2-chloro-1-propanol is converted by potassium hydroxide into (*R*)-1,2-epoxypropane without any loss in optical purity (316).

In particular cases, when a neighboring group facilitates epimerization, the reaction may not be entirely stereospecific: thus whereas the erythro form of *500* gives with base only *trans*-epoxide (*501*), *threo*-*500* forms a mixture of *cis*- and *trans*-*501* (317), and *cis*-*502*, on treatment with NBS followed by base, produces 70% *cis*- and 30% *trans*-*503* (318). Epimerization does not

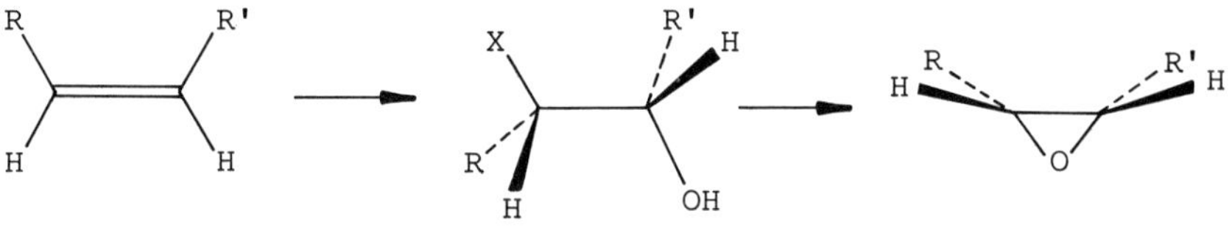

498 *499*

take place on the epoxide, but rather at the chlorohydrin stage for *500* and at the enone stage for *502*.

$PhCHClCH(OH)CONEt_2$

500

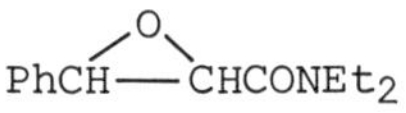

PhCH—O—CHCONEt_2

501

MeCH=CHCOMe

502

MeCH—O—CHCOMe

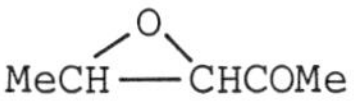

503

The alternative use of the peroxyacid or of the halohydrin route may often be a matter of choice, but in several cases only the latter method can be applied, e.g., for the preparation of epoxides containing amino functional groups. Stilbazole oxides, such as *507*, cannot be prepared from stilbazoles (*504*) by direct epoxidation, since the peroxyacid oxidizes much faster at nitrogen than at the double bond to give the *N*-oxide, which is resistant to epoxidation because of the deactivating effect of the positive nitrogen. On the other hand, *507*, its diastereoisomer, and the corresponding 3- and 4-stilbazole oxides can be easily obtained in a stereospecific way through the halohydrin route. The *threo*-chlorohydrin *505* is also available through stereoselective reduction of the chloroketone *506* with LAH (319,320). Similarly, the synthesis of the epoxide *508* succeeds via the halohydrin, but not through the treatment of the corresponding unsaturated amine with peroxyacids (321).

Halohydrins can also be obtained by reaction of epoxides with hydrogen halides. This reaction is of no preparative use for the synthesis of aliphatic epoxides, since the ring opening of the oxirane and the ring closure of the halohydrins both involve inversion of configuration, and therefore one gets back to the starting epoxide. However, in the case of aryl substituted epoxides ring opening takes place very often with retention of configuration, and can provide a useful route for the

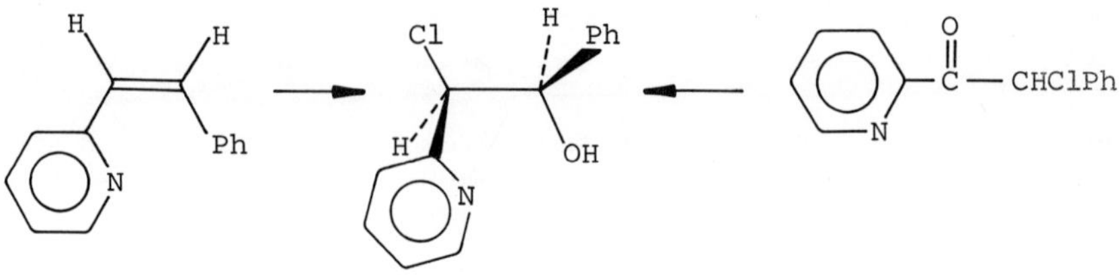

504 505 506

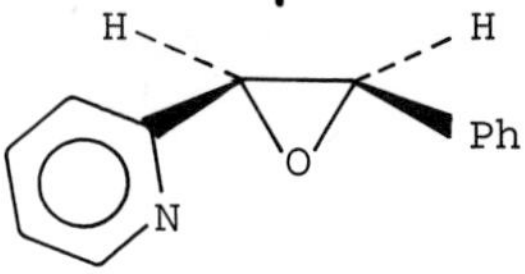

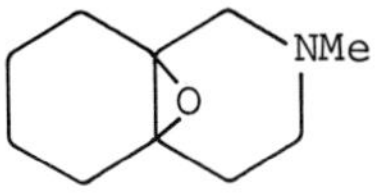

507 508

conversion of an epoxide into its diastereoisomer. Thus *trans*-stilbene oxide (*509*) reacts with hydrogen chloride in chloroform or benzene to form exclusively the *threo*-chlorohydrin *511*; the stabilizing influence of the phenyl group on the developing positive charge favors the formation of the ion pair *510*, and the direction of attack by Cl^- is determined by the relative position of the two ions in this ion pair. Recyclization of *511* with base produces the *cis*-epoxide *512* in better than 90% overall yield from *509* (322).

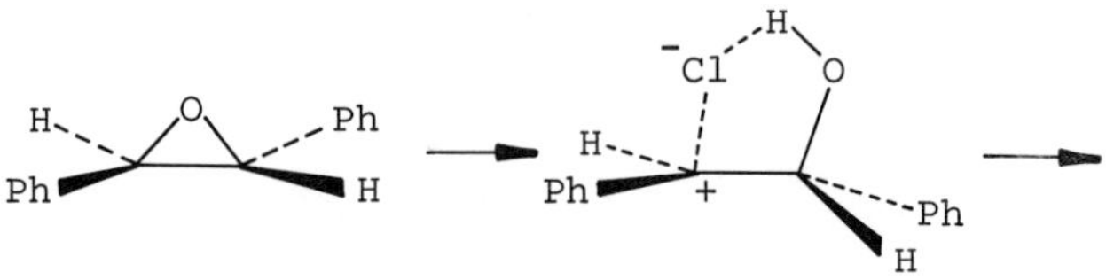

509 510

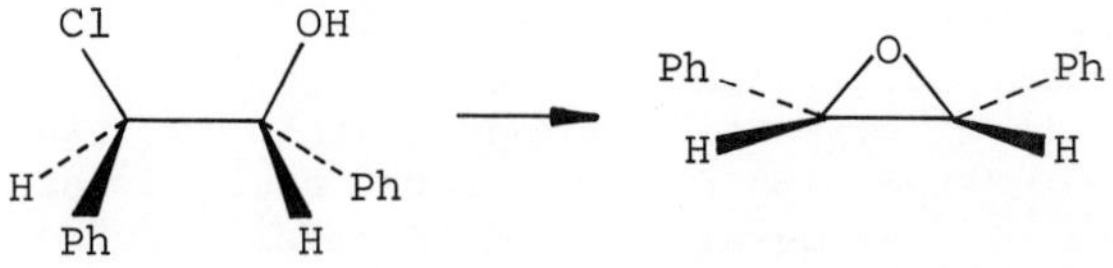

511 512

The preparation of halohydrins from α-haloketones by reduction with mixed hydrides is usually possible, provided one avoids an excess of the reducing agent to prevent halogen hydrogenolysis; several phenacyl bromides (*513*) have thus been converted into bromohydrins (*514*) and epoxides (323). Similarly, halohydrins with tertiary hydroxy groups can be obtained from α-haloketones with Grignard reagents. This reaction can exhibit a rather pronounced stereoselectivity if carried out at -70°C, and its steric course can be interpreted on the basis of an addition from the least hindered side to the preferred *anti*-parallel conformation *515* of the chloroketone, with formation of *516*. It was thus possible to prepare a mixture of epoxides containing 80-85% of the diastereoisomer *517* (R' = R" = Me, R''' = Et) starting from 3-chlorobutanone and ethylmagnesium bromide (324). This method was also used for the synthesis of all-*trans*-squalene (325).

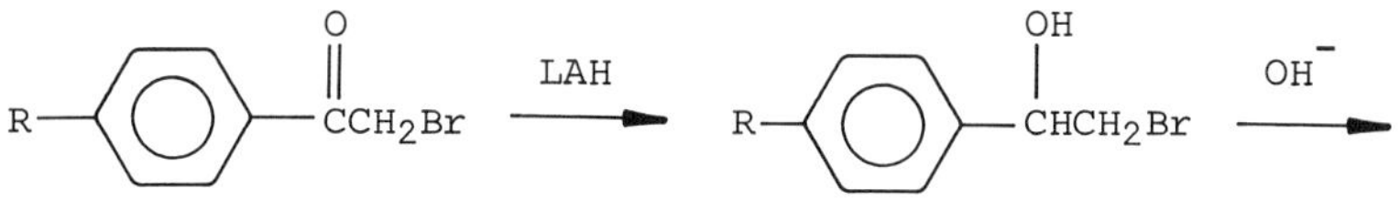

R—C₆H₄—C(=O)CH₂Br —LAH→ R—C₆H₄—CH(OH)CH₂Br —OH⁻→ R—C₆H₄—(epoxide)

513 *514*

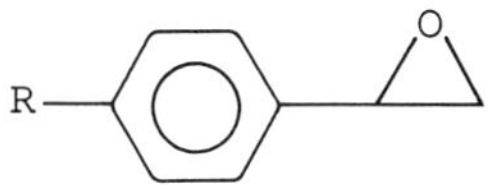

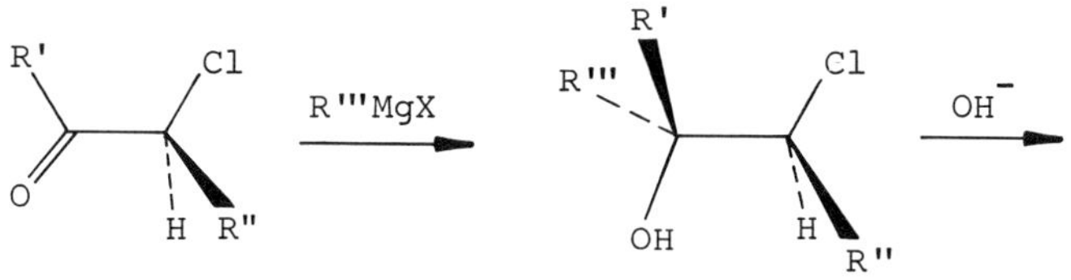

R' Cl O H R" —R'''MgX→ R' R''' Cl OH H R" —OH⁻→

515 *516*

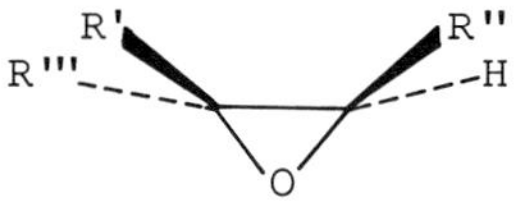

R' R''' R" H O

517

The Grignard method is not being used much at present for the preparation of halohydrins, because it is subject to many side reactions that can depress yields considerably. It can, however, be very useful in special cases, such as for the preparation of acetylenic epoxides, as first discovered by Kohler and Tishler (326), who obtained the epoxide *519* from the ketone *518*. The haloketone approach to halohydrins and epoxides, and its complications, which gave rise to considerable discussion in the past, are not treated here, since excellent reviews are available on this topic (327,328).

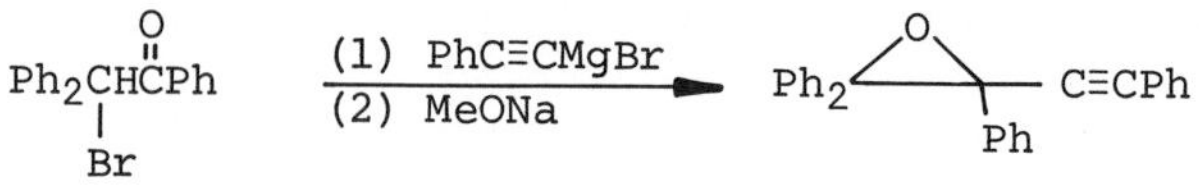

$$Ph_2CH(Br)C(=O)Ph \xrightarrow[(2)\ MeONa]{(1)\ PhC{\equiv}CMgBr} Ph_2\text{-epoxide-}C(Ph)C{\equiv}CPh$$

518 *519*

A particularly useful application of the bromohydrin route was discovered by van Tamelen (314,329,330) in the selective epoxidation of the terminal double bond of squalene, which opened the road to the investigation of the role of squalene monoepoxide (*522*) as an intermediate in the biosynthetic pathway from acetate to triterpenoids and steroids. The treatment of squalene (*520*) with one equivalent of NBS in aqueous glyme produces 96% addition to one of the two terminal double bonds to give *521*, which is cyclized to *522*; only 4% of the addition occurs on the four internal double bonds. This remarkable selectivity, observed also in other long-chain nonconjugated polyenes, decreases when less polar solvents are employed, and cannot be attributed to electronic factors, since all the double bonds of *520* are trisubstituted. It is assumed rather that the apolar molecule takes a coiled conformation in the polar solvent, so that the

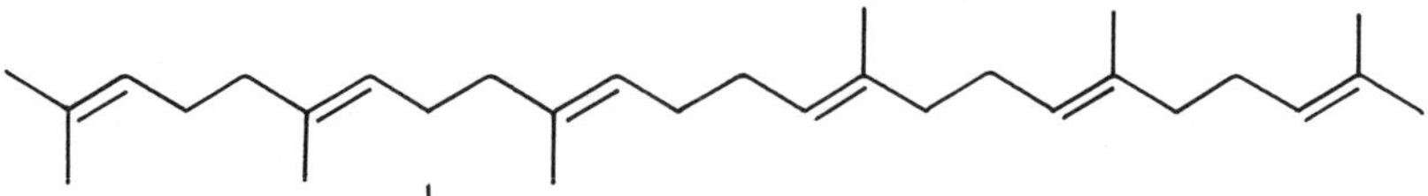

520

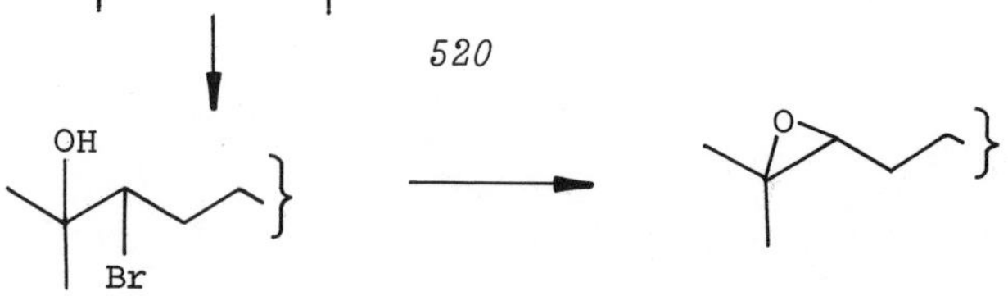

OH

Br

521 *522*

internal double bonds are sterically shielded and less readily available for the attack by the reagent. Solvent clustering at the reacting centers in the transition state may further contribute in preventing attack on the internal double bonds.

The cyclization of sulfonate esters of 1,2-diols can provide a useful alternative to the halohydrin route, since it takes place under very mild alkaline conditions, because of the excellent leaving group properties of the sulfonate anions. The reaction is usually carried out by selective esterification of one of the two hydroxyls of a glycol with tosyl or mesyl chloride to give *523*, followed by treatment with a base. This reaction has the same mechanism and steric course as the halohydrin method. For instance, *meso*-butane-1,2-diol is converted stereospecifically into *trans*-2,3-epoxybutane, and optically active styrene oxide (*524*) can be obtained from active styrene glycol, easily available by the LAH reduction of mandelic acid (331). Optically active α-methyl- and α-ethylstyrene oxides have been prepared in a similar way (332).

HO–CH₂–CH₂–OSO_2R ⇌ (base) $^-$O–CH₂–CH₂–OSO_2R → epoxide + RSO_3^-

523

TsO–CH₂–CH(OH)(Ph)(H) → epoxide (Ph, H)

524

cis- and *trans*-*526* can be obtained from *d,l*- and *meso*-*525* by conversion into the disodium salt with sodium hydride, followed by treatment with one equivalent of tosyl chloride; the monotosylate alkoxy anions are converted directly into the epoxides (333).

CH_2=CH-CH(OH)-CH(OH)-CH=CH_2 → CH_2=CH-CH—CH-CH=CH_2 (bridged by O)

525 *526*

The monotosylation can be very selective if differently hindered hydroxy groups are present in the molecule; for instance, the tetraol *527*, which has one primary, two secondary, and one tertiary hydroxyls, is esterified only at the primary group to give a monoester which cyclizes to epoxide *528* (334).

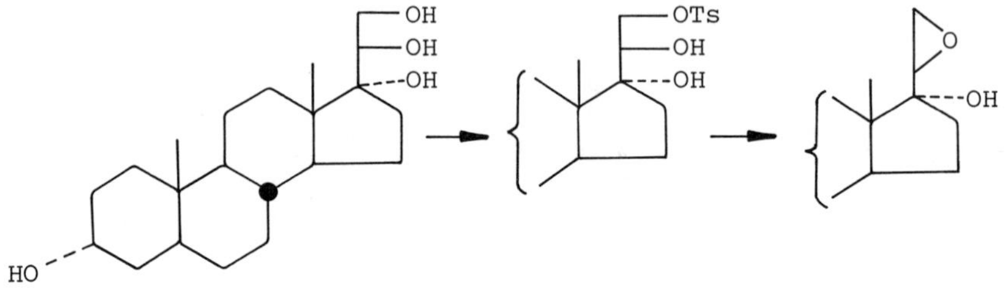

527 *528*

2. *Cyclization of Alicyclic Bifunctional Compounds*

The formation of epoxide rings on cyclic systems is subject to some limitations that are absent in the case of acyclic compounds, the main one being that the hydroxy function and leaving group are not always capable of reaching the *anti*-coplanar conformation. Epoxides can therefore be obtained through elimination reactions only from *trans*-1,2-disubstituted cycloalkanes (with the exception of macrocyclics), and the reaction is much easier with diaxial than with diequatorial derivatives. This is illustrated by the diastereoisomeric halohydrins *529* and *530*, which are both converted into epoxides by potassium *t*-butoxide, but at greatly different rates (335). The importance of the transition-state geometry is also stressed by the comparison of the behavior with base of the two similar bromolactones *531* and *533*; whereas the latter forms the epoxide *534*, the former gives only the ketone *532*, probably because the *anti*-coplanar transition state involves a high degree of torsional strain in the bicycloheptane system of *531* (336).

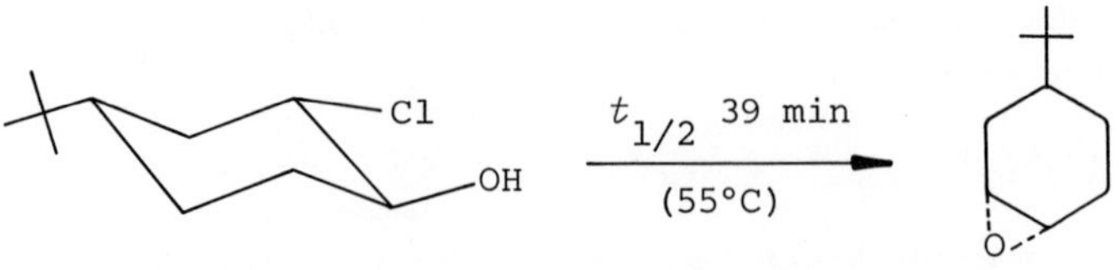

$t_{1/2}$ 39 min (55°C)

529

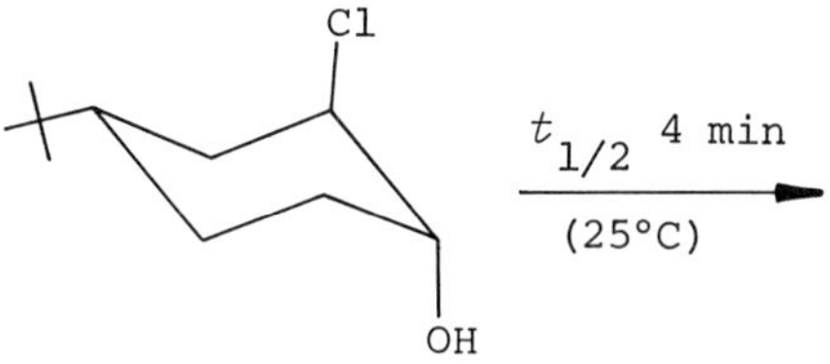

$t_{1/2}$ 4 min (25°C)

530

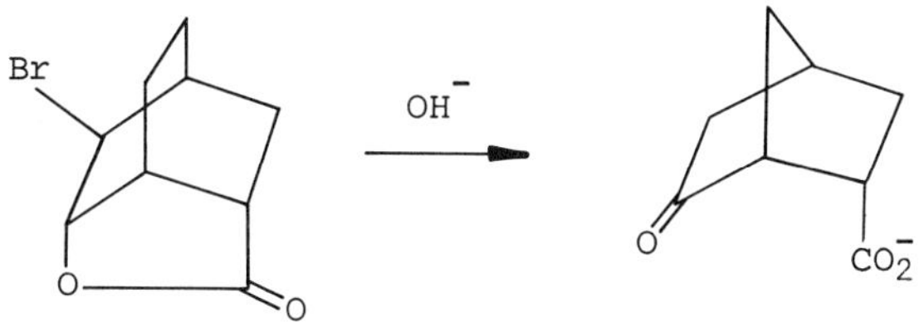

531 532

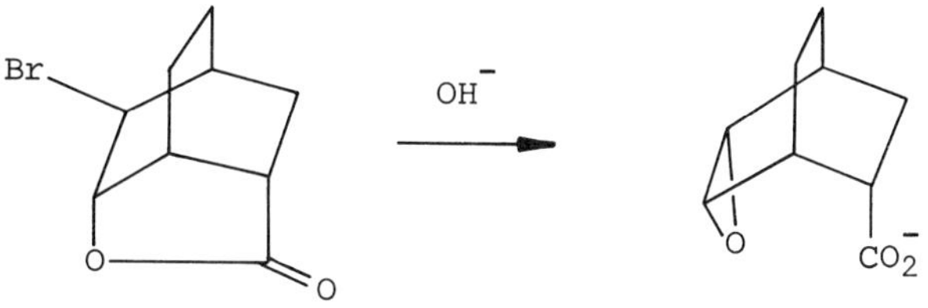

533 534

The only notable exception to the rule of anti elimination appears to be in the recent report (337) about the high-yield conversion of the *cis*-tosylate *535* into the *exo*-epoxide. One possible explanation for this very unusual reaction may be sought in the exceptional nature of the bornane system that could favor the heterolytic fission to tosylate anion and non-classical carbonium ion, followed by cyclization through nucleophilic attack by the hydroxyl. A report on the formation of *exo*-2,3-epoxy-2-*p*-anisylnorbornane by treatment of the *exo-cis*-diol with acids (338) was shown to be wrong, a 1,4-dioxane derivative being formed instead (339).

A useful application of the halohydrin method is in the synthesis of hindered epoxides that are diastereoisomeric to those prepared by peroxyacid oxidation. This is possible since the formation of the halohydrin involves electrophilic attack by the halogen from the less hindered side, and nucleophilic attack by OH^- or H_2O from the more hindered side of the double bond; and the latter determines the configuration of the epoxide. This can be seen in the synthesis of the carene α- and β-oxides *537* and *538* from carene (*536*) (340). Similarly, the

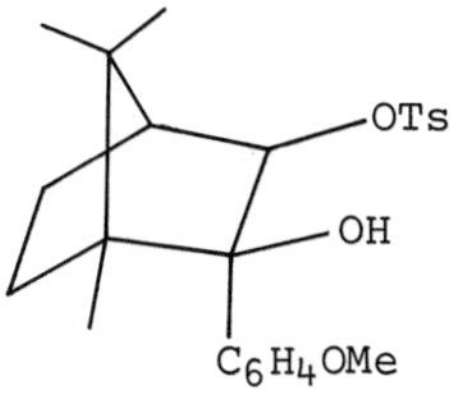

OTs OH C_6H_4OMe *535*

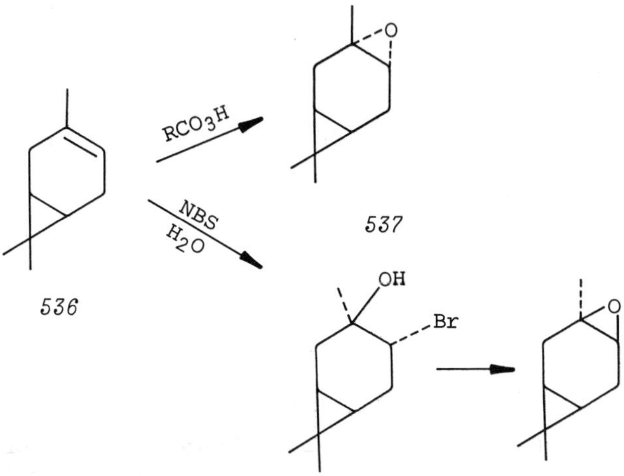

536 RCO_3H *537* NBS H_2O OH Br O *538*

exocyclic epoxide *540* can be obtained in 80% yield and 92% diastereoisomeric purity from *539*, by reaction with NBA followed by cyclization (341), whereas peroxyacid gives an excess of its diastereoisomer (Sect. II-A-6). This approach has been extensively employed in the steroid field to introduce the β-epoxy function, e.g., in the 4,5- (271), 5,6- (342), and 14,15-positions (343). The Δ^{15}-19-norsteroid olefins are converted by the halohydrin route into the α-epoxides (195), and by peroxyacid oxidation into the β isomers. 5,6β-Epoxy-6α-methylcholestanol can be prepared through the route *541* → *542* (344).

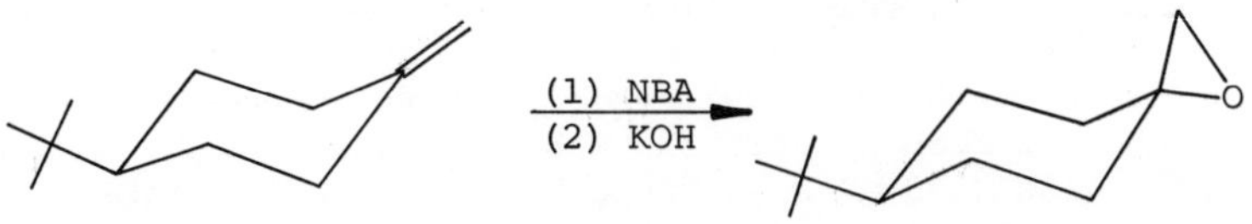

(1) NBA (2) KOH

539 *540*

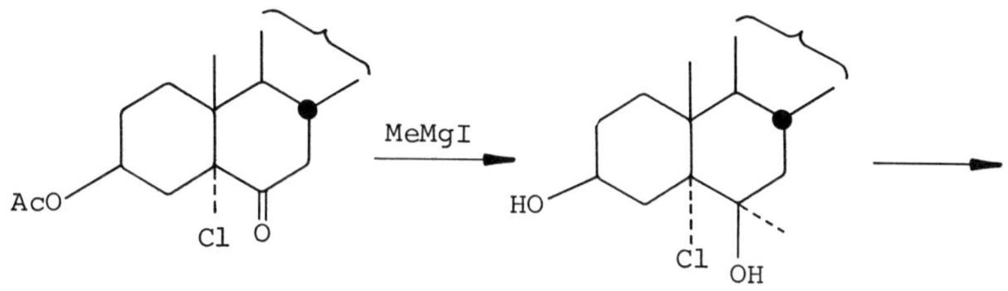

AcO

Cl O

MeMgI

HO

Cl OH

541

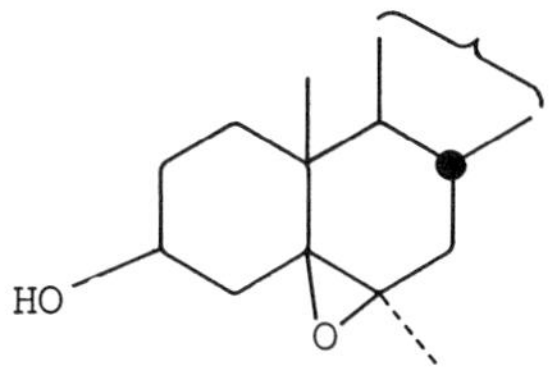

HO

O

542

When an axial allylic hydroxyl group is present in the substrate (*543*) aqueous chlorine can produce a chloroepoxide (*544*), probably through intramolecular attack by the hydroxyl on the intermediate chloronium or chlorocarbonium ion (345).

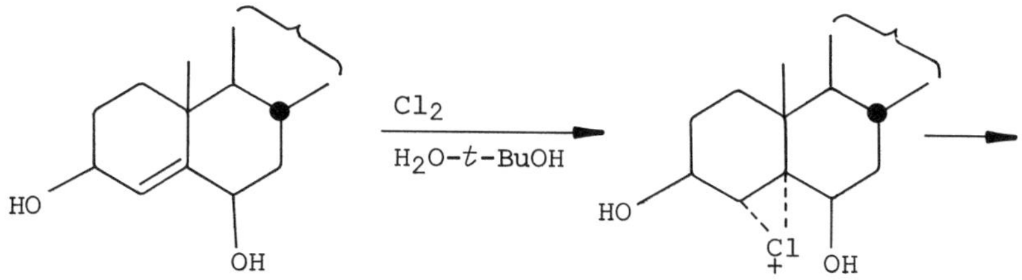

Cl_2

H_2O-*t*-BuOH

HO

OH

HO

Cl
+

OH

543

HO

O

Cl

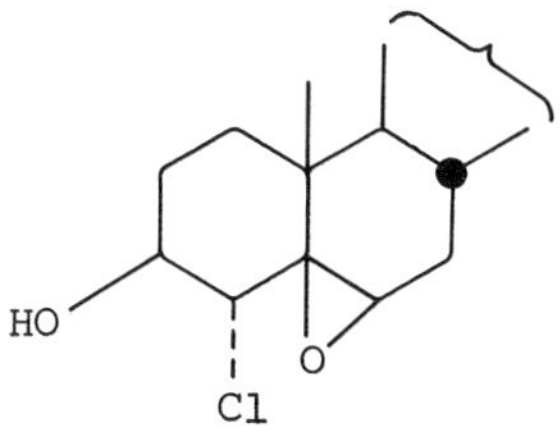

544

Acyl hypohalites can be used to produce halohydrin esters. The addition of acetyl hypobromite to 4-methylcyclopentene gives the bromoester *545*, which is cyclized by base to the *cis*-epoxide *546*. The stereospecificity is higher than in the PLA epoxidation, which gives an excess of the *trans*-epoxide (36,48,212). 2-Cholestene is analogously converted into the β-epoxide with trifluoroacetyl hypoiodite (346).

When an alkene such as *547* is treated with bromine or NBS in the presence of dimethylformamide and silver perchlorate, the product is the monoformate *549*, formed by hydrolysis of the intermediate *548*; cyclization of *549* with base gives the β-epoxide in high yield (347).

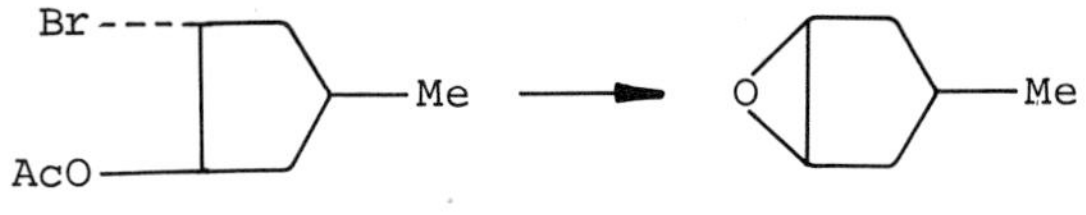

Br AcO Me → O Me

545 *546*

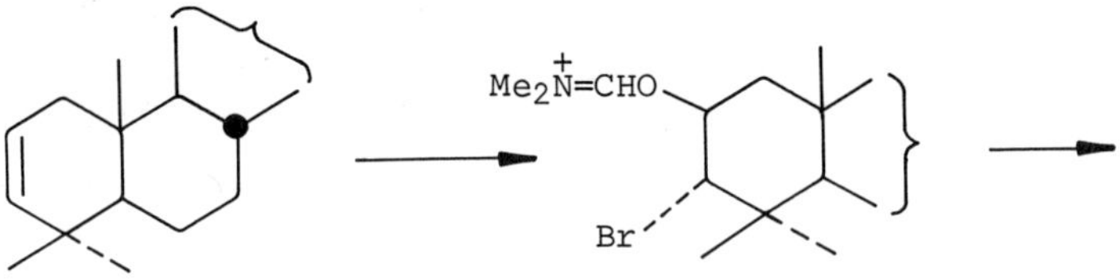

Me$_2\overset{+}{N}$=CHO Br

547 *548*

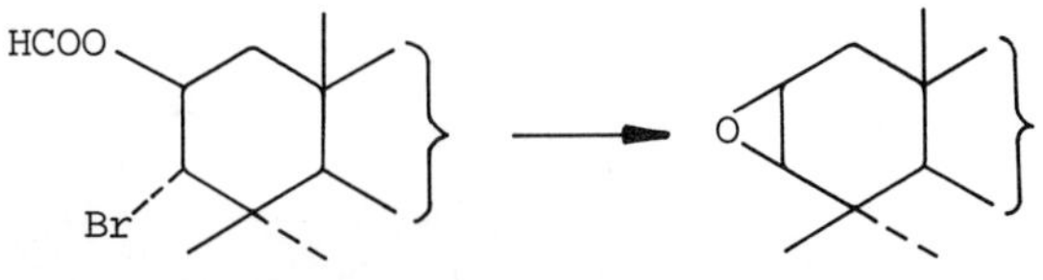

HCOO Br → O

549

The tosylate route can be very useful for the stereospecific synthesis of cyclohexene oxides. Thus the equimolar mixture of epoxides obtained in the peroxyacid oxidation of limonene is opened by aqueous acid to give almost exclusively the diaxial diol *550*, which is monotosylated to *551*; cyclization with base gives the pure *trans*-epoxide *552* (43,348). Reaction of *552* with acetic acid yields 90% of the acetate *553*, conversion of which into the mesylate, followed by treatment with base (which hydrolyzes the secondary ester group and catalyzes the cyclization), provides the *cis*-epoxide *554* (42). Through a similar acetolysis, mesylation, elimination sequence *trans*-2,3-epoxy-9-methyldecalin (*41*) was converted into the cis isomer (52).

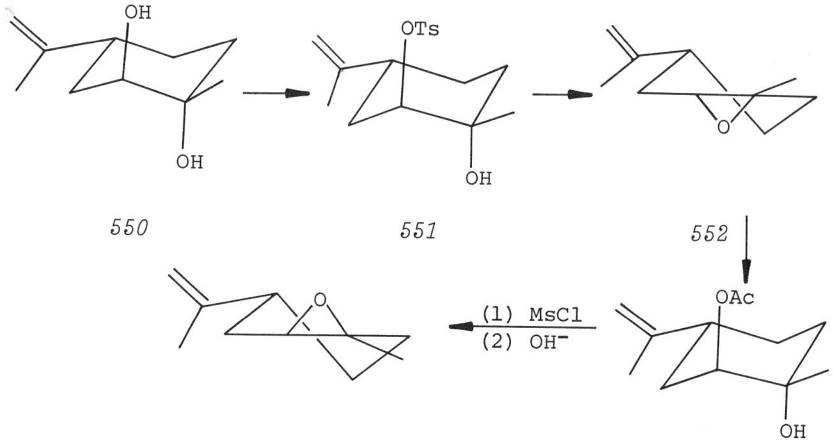

550 *551* *552*

(1) MsCl
(2) OH^-

554 *553*

In the 4-*t*-butylcyclohexane series, the *trans*-epoxide (*559*) can be prepared from the mixture *555* by diaxial opening with hydrogen chloride in the presence of *p*-nitrobenzoyl chloride, separation of the less soluble isomeric chloroester (*556*), and cyclization (349). Alternatively, the chloroketone *557* is reduced with borohydride to 50% *558*, the cyclization of which gives *559*, which can in turn be further transformed into the *cis*-epoxide *561* through *560* (335). Analogous sequences give the diastereoisomeric 1-methyl-4-*t*-butyl-1,2-epoxycyclohexanes (44) and β-carene oxide (*538*) (73,350,351).

An alternative method for isolating individual epoxides from the mixtures obtained by peroxyacid oxidation is based on differences in the rates of the basic or acidic hydrolysis of the diastereoisomers. For instance, of the 1-phenyl-4-*t*-butylcyclohexene oxides, the trans isomer *562* is converted into the diaxial glycol *563* by heating with sodium hydroxide in DMSO, under conditions that leave the cis isomer *564* almost unaffected (352). This difference is due to the fact that diaxial opening involves a nucleophilic attack by base which is highly sensitive to steric effects and takes place much faster with *562* (attack on

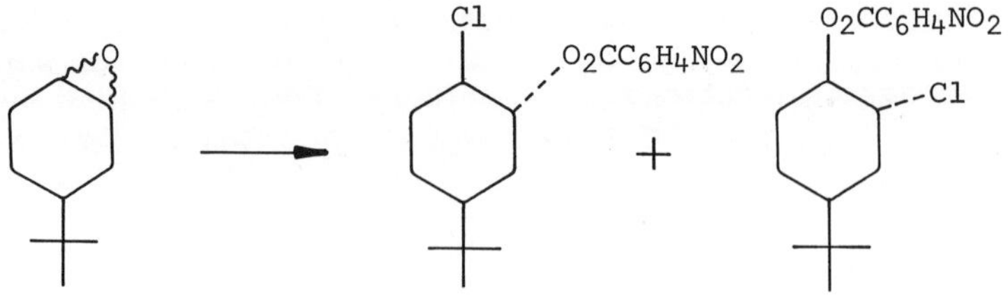

Cl $O_2CC_6H_4NO_2$ $O_2CC_6H_4NO_2$ Cl

O +

555 *556*

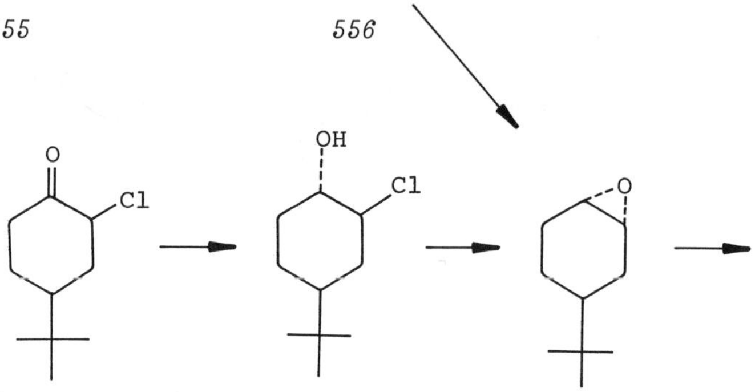

O OH Cl Cl O

557 *558* *559*

OAc OH O

(1) TsCl
(2) MeO^-

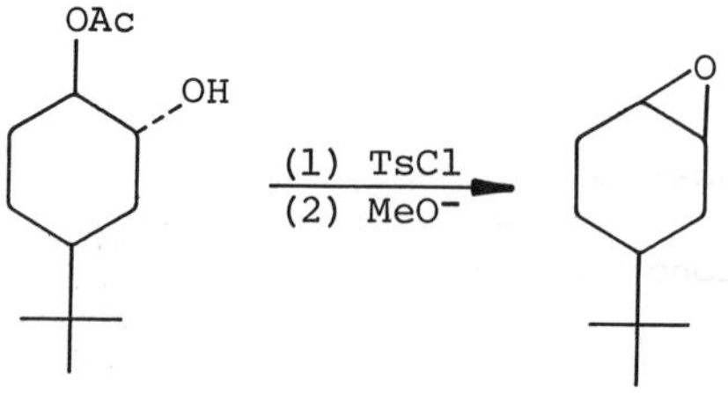

560 *561*

secondary carbon) than with *564* (attack on tertiary carbon); the diol *563* can then be reconverted into *562* through the tosylate route. Under acidic conditions the rates are reversed, since the transition state has more carbonium ion character and attack is easier on the tertiary than on the secondary position; therefore the *cis*-epoxide reacts faster than the trans isomer. A patent describes the separation of *trans*-limonene oxide (*552*) from the epoxidation mixture by preferential hydrolysis of the *cis*-epoxide (*554*) under mildly acidic conditions (353). A further example of selective hydrolysis is that of the separation of *565*, by preferential cleavage of its diastereoisomer, which is attacked more easily by sodium hydroxide in DMSO, because of the equatorial nature of the exocyclic methylene group (106). The separation of the unreacted epoxides from the diols is always easy because of the much lower solubility and higher polarity of the latter.

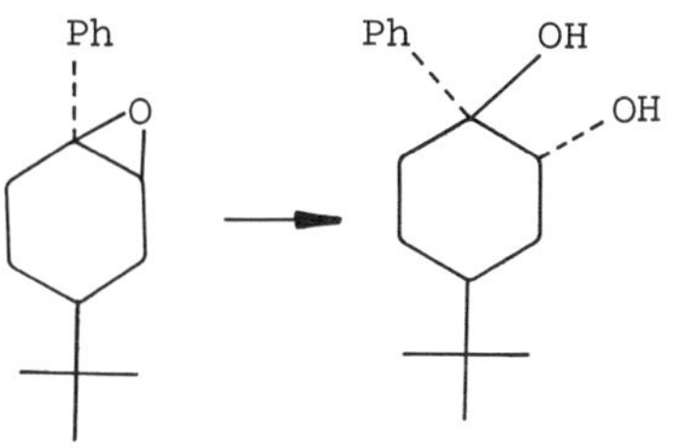

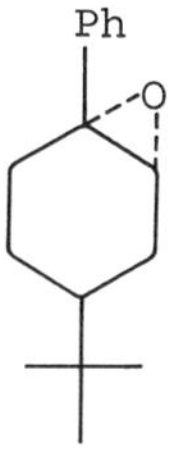

562 *563* *564*

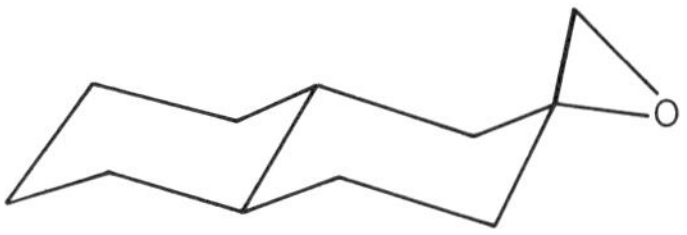

565

The tosylate route has also been employed for the synthesis of sugar epoxides; this aspect has been extensively reviewed (354,355) and will not be treated here, since relatively little work has been done recently in this field.

A carboxylate anion can function as a leaving group for epoxide synthesis under favorable conditions. This approach has been employed mainly for the synthesis of 5,6β-epoxysteroids, on the basis of the observation by Davis and Petrow (356) that 5α,6β-diacetoxy derivatives *566* are converted by ethanolic potassium hydroxide into the β-epoxides *567*. The reaction involves hydrolysis of the more easily accessible secondary ester function, followed by intramolecular displacement of the tertiary acetoxy group by the anionic oxygen (357,358). The conversion of the triterpene diol *558* into the epoxide *570* (359) through treatment with ethyl carbonate and ethoxide provides a different approach to the same type of reaction: formation of the anion by selective transesterification and cyclization with expulsion of the $EtOCO_2^-$ anion. All these reactions require a diaxial disposition and different reactivities of the two hydroxyl groups.

Vicinal aminoalcohols can be converted into epoxides by two different routes: Hofmann-type elimination on the quaternary

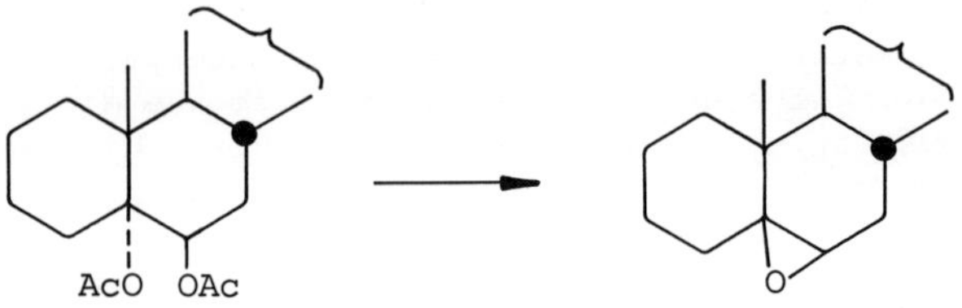

AcO OAc

566

567

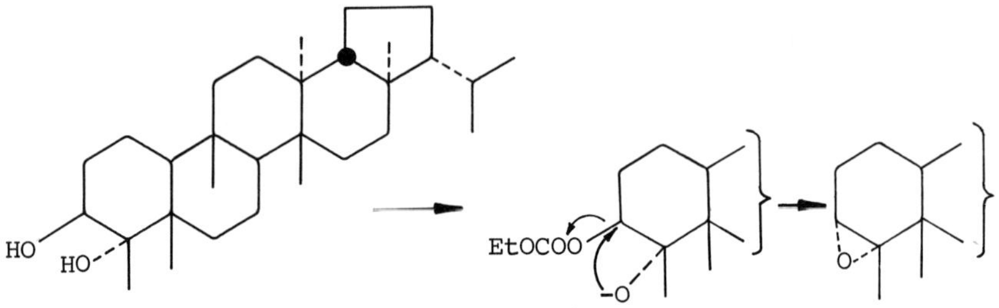

HO HO

EtOCOO

-O

O

568

569

570

hydroxides (*571*), or nitrous acid deamination of the primary amines (*572*). Both reactions normally exhibit the usual requirement of *anti*-coplanar approach to the transition state, although a route through a free carbonium ion could provide an alternative, particularly in the case of the nitrous acid reaction. The only reported case of the formation of an epoxide from a *cis*-aminoalcohol is that shown in *573* → *574* (360). The fact that it appears to be unique so far may suggest some further investigation of this point.

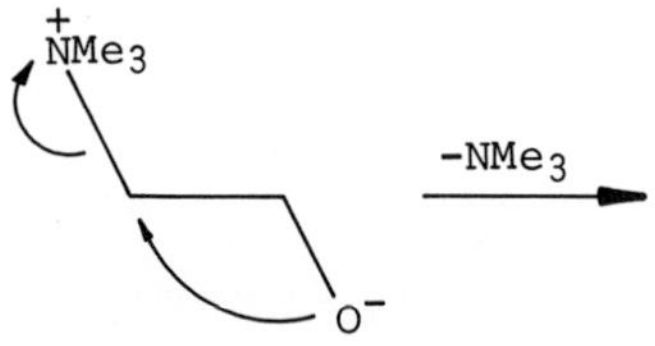

$\overset{+}{N}Me_3$

$-NMe_3$

O^-

571

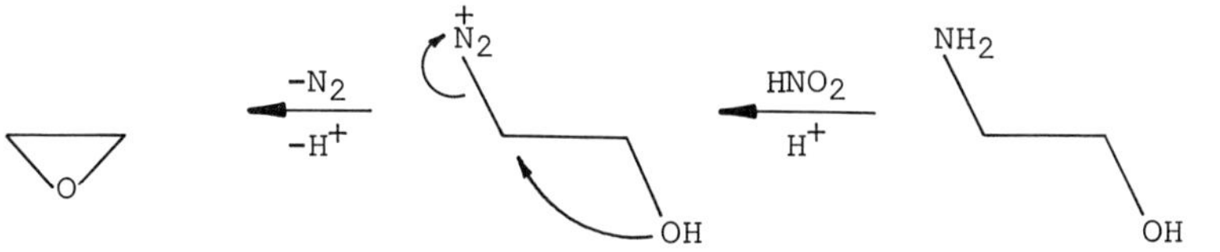

572

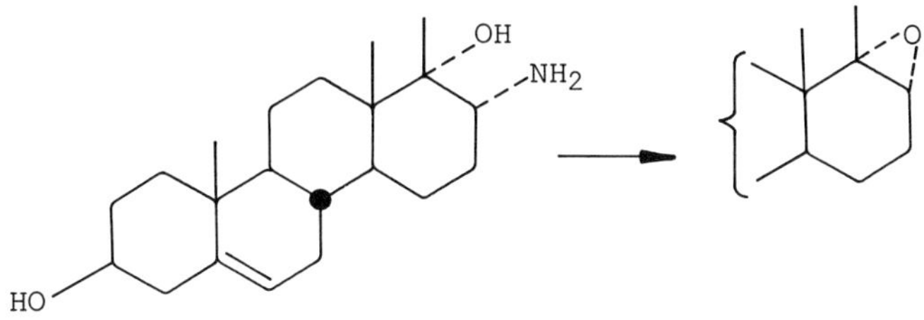

573

574

Examples of the use of the nitrous acid method are rather few, and the usefulness of this reaction appears to be limited in most cases to compounds with axial amino and hydroxyl groups: *575* gives the epoxide *576*, but the diequatorial isomer *577* forms exclusively the ring-contracted aldehyde *578* (361). The diaxial aminoalcohol *579* produces a quantitative yield of epoxide, but the mobile *trans*-2-aminocyclohexanol, which certainly exhibits a high preference for the diequatorial conformation, gives only cyclopentanecarboxaldehyde with nitrous acid (362).

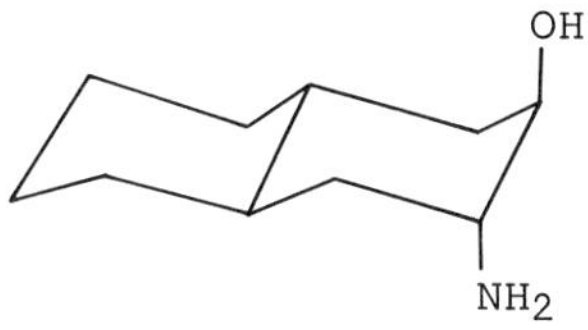

575

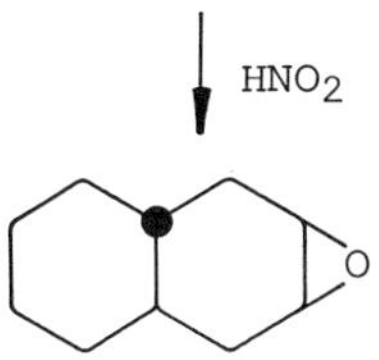

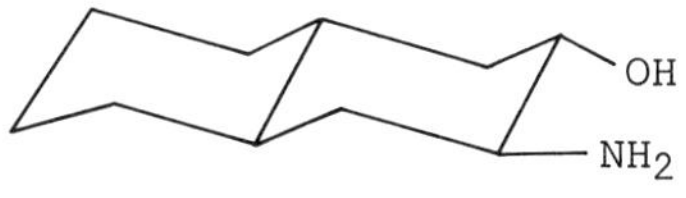

577

576

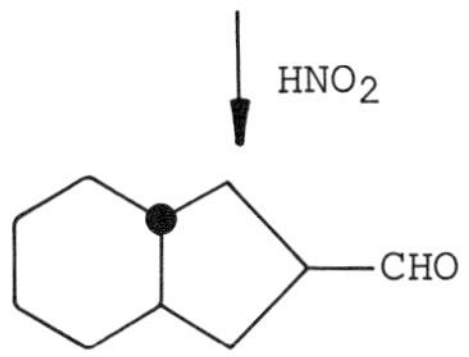

578

OH

NH_2

579

The Hofmann route to epoxides is more useful from the preparative point of view (363), particularly for obtaining optically active epoxides from the easily resolvable aminoalcohols. (-)-Ephedrine (*580*) and (+)-pseudoephedrine (*581*) have been converted into (+)-*trans*- (*582*) and (+)-*cis*-1,2-epoxy-1-phenylpropane (*583*) (364), respectively, and (-)-*erythro*-1,2-diphenylethanolamine into (+)-*trans*-stilbene oxide (365). In contrast with the nitrous acid deamination, the reaction can be carried out with diequatorial aminoalcohols, as shown by the conversion of *584* into *576* (361). The quaternary hydroxide from *cis*-2-aminocyclodecanol (*585*, n = 6), as well as those deriving from *cis*-aminoalcohols with smaller rings, do not form epoxides, but the cyclododecanol derivative (*585*, n = 8) gives the *trans*-epoxide and cyclododecanone in a 3:2 ratio, and the cyclotridecanol derivative (n = 9) only the *trans*-epoxide. This behavior can clearly be related with the different strain involved in reaching the *anti*-coplanar transition state (366).

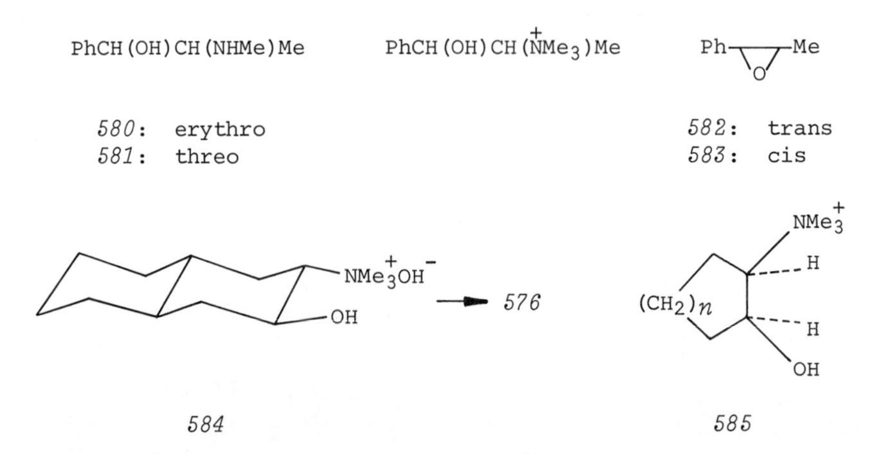

Among the few side reactions that can occur in the Hofmann-type epoxide synthesis is a rearrangement. The quaternary ion

586 (n = 2) forms with sodium amide, besides the epoxide *587* (64%), 13% of the rearranged aminoalcohol *588*, whereas the analogous cyclopentane derivative (*586*, n = 1) gives under the same conditions 77% of *587* and no epoxide (367).

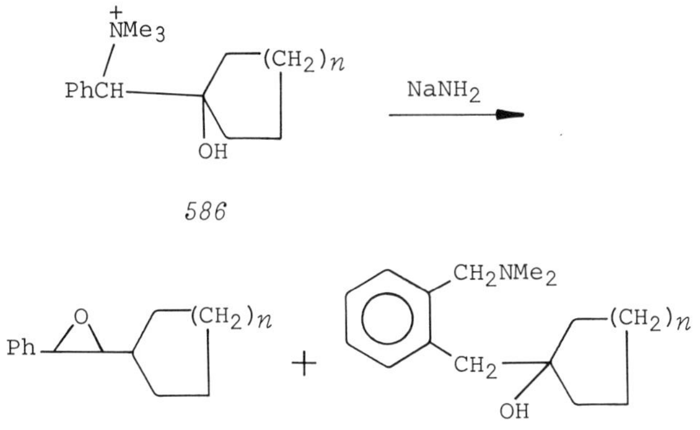

586

587 *588*

The Hofmann route, too, has been applied to the preparation of pure diastereiosomeric cyclohexene oxides from their mixtures. The mixed limonene oxides *589* and *591* react with dimethylamine to give the amino alcohols *590* and *592*, which can be easily separated by fractional crystallization of their picrates. The individual aminoalcohols are then reconverted into the pure epoxides, through quaternization-elimination (41). *Cis*-4-Butyl-

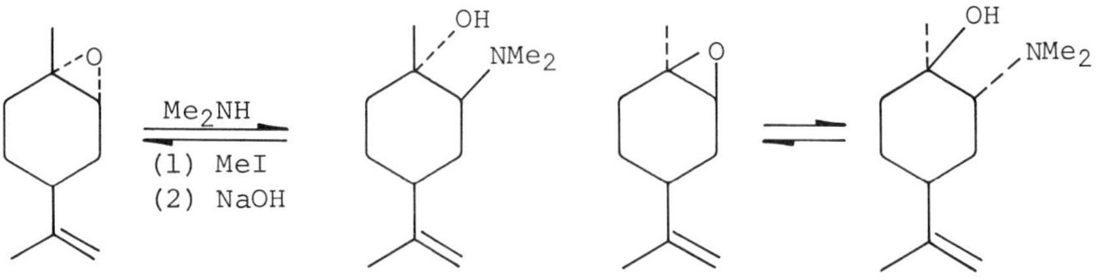

589 *590* *591* *592*

cyclohexene oxide (*561*) can be similarly prepared (368). Resolution of the intermediate aminoalcohol with tartaric acid gives (+)-*561* (369). Pure α- and β-1,2-epoxy-*trans*-decalins were prepared analogously in racemic and optically active forms (54). This approach proved to be particularly useful for the synthesis of optically active 1-phenylcyclohexene oxide. Racemic *593* reacts with dimethylamine to give a mixture of (±)-*594* and (±)-*595*, which can be separated and resolved at the same time by a treatment with (+)-tartaric acid, to yield the (+)-tartrate of (-)-*594*, followed by one with (-)-dibenzoyltartaric acid, which gives the salt of (+)-*595*; other treatments on the mother liquors with (-)-tartaric and (+)-dibenzoyltartaric acid yield the salts of (+)-*594* and (-)-*595*. The individual aminoalcohols are reconverted separately into (+)- and (-)-*593* in high optical, and satisfactory chemical, yields (370).

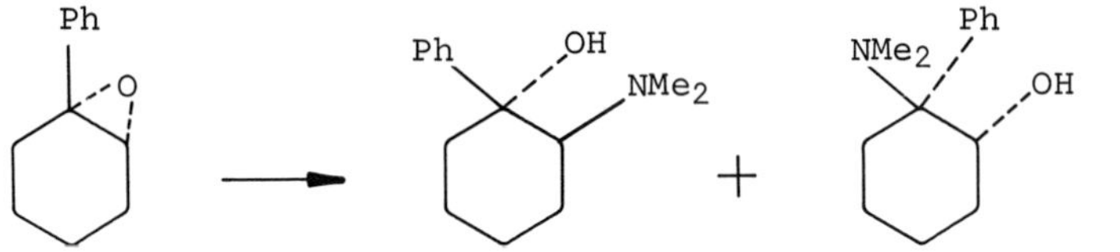

Ph O Ph OH NMe_2 + NMe_2 Ph OH

593 *594* *595*

In particular cases epoxides can be formed by dehydration of 1,2-glycols under acidic conditions. Pocker and Ronald (371) have provided evidence that epoxides are intermediates in the pinacolic rearrangements of tetraarylethylene glycols. The diol *596* is converted into the epoxide *597* on heating with potassium hydrogen sulfate and acetic anhydride, a reaction probably involving the formation of a sulfate ester in position 6, followed by the usual intramolecular displacement (372). The same type of conversion can be achieved by treatment of diaxial diols with thionyl chloride-pyridine (373-375), or with phosphorus trichloride-pyridine (376). The unusual diadamantyl glycol *598* is transformed by hydrogen chloride into the corresponding epoxide, which exhibits an astonishing resistance to attack by acids (377).

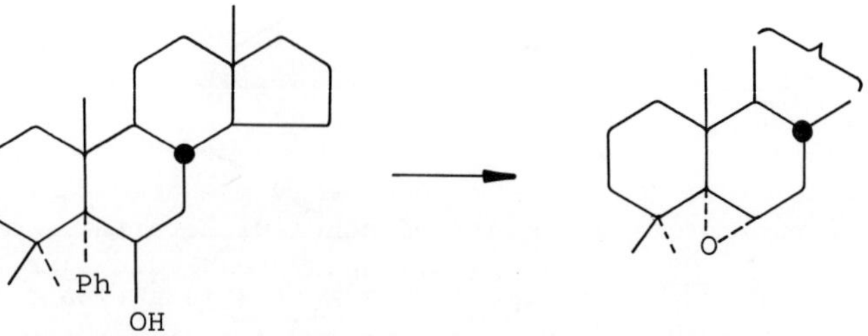

Ph OH O

596 *597*

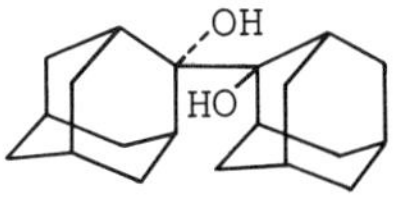

598

3. *Reactions Involving Addition-Elimination on Halocarbonyl Compounds*

Epoxides can be obtained without isolation of the intermediate halohydrin in the reaction of halocarbonyl derivatives with an appropriate base; the intermediate anion *599* is immediately cyclized to epoxide. Such reactions have been observed in the treatment of some ketones having an axial α-bromine, such as *600*, with one equivalent of LAH; epoxide formation probably occurs through an intermediate *599*, where X is hydrogen (378).

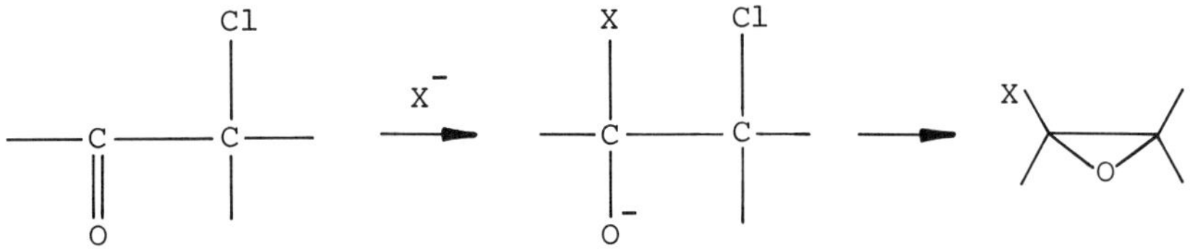

599

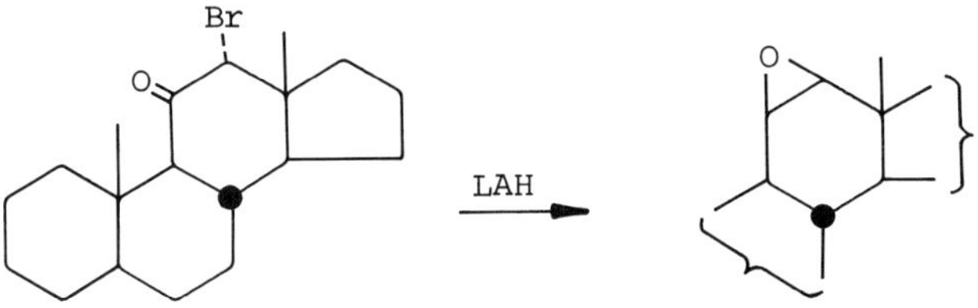

600

This type of reaction provides the best approach to epoxy ethers when X^- is an alkoxide ion (379). The stereochemistry of this reaction has been investigated for a series of α-bromoaldehydes *601* (380), which on treatment with methoxide give 80% or more of the *trans*-epoxy ethers *602*. This has been interpreted on the basis of an approach by the base to the least hindered side of the aldehyde in the conformation *603*, in which the polar groups are *anti*-parallel. Such a model can not, however, account for the exclusive conversion of desyl chloride (*604*) into the *trans*-epoxide *605* (381). The steric course of the latter reaction may be better explained in terms of a reversible first step and of the preferential cyclization of the diastereoisomer involving a more favorable transition state.

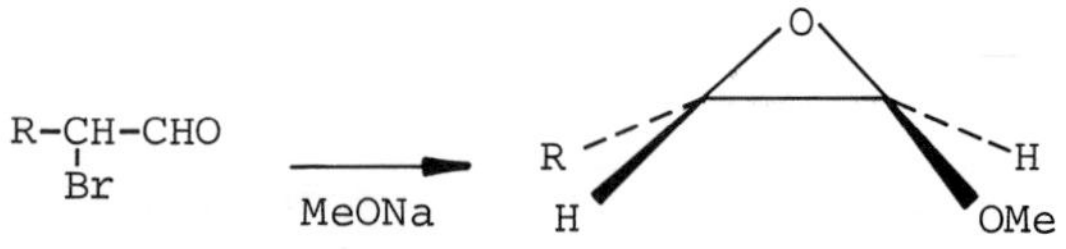

R-CH(Br)-CHO —MeONa→ (R, H, H, OMe epoxide)

601 *602*

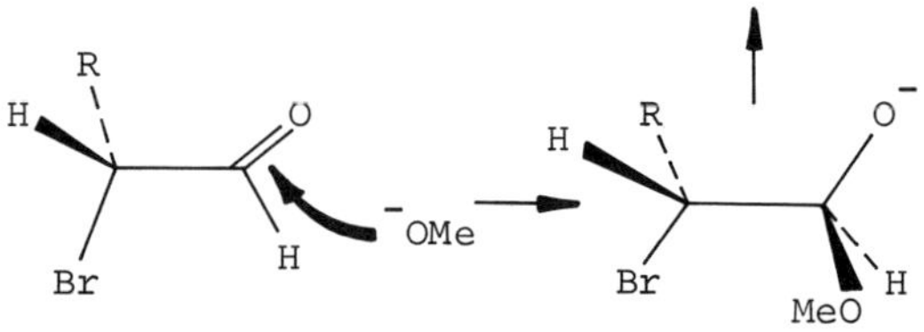

H, R, Br, O, H, $^{-}$OMe → H, R, Br, O^{-}, MeO, H

603

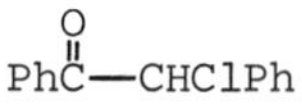

PhC(=O)—CHClPh

604

↓ MeONa

Ph, OMe, H, Ph (epoxide)

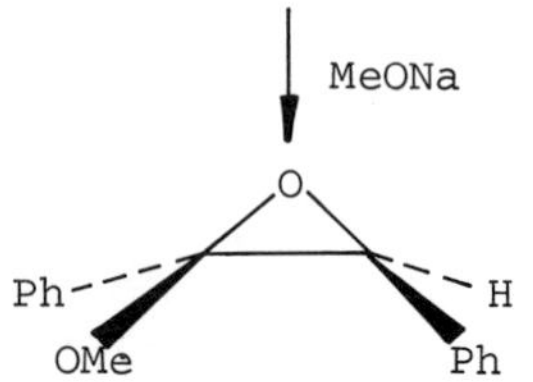

605

A further example of this type of reaction is given by the easy formation of epoxynitriles in the reaction of cyanide ion with haloketones. For instance, the bromochloroketone *606* is converted into the chloroepoxynitrile *607*, bromine being displaced more easily than chlorine; *607* can be further transformed into the epoxydinitrile *609*, through rearrangement to *608* under the action of pyridine and repetition of the cyanide cyclization

(382). Both *607* and *609* contain 65% trans and 35% cis isomer, which corresponds to an excess of the less hindered product. The 2-bromo-4-*t*-butylcyclohexanones *610* and *612*, which are equilibrated in the presence of cyanide, can be converted almost completely either into *611* or into *613* by simply changing the amount of cyanide (383). For other recent examples of applications of this reaction, see (384,385).

$EtCBrClC(=O)Me \xrightarrow{CN^-} EtCCl\text{—}(O)\text{—}C(CN)Me \xrightarrow{C_5H_5N}$

606 *607*

$EtC(=O)CCl(CN)Me \xrightarrow{CN^-} Et(CN)C\text{—}(O)\text{—}C(CN)Me$

608 *609*

O Br $\xrightarrow{CN^-}$ CN O

610 *611*

⇅

O Br $\xrightarrow{CN^-}$ CN O

612 *613*

F. Reactions of Aldehydes and Ketones Involving the Formation of the C-C Bond

This section deals with an important group of methods in

which the synthesis of the oxirane ring is brought about by formation of the C-C bond through attack on the carbonyl group of an aldehyde or ketone by an appropriate nucleophile *614* containing a good leaving group X, in order that the intermediate *615* can immediately cyclize to the epoxide by the usual intramolecular displacement.

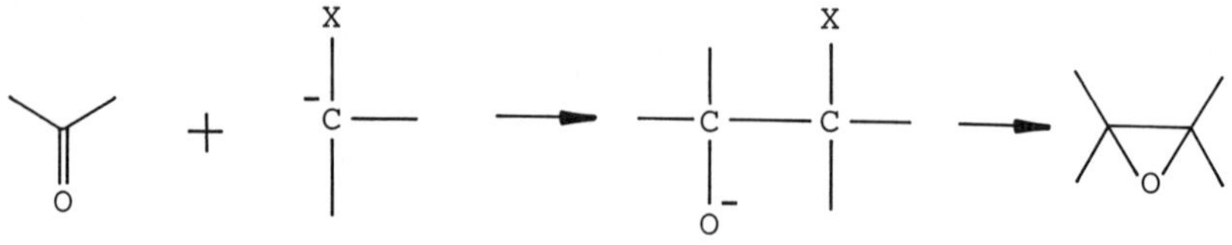

614 *615*

1. *Darzens Reaction*

This is one of the most widely used methods for the synthesis of epoxides carrying electronegative substituents, and is based on the condensation of an aldehyde or ketone with a halide activated by an electron-withdrawing group such as carbalkoxy, cyano, keto, sulfonyl, etc. This reaction has been reviewed extensively (386-388), and reference is made to these reviews for a discussion of the early mechanistic proposals, some of which, such as the one assuming carbenoid intermediates, have by now been disproved (389). It is now generally accepted that the Darzens reaction goes through the mechanism shown in Scheme 2, involving the formation of a carbanion *616*, which attacks the carbonylic component, giving rise to the *threo*- and *erythro*-halohydrins *617* and *618*, which then cyclize to the *cis*- and *trans*-epoxides *619* and *620*, respectively.

Although the steric course of the Darzens reaction has been the subject of many investigations, the present situation in this field is all but clear. The diastereoisomer ratio can vary greatly, depending not only on the structures of the reactants, but also on solvent and base. Since the reacting centers are not chiral, the steric course will be determined by the configurations of the intermediate halohydrins and controlled by kinetic or thermodynamic factors, depending on whether the rates of cyclization (k_T' and k_E') are greater or smaller than those of retroaldolization (k_{-T} and k_{-E}). The situation can be further complicated by the possibility that the glycidic derivatives epimerize in the reaction medium, as mentioned before (p. 167). Many cases of a lack of stereoselectivity in the Darzens reaction have been reported; for instance, the condensation of benzaldehyde with several α-chloro esters gives

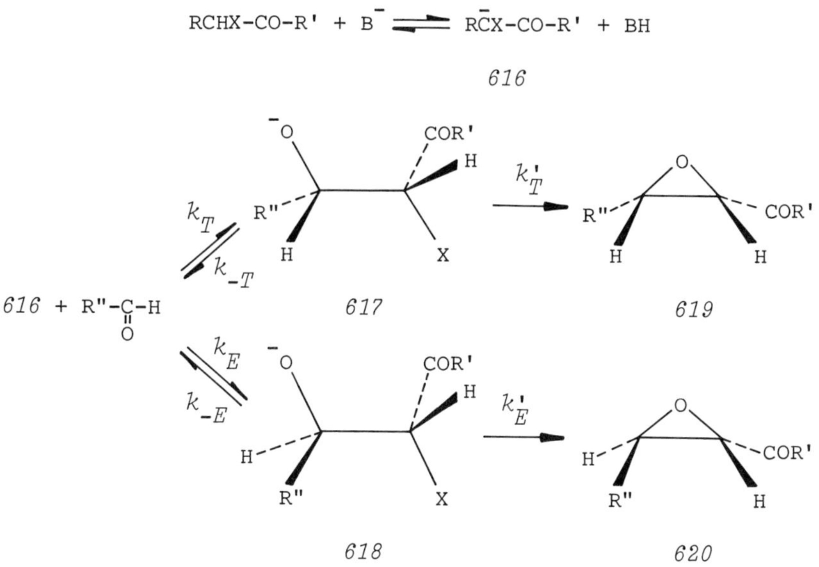

$$RCHX\text{-}CO\text{-}R' + B^- \rightleftharpoons R\bar{C}X\text{-}CO\text{-}R' + BH$$

616

616 + R"-C(=O)-H ⇌ (k_T, k_{-T}) 617 → (k'_T) 619

616 + R"-C(=O)-H ⇌ (k_E, k_{-E}) 618 → (k'_E) 620

Scheme 2

mixtures of *cis*- and *trans*-epoxides *621* (390). The formation of about equimolar amounts of diastereoisomeric glycidamides *622* from benzaldehydes and α-chloroacetamides with *t*-butoxide was interpreted as being due to a slow aldolization, followed by a rapid cyclization (391).

The dependence of the cis/trans ratio on the type of starting halogen derivative and solvent and base is illustrated in Table 12, which refers to the reaction between *623* and *624*. The most striking differences are observed in the case of the reactions of the esters in HMPT and in benzene. Information about the reasons for these differences were obtained by subjecting

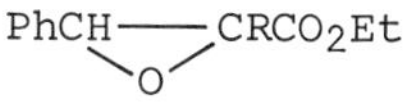

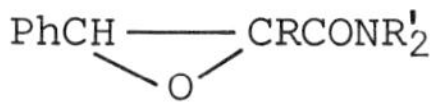

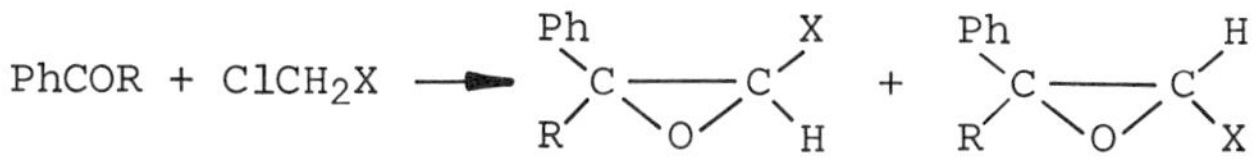

PhCH—CRCO$_2$Et (epoxide) *621*; PhCH—CRCONR'$_2$ (epoxide) *622*

PhCOR + ClCH$_2$X → *625* + *626*

623 *624* *625* *626*

Table 12 Darzen Reactions

623 R	*624* X	Solvent	Base	*625/626* ratio	Ref.
H	CO_2Et	HMPT	NaH	1.0	392
		HMPT	EtONa	0.65	393
		C_6H_6	NaH	0.11	392
H	CN	HMPT	NaH	1.0	394
		t-BuOH	*t*-BuOK	0.8	393
		C_6H_6	NaH	0.8	394
H	$CONH_2$	HMPT	NaH	1.1	393
		C_6H_6	NaH	1.2	393
Me	CN	HMPT	NaH	1.2	395
		C_6H_6	$NaNH_2$	0.47	395

the *threo*- and *erythro*-halohydrins *627*, corresponding to the intermediates in the condensation (392), to the same conditions of the Darzens reaction. *erythro*-*627* when treated with sodium hydride in either HMPT or benzene gives only the *trans*-glycidic ester *626*, and does not incorporate deuterium if the reaction is carried out in the presence of C_6H_5CDO. Therefore the aldolization step is not reversible. *threo*-*627* forms only the cis derivative *625* in HMPT, even if the rate of cyclization is slower than that of the threo isomer, and does not exchange with C_6H_5CDO, but yields a 20:80 mixture of *625* and *626* and exchanges if the reaction is conducted in benzene or ethanol: the aldolization is reversible under these conditions. These data agree with those obtained in the actual Darzens condensation and can be

$$PhCH(OH)CHClCO_2Et$$

627

interpreted by assuming the following relations between the rate constants of Scheme 2 (392):

in HMPT: $k'_E > k'_T \gg k_{-E}, k_{-T}; \quad k_E = k_T$

in C_6H_6 and EtOH: $k'_E > k_{-E}; \quad k'_E > k_{-T} > k'_T$

These assumptions appear reasonable, since the transition state for epoxide formation is reached more easily from the *erythro*-

(*628*) than from the *threo*-chloroalkoxide (*629*). On the other hand, the well-known accelerating properties of HMPT for nucleophilic displacements should increase very much the rate of cyclization, even in the less favorable threo case, and make it competitive with retroaldolization. Such effects are much

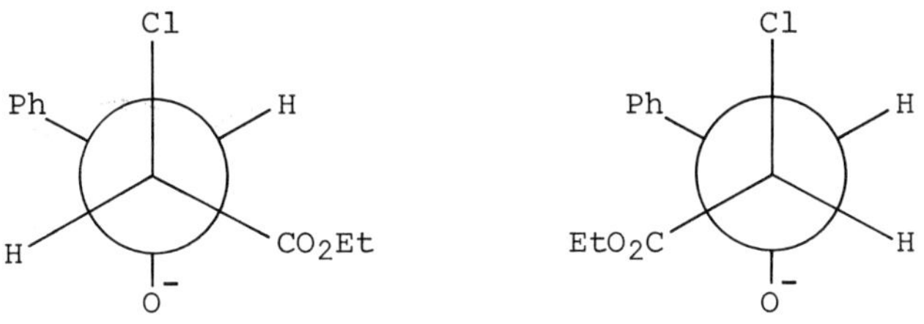

628 *629*

smaller with the nitriles corresponding to *627*, in which retroaldolization occurs in a very minor proportion. This is also true for the threo isomer because of the smaller steric requirements of the CN group (394), which reduces the energy difference between the transition states corresponding to *628* and *629*. Therefore the solvent effect on the steric course of the Darzens synthesis of the nitriles *625* and *626* (R = H, X = CN) is relatively small.

The exclusive formation of the thermodynamically less stable *cis*-diphenylglycidic ester *630* has been interpreted by Zimmermann and Ahramjian (389) by assuming that the steric course of this reaction is controlled by the tendency toward maximum overlap of the carbonyl π orbital of the ester and the developing *p* orbital at the α carbon atom in the transition state for the final cyclization (*631*). The carbethoxy group therefore prefers to be in the less hindered position anti to the phenyl, even if this places the two phenyl groups syn to each other; such a transition state leads to the epoxide *630*. This explanation is similar to that proposed to account for the formation of the thermodynamically less stable diastereoisomer in the Weitz-Scheffer oxidation (Sect. II-C, formulas *388* and *389*). It

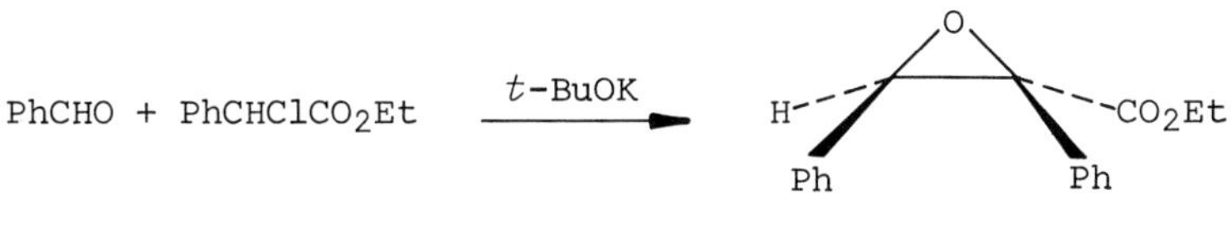

$PhCHO + PhCHClCO_2Et \xrightarrow{t\text{-BuOK}}$

630

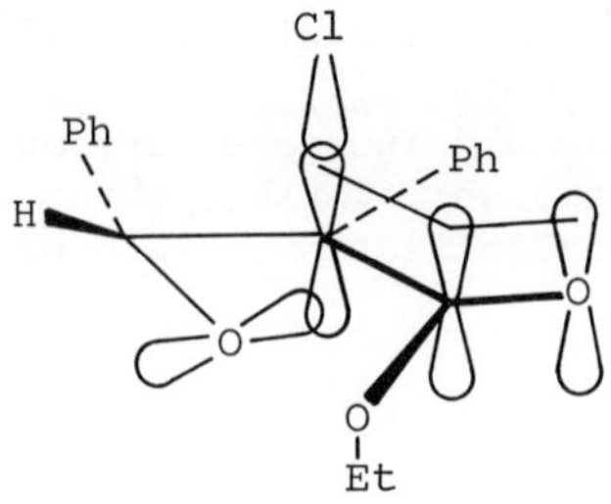

631

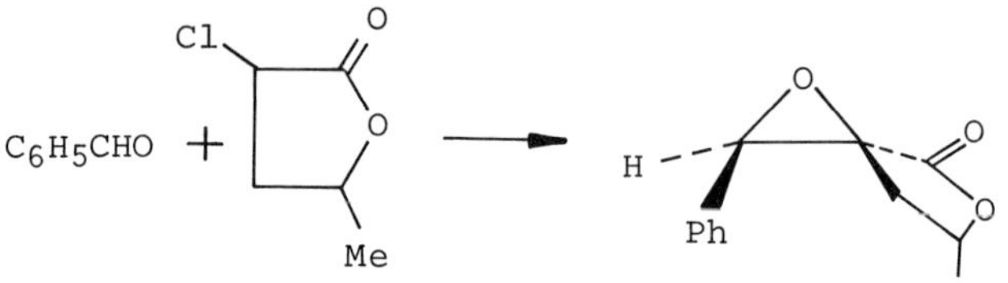

C_6H_5CHO + *632* → *633*

appears somewhat surprising that the steric courses of the two reactions should be rationalized on the basis of analogous assumptions, since the cyclization step involves a nucleophilic attack of oxygen on carbon in the Darzens reaction, but one of carbon on oxygen in the Weitz-Scheffer reaction.

The condensation of benzaldehyde with the chlorolactone *632* was reported to give exclusively the diastereoisomer *633*, in which again the carbonyl group is anti to the phenyl group (396), in accordance with the Zimmermann theory, but the fact that in several other cases the glycidic derivative with the carbalkoxy group cis to the larger substituent is the main product indicates that orbital overlap may be one of the factors, but certainly not the only one, in determining the steric course.

Bachelor and Bansal (397) have investigated the influence of the sizes of a β-alkyl (R) and of the ester alkyl group (R') on the cis/trans ratio in the Darzens synthesis of phenylglycidic esters (*634*). It can be seen from Table 13 that the amount of *cis*-epoxy ester (cis refers to phenyl and carbalkoxy) increases with increasing sizes of both R and R'. These effects cannot be explained on the basis of steric effects in the cyclization step alone, and it was assumed (397) that the configuration of the final product was determined in the aldolization step, based on the rather novel hypothesis that when R and R' are small the preferred transition state leading to the intermediate chloroalkoxide has the eclipsed conformation *635*, where the electron

Table 13 Variations in cis/trans Ratios in the Synthesis of *634*

R	R'	cis/trans ratio
H	Me	0.27
	Et	0.35
	i-Pr	0.49
	t-Bu	1.2
Me	Me	0.75
	Et	0.96
	i-Pr	1.5
	t-Bu	1.6
i-Pr	Me	2.7
	Et	3.0
	i-Pr	3.7
	t-Bu	3.0

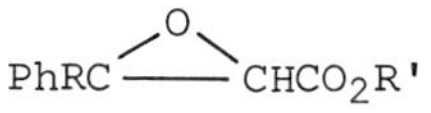

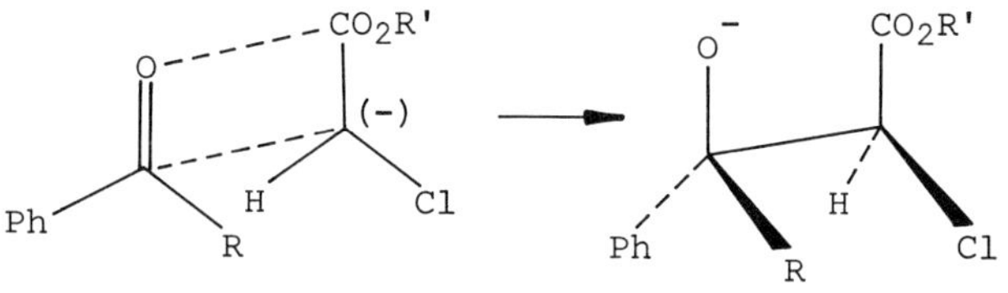

634 *635* *636*

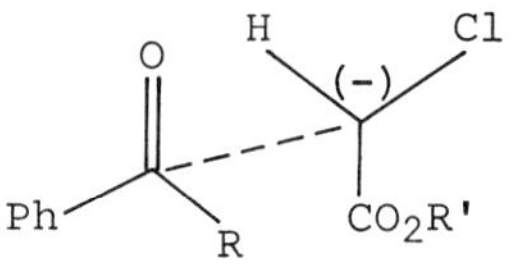

637

pair on the developing oxyanion can overlap with the π orbital of the ester carbonyl; *635* would give *636*, the precursor of the *trans*-epoxide. An increase in the size of R or R' would increase the repulsive energy and counterbalance the overlap energy, making a staggered conformation, such as *637* (leading to *threo*-chlorohydrin and *cis*-epoxide) competitive.

The above interpretation must be accepted with some caution, since more recent work (398) on the reaction between isobutyraldehyde and chloroacetic esters in the presence of *t*-butoxide at room temperature has given results that could be consistent with those in Table 13, with a strong increase in the cis/trans

$$Me_2CHCHO + ClCH_2CO_2R \longrightarrow Me_2CHCH(OH)CHClCO_2R$$

638

$$\downarrow$$

$$Me_2CHCH\text{—}CHCO_2R \text{ (epoxide, O bridge)}$$

639

ratio for *639*, in going from the methyl to the *t*-butyl ester (Table 14). However, when the reaction was conducted at -78°C

Table 14 Variation in cis/trans Ratio in the Synthesis of *639*

R	*cis-639*/*trans-639* at room temp.	(*cis-639* + *threo-638*)/(*trans-639* + *erythro-638*) at -78°C
Et	0.19	2.56
i-Pr	0.25	2.70
t-Bu	2.30	2.30

in order to reduce the rates of cyclization and retroaldolization, mixtures of chlorohydrins *638* and epoxides *639* were obtained, in which the ratios of *threo-638* + *cis-639* to *erythro-638* + *trans-639* were about constant at around 7:3, independent of the size of R. It therefore appears that, at least in this case, the product ratios of the glycidic esters formed at room temperature do not provide a true picture of the primary steric course of the reaction; the rate-determining step is given by the cyclization for the ethyl and isopropyl esters, and by the aldolization for the *t*-butyl ester. This could perhaps hold as well for the data in Table 13, which were obtained at room temperature, and a reappraisal of the mechanistic proposals may be necessary. The same applies for most of the older literature data, and on the whole it can be said that at present the interpretation of the steric course of the Darzens reaction is still wide open.

In the case of the Darzens reaction with α,α-dichloro esters, leading to α-chloroglycidic esters (*640*), two rotamers of a single chlorohydrin intermediate *641* can give rise to either of the diastereoisomeric forms of *640*. The only results so far reported on this type of reaction are given in Table 15 (399). The compound in which the larger alkyl R and the carboisopropoxy group are trans to each other is formed preferentially, and in accordance with a transition state deriving from *642*, the preference is higher, the greater the difference in size between R and R'.

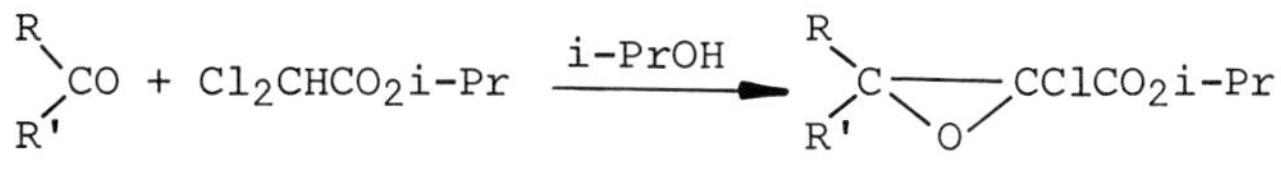

Table 15 Steric Course of the Synthesis of *640*

R	R'	% trans	% cis
Me	H	100	0
Et	Me	75	25
i-Pr	Me	95	5

$$RR'CO + Cl_2CHCO_2i\text{-}Pr \xrightarrow{i\text{-}PrOH} RR'C\overset{O}{—}CClCO_2i\text{-}Pr$$

640

$RR'C(OH)CCl_2CO_2i\text{-}Pr$

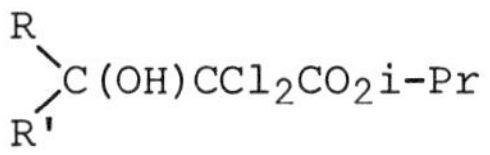

641

O⁻, i-PrO₂C, Cl, R', R, Cl

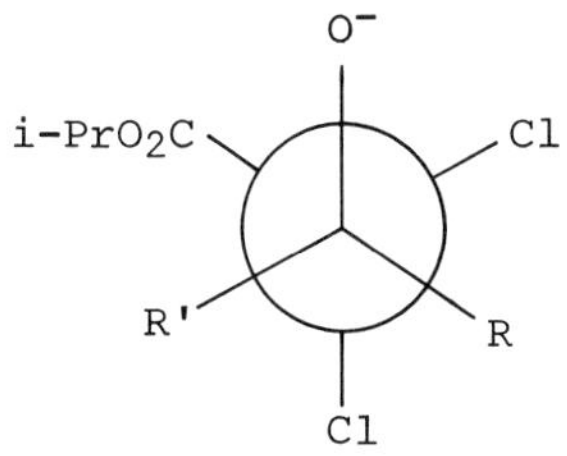

642

The Darzens reaction between tosyl chloromethyl sulfone and benzaldehyde in the presence of potassium *t*-butoxide (400) or sodium hydride (401) gives exclusively the *trans*-epoxysulfone *643*. The analogous reaction between the corresponding sulfoxide and benzaldehyde produces instead predominantly the *cis*-epoxy-sulfoxide *644* (402). Since it is not likely that the two reactions differ considerably in mechanism, the isolation of *644* is probably due to an easy equilibration between diastereoisomers, and to the preferential precipitation of *644* from the reaction medium, because of its lower solubility.

$$p\text{-MeC}_6\text{H}_4\text{SO}_2\text{CH}_2\text{Cl} + \text{PhCHO} \longrightarrow p\text{-MeC}_6\text{H}_4\text{SO}_2\text{(H)}\overset{\text{O}}{\text{C—C}}\text{(H)(Ph)}$$

643

$$p\text{-MeC}_6\text{H}_4\text{SOCH}_2\text{Cl} + \text{PhCHO} \longrightarrow p\text{-MeC}_6\text{H}_4\text{SO(H)}\overset{\text{O}}{\text{C—C}}\text{(Ph)(H)}$$

644

2. *Reactions with Diazoalkanes*

Diazoalkanes react with aldehydes and ketones to give mixtures of epoxides and homologous carbonyl compounds (403-406). Although epoxides are usually the minor products, the reaction can be of preparative interest in particular cases.

The usually accepted mechanism involves nucleophilic attack of the diazoalkane on the carbonyl group to form an intermediate *645*, which can either cyclize to epoxide through intramolecular displacement of nitrogen by the alkoxy anion (this is similar to the final step in the conversion of amino alcohols into epoxides, p. 202), or undergo a 1,2-shift, producing a homologous ketone. Recent kinetic work has provided evidence for a second intermediate *646*, formed by electrophilic attack of diazomethane on the carbonyl oxygen, which could give only epoxide (407,408).

$$\text{RR'}\overset{+}{\text{C}}\text{—O—CHR''}\overset{-}{\text{N}_2} \xrightarrow{-\text{N}_2} \text{RR'C}\overset{\text{O}}{\text{—}}\text{CHR''}$$

646

$$\text{RR'C=O} + \text{R''CHN}_2 \longrightarrow \text{RR'C(O}^-\text{)—CHR''}\overset{+}{\text{N}_2} \xrightarrow{-\text{N}_2} \text{R-CO-CHR'R''}$$

645

(*645* $\xrightarrow{-\text{N}_2}$ epoxide; $\text{RR'C=O} + \text{R''CHN}_2 \longrightarrow$ *646*)

The formation of epoxides can predominate over homologation if electron-withdrawing groups are present α to the carbonyl group (404), but other reactions can also take place: the main product formed when an excess of chloral is reacted with diazomethane is the epoxide *647* (409); this reaction exhibits a remarkable stereoselectivity.

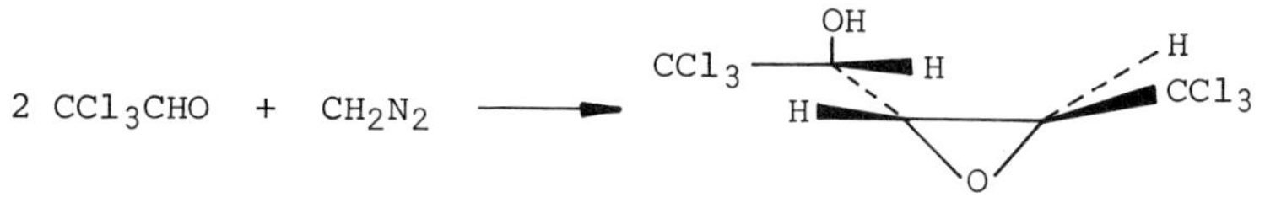

$$2\ CCl_3CHO + CH_2N_2 \longrightarrow$$

647

Acyclic ketones produce more epoxide than homologues, 2-dodecanone having been reported to give only the former (403), whereas yields of epoxides from alicyclic ketones are usually low; only cycloalkanones with six- or more-membered rings form them, and in moderate yields. Also 2-norbornanone does not yield any epoxide with diazomethane (410). Diastereoisomeric carbonyl compounds may behave differently: the 17β-formyl derivative *648* gives 33% epoxide *649* with diazomethane, whereas the α-formyl epimer *650* is converted entirely into the ketone *651* (411).

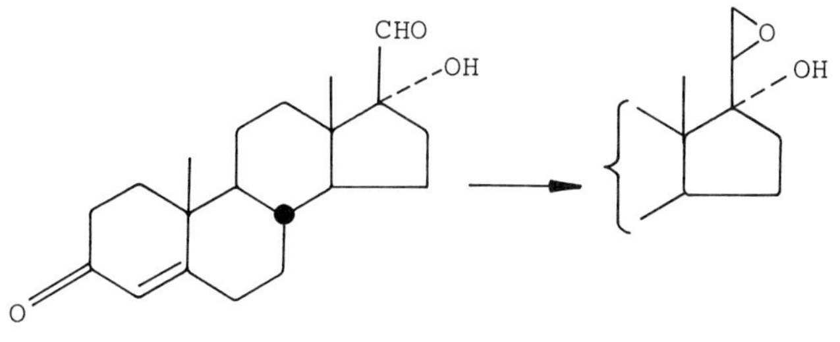

648 *649*

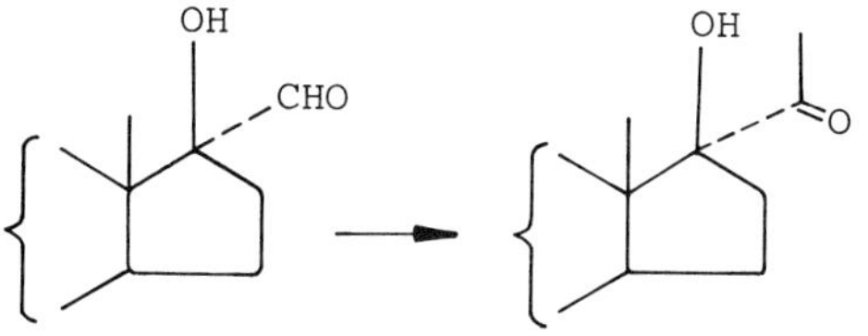

650 *651*

Relatively little is known about the steric course of the diazoalkane attack. Marshall and Partridge (412) have investigated the reaction of some 4-alkylcyclohexanones with diazoethane. The 4-methyl derivative gives 12% of a 1:1 mixture of epoxides *652* and *653*, the 4-isopropyl and 4-*t*-butyl derivatives 18 and 14%, respectively, of a single isomer, probably *652*. This has been interpreted on the assumption that only with an equatorial diazoalkyl group can the *anti*-coplanar disposition *654* necessary for the cyclization be reached, because of the strong repulsive interaction between N_2^+ and *syn*-axial hydrogens

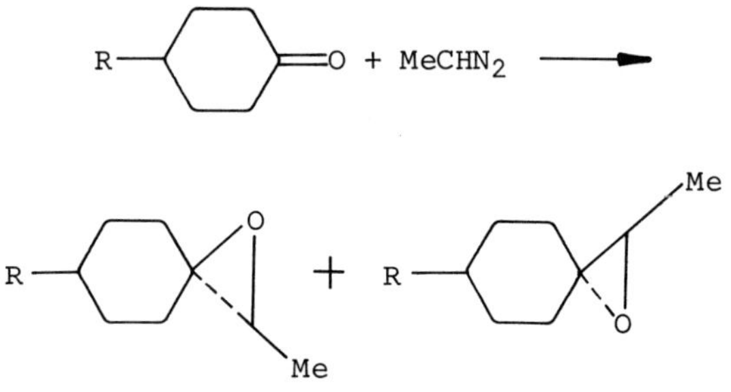

652 *653*

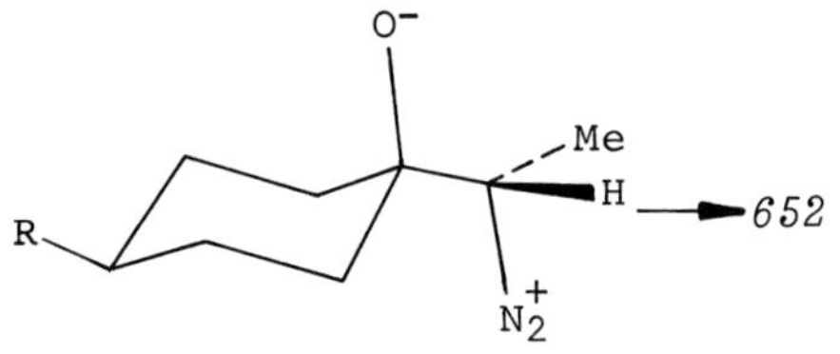

654 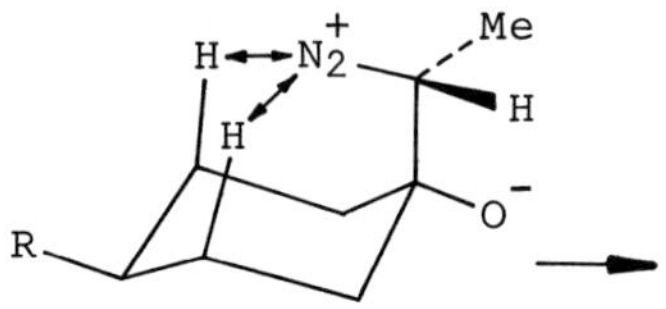 *655*

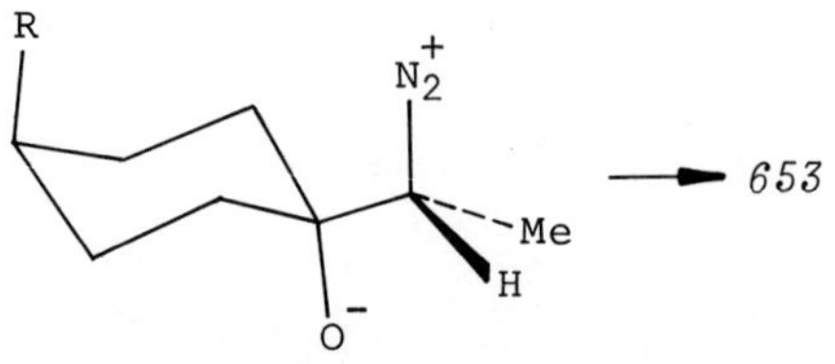

656

in the axial conformation *655*. The *cis*-epoxides *653* must therefore be formed from conformation *656*, which is the less stable, the greater the size of R.

Further evidence for a preferential equatorial attack by diazomethane is given by the fact that 2-decalone forms the epoxides *657* and *658* in a ratio of 4:1 (413), and that only *659* is obtained in 50% yield from the reaction of 5α-androstan-17β-ol-3-one with diazomethane in the presence of alumina (414).

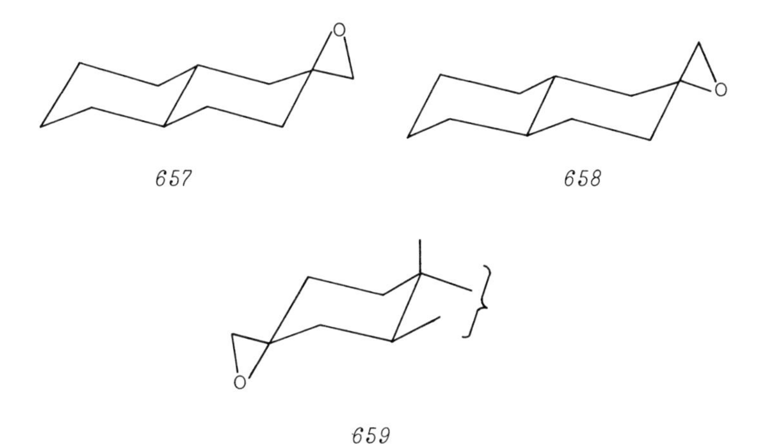

657 *658*

659

Norbornen-7-one (*660*) gives a 34% yield of the epoxide *662*, with the methylene group syn to the double bond, and none of the diastereoisomeric epoxide. The betaine deriving from syn attack (*661*) can more easily reach the conformation for intramolecular displacement, because of the lesser shielding from this side; charge delocalization from N_2^+ to the double bond may also contribute to stabilize *661* (415).

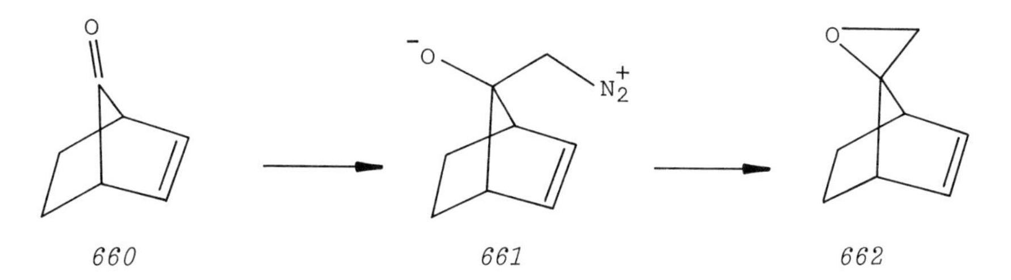

660 *661* *662*

3. *Reactions with Sulfonium Ylides*

Aldehydes and ketones can be converted, often in better than 90% yields, into epoxides by reaction with certain sulfur ylides. After a first report by Johnson and La Count (416) that the ylide *663* reacts with substituted benzaldehydes to give, among other products, the epoxide *664*, the method was developed to its full synthetic potential by the use of dimethylsulfonium methylide (*668*) (417,418), and particularly dimethyloxosulfonium methylide (*669*) (419-421). The early developments in this field have been reviewed (422).

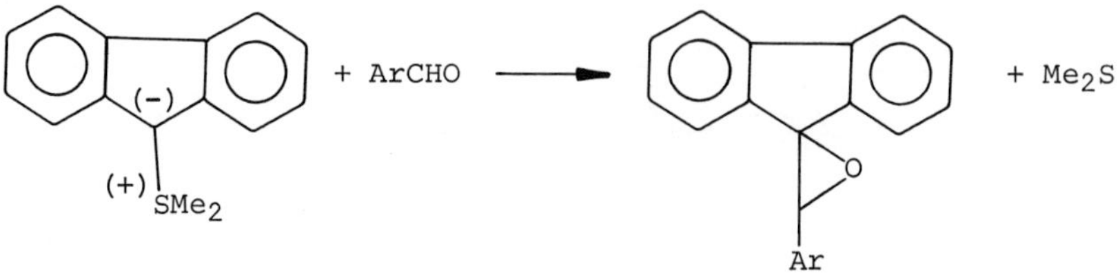

(-) (+) SMe_2 + ArCHO → O Ar + Me_2S

663 *664*

It is well known that ylides are very useful reagents for attack on π systems. Their reactions with carbonyl functions involve as the first step a nucleophilic attack by the anionic carbon on the carbonyl carbon to give betaines of type *665*, but the further course of the reaction depends on the nature of the cationic part of the ylide (Y). If it is a phosphonium group, the betain collapses into an olefin and a phosphonium oxide (path *a*, Wittig reaction), whereas if it is a sulfonium group, displacement of Y by the anionic oxygen with formation of an epoxide (path *b*) is the normal course. Arsonium ylides show an intermediate behavior since they can give both olefins and epoxides, as found for the reaction between *666* and *667* (423). For a discussion of the possible reasons of these differences, see (422).

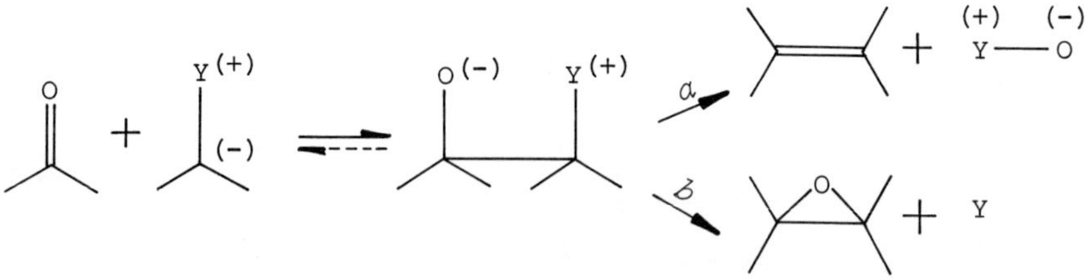

O + Y(+) (-) ⇌ O(-) Y(+) —*a*→ + (+)Y—O(-) ; —*b*→ O + Y

665

$$\overset{(+)\ (-)}{Ph_3As\text{-}CHPh} + p\text{-}NO_2C_6H_4CHO \longrightarrow \begin{cases} PhCH{=}CHC_6H_4NO_2 + Ph_3AsO \\ PhCH\overset{O}{\text{—}}CHC_6H_4NO_2 + Ph_3As \end{cases}$$

666 *667*

Dimethylsulfonium methylide (*668*) and dimethyloxosulfonium methylide (*669*) are easily available through the treatment of the corresponding trimethylsulfonium salts with sodium hydride or butyllithium in DMSO or THF. When the formation of diastereoisomers is not involved, *669* is the reagent of choice, since it is much more stable than *668*, and its solutions can be stored several days at room temperature without appreciable decomposition. However, when diastereoisomeric epoxides can be formed, the two reagents may show different stereoselectivities, and this is a further advantage of the method: *668* usually attacks

$$\rangle{=}O + \overset{(+)\ (-)}{Me_2S{-}CH_2} \longrightarrow \text{(epoxide)} + Me_2S$$

668

$$\rangle{=}O + \overset{(+)\ (-)}{Me_2S(\to O){-}CH_2} \longrightarrow \text{(epoxide)} + Me_2SO$$

669

the carbonyl group from the more hindered, *669* from the less hindered side. Thus in the case of 4-*t*-butylcyclohexanone, *668* yields an 83:17 mixture of *670* and *671*; *669* yields only *671*. Similarly, although *673* is the only product of the reaction of 3,3,5-trimethylcyclohexanone with *669*, a 55:45 mixture of *672* and *673* is obtained with *668* (420).

Analogous stereoselectivities are found for *trans*-2-decalone (106), 3-, and 6-oxosteroids (118,119,424). The formation of an

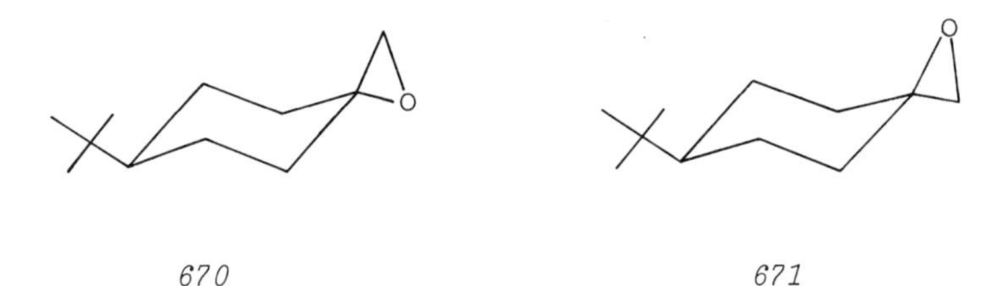

670 *671*

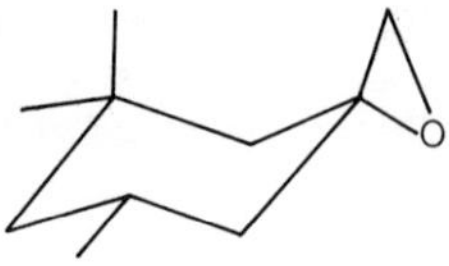

672 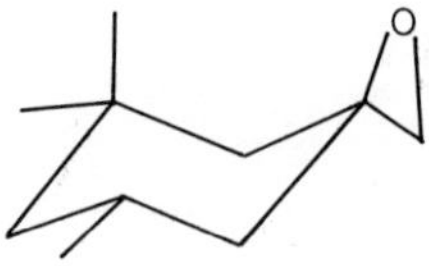 *673*

excess of the kinetic product with *668* can be explained by its higher reactivity and smaller size. As in other nucleophilic reactions (p. 118) preferential attack occurs from the axial side, and the rate of cyclization of the intermediate betaine *674* is probably much faster than that of its reversion to the initial reagents. In the case of the reaction with *669*, it is likely that the intermediates *675* and *676* can equilibrate before cyclization, because of the reversibility of the first step, and because of the greater stability of *675* and *676* toward cyclization. A severe interference between the oxosulfonium group and the *syn*-axial hydrogens makes the axial *anti*-coplanar conformation *675*, leading to the epoxide *670*, highly unfavorable with respect to the equatorial one *676*, leading to *671*.

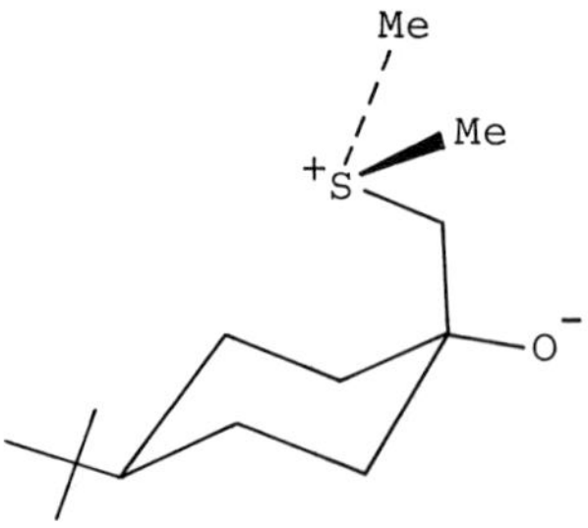

674

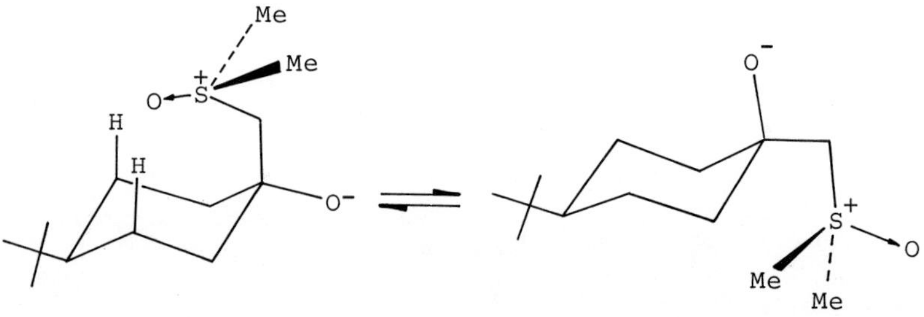

675 *676*

Although these stereoselectivity rules appear to be quite valid generally for the reactions of cyclohexanones with *668* and *669*, very little is known about the steric course of the reactions with other types of cyclic and acyclic ketones. In the case of 17-oxosteroids, whereas *668* gives only the β-epoxide (*677*), *669* produces a mixture of *677* and *678* (424,425); this is just the opposite of what one would expect on the basis of a steric effect. Important steric hindrance can lead to the exclusive formation of the less hindered epoxide with *668* also. This is the case of the bicyclic ketone *679*, which forms only *680*, the product of attack anti to the bulkier bridge (156). Norcamphor gives a large excess of *681* with both *668* and *669*,

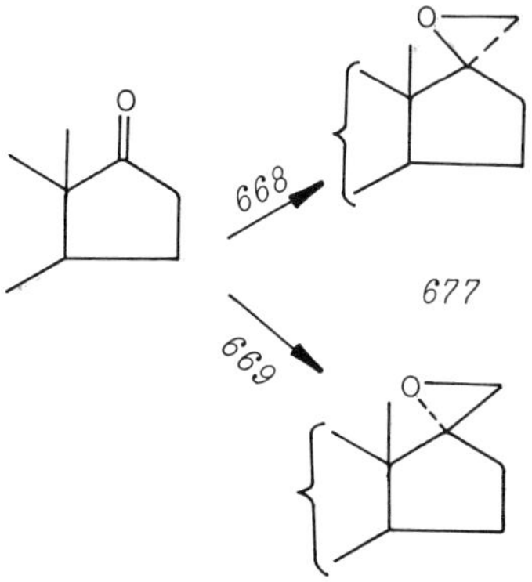

668

669

677

678

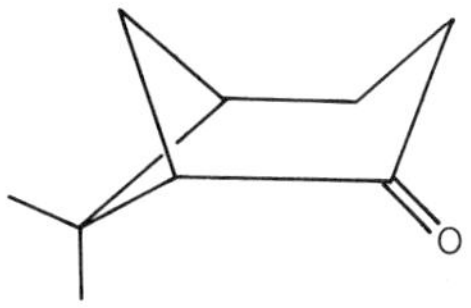

679

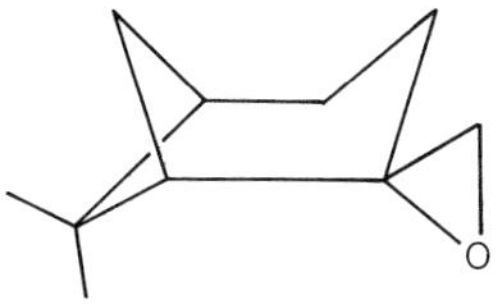

680

in accordance with the very unfavorable character of endo attack (426). 7-Norbornenone forms exclusively *682* with either ylide. That this is due not only to a smaller steric hindrance to ylide attack syn to the double bond is indicated by the formation of 67% of the endo attack product *683* from 5,6-dihydronorcamphor. A participation of the π electrons of the double bond (*684*), favoring the anti conformation of the betaine may explain these results, as in the case of the diazomethane reaction (Sect. II-F-2).

One limitation in the use of *668* and *669* is the fact that they can give only epoxides with one unsubstituted methylene group in the oxirane ring. By a suitable choice of reagents it is possible to obtain tri- and tetrasubstituted epoxides also. Diphenylsulfonium benzylide (*685*, R = Ph) can easily be prepared from diphenylbenzylsulfonium fluoborate and allowed to react

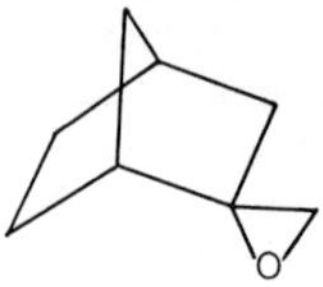

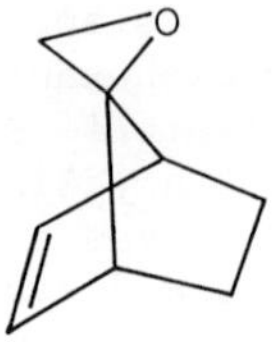

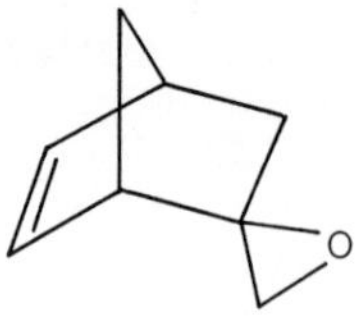

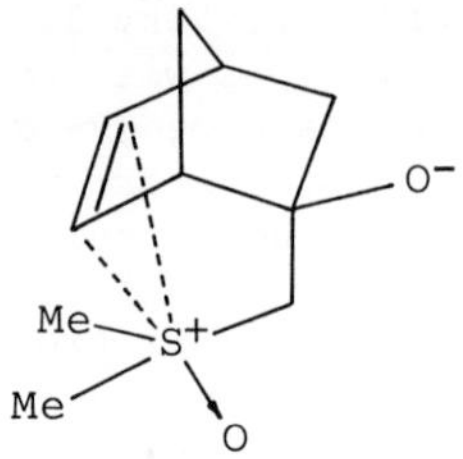

681 682 683 684

with benzaldehydes to yield 60 to 75% *trans*-stilbene oxides free of cis isomers (427). Even better yields of 2-phenyl-3-alkyloxiranes are obtained through the use of the ylides *685* (R = Me or Pr), but the reaction exhibits little or no selectivity (428). The reason for this variation in the steric course of the epoxidations with aryl- or alkyl-substituted *685* may be sought in the difference in the sizes of the R groups. When R is phenyl the threo conformation *686*, leading to the *cis*-epoxide, is much more hindered than the erythro one (*687*); if the betaine formation is reversible and the cyclization slow, a high preference for the formation of the *trans*-epoxide can be observed. The

$$Ph_2\overset{+}{S}-\overset{-}{C}HR + ArCHO \longrightarrow ArCH\overset{O}{—}CHR + Ph_2S$$

685

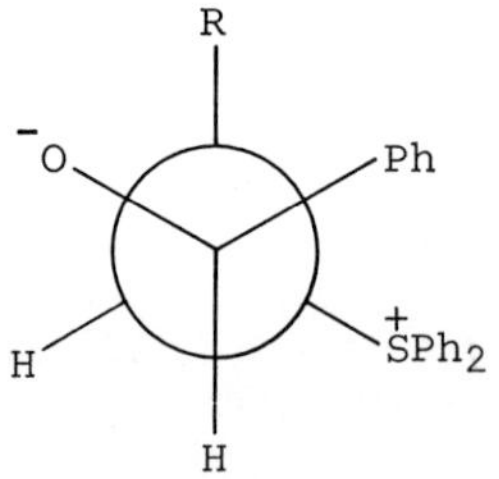

686 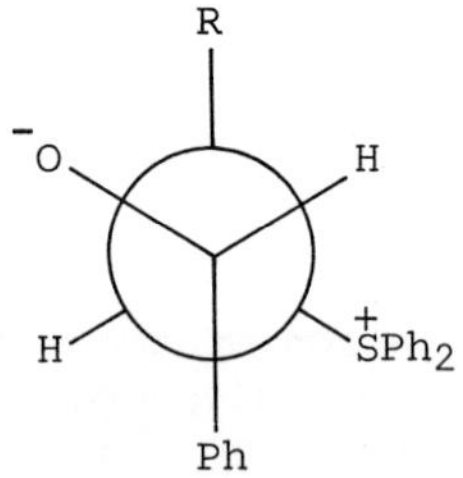 *687*

effect gets less pronounced as R gets smaller: with R = Me the *cis*- and *trans*-epoxides are formed in about equal amounts; with R = Pr their ratio is 1:1.5.

An approach to tri- and tetrasubstituted epoxides is provided by the alkylation of *685* to give *688*, which is then converted into the ylide and treated with the carbonyl partner. This reaction is nonstereoselective (429). The combination lithium diisopropylamide-methylene chloride in dimethoxyethane, which is equivalent to dichloromethyllithium, gives the best results in these reactions. The ylides derived from dimethylthetin (*689*) (430) and from the allyldiphenylsulfonium ion (*690*) (431) also give satisfactory yields of trisubstituted epoxides; they attack 4-*t*-butylcyclohexanone preferentially from the equatorial side to form *691* and *692*, respectively.

$$685 \xrightarrow[(2)\ \mathrm{MeI}]{(1)\ (i\text{-}Pr)_2NLi/CH_2Cl_2}$$

$$\underset{688}{Ph_2\overset{+}{S}CHMeR} \xrightarrow[(2)\ R'COR'']{(1)\ (i\text{-}Pr)_2NLi/CH_2Cl_2} R'R''C\overset{O}{—}C(Me)R$$

$$\underset{689}{Me_2\overset{+}{S}—\overset{-}{C}HCO_2^-} \qquad \underset{690}{CH_2=CH\overset{-}{C}H-\overset{+}{S}Ph_2}$$

691: R = COOH
692: R = $CH=CH_2$

Ylides can also be formed in strongly basic aqueous solutions. Under these conditions triethylsulfonium bromide and benzaldehyde give an 80% yield of a mixture of *cis*- and *trans*-propenylbenzene oxides, and allyldimethylsulfonium bromide gives a 60% yield of the diastereoisomers of *693* (432).

$$CH_2=CHCH_2(Me)_2\overset{+}{S}\,\overset{-}{Br} + PhCHO \xrightarrow[H_2O]{NaOH} \underset{693}{PhCH\overset{O}{—}CHCH=CH_2}$$

Dimethylaminooxosulfonium ylides can also be employed for the preparation of epoxides (433); *694* is obtained in optically active form from active methyl *p*-tolyl sulfoxide, and can in turn be used for an asymmetric synthesis of styrene oxide (*696*); the optical yield is only 5%, whereas the corresponding reaction of *694* with α,β-unsaturated ketones and esters produces substituted cyclopropanes of up to 35% optical purity (434). The sulfinamide *695*, formed as the second product, can be reconverted into the starting active sulfoxide with an overall loss of optical purity of only 20%.

The tosylimino ylide *697* converts 4-*t*-butylcyclohexanone into *671* with an 84% chemical and a quantitative diastereoisomeric yield (435).

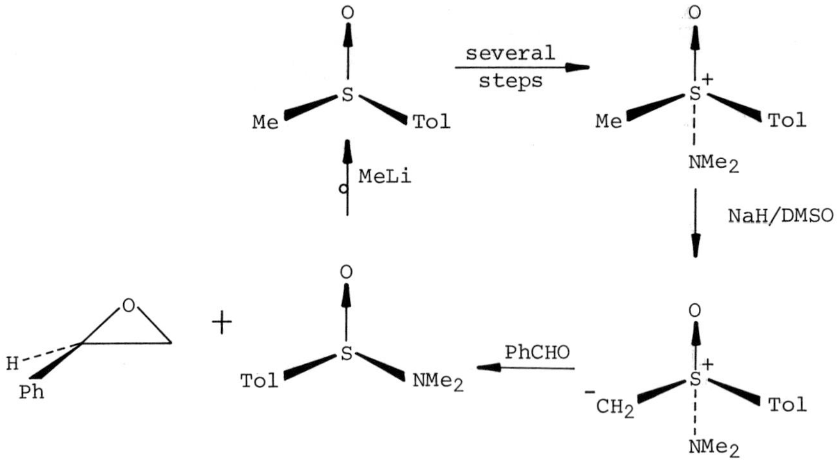

O, S, Me, Tol — several steps → O, S+, Me, Tol, NMe2

MeLi; NaH/DMSO

H, Ph, O + O, S, Tol, NMe2 ← PhCHO — O, S+, $^-CH_2$, Tol, NMe2

696 *695* *694*

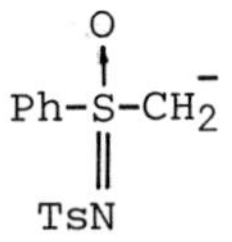

O, Ph-S-CH_2^-, TsN

697

4. *Reactions of Aldehydes with Trivalent Phosphorus Derivatives*

Mark (436) found that diaryloxiranes can be obtained, often in high yields, from the reaction of two moles of an aromatic or heterocyclic aldehyde with one of tris(dimethylamino)phosphine:

$$2\ ArCHO + (Me_2N)_3P \longrightarrow ArCH\text{—}CHAr\ (\text{epoxide, O bridge}) + (Me_2N)_3PO$$

A mechanism was proposed involving the nucleophilic attack of the phosphorus derivative on the carbonyl carbon to give an intermediate *698*, which reacts with a second molecule of aldehyde to produce the *cis*- and *trans*-epoxides, through *699* and *700*. The actual isolation of compounds supposed to have structure *698* from the reactions of aldehydes having no electron-withdrawing substituents on the ring lent support to this mechanism. However, Ramirez and co-workers (437) offered a different interpretation for the reaction, which provided a more satisfactory picture, at least for aromatic aldehydes with electron-withdrawing substituents. It is by now firmly established that in these cases trivalent phosphorus attacks nucleophilically the oxygen, and not the carbon, of the carbonyl group. Such an attack gives rise to an intermediate *701*, which reacts with a second molecule of aldehyde to form the diastereoisomeric dioxaphospholanes *702* and *703*. These are converted into the epoxides through ring opening, followed by displacement of hexamethylphosphoramide, with inversion on one of the chiral carbons. The intermediates *702* and *703* are too labile to be separated, but the analogues *704* are stable enough for isolation and characterization; they are converted into the dinitrostilbene oxides stereospecifically with inversion by heating at 80°C.

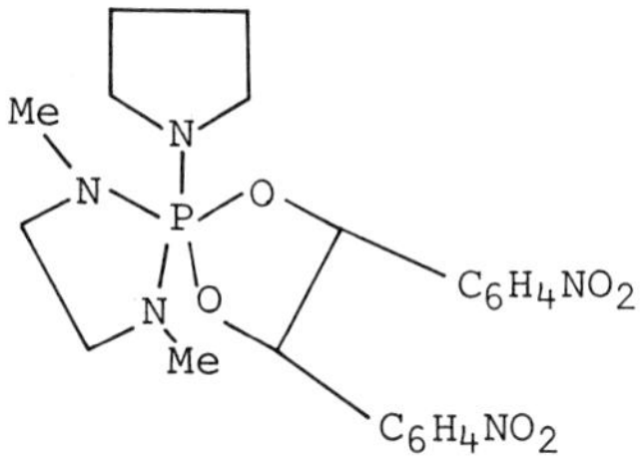

704

Compounds containing two aldehydo groups in a suitable position can react intramolecularly with tris(dimethylamino)phosphine. It is possible to so prepare phenanthrene-9,10-oxide (*706*) from phenanthrene, through the dialdehyde *705* (438); *706* cannot be prepared directly from phenanthrene by peroxyacid oxidation.

A reaction that probably falls into the same mechanistic pattern is the condensation of benzoyl cyanide with triethyl phosphite, giving the adduct *707*, the thermolysis of which at 140°C produces the *cis*-epoxide *708* (439,440); *707* must have the trans configuration if the mechanism of its conversion into *708*

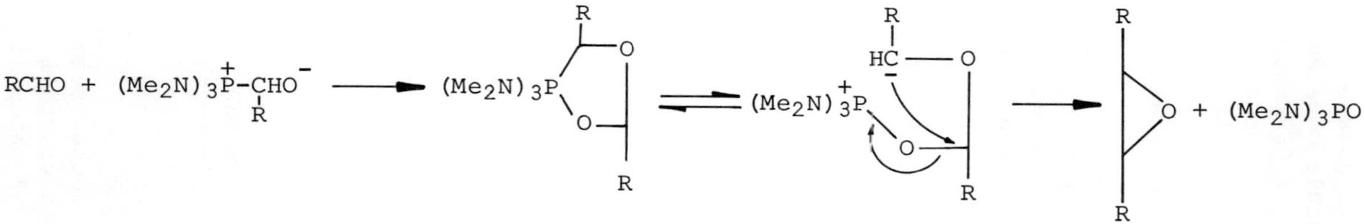

698 699 700

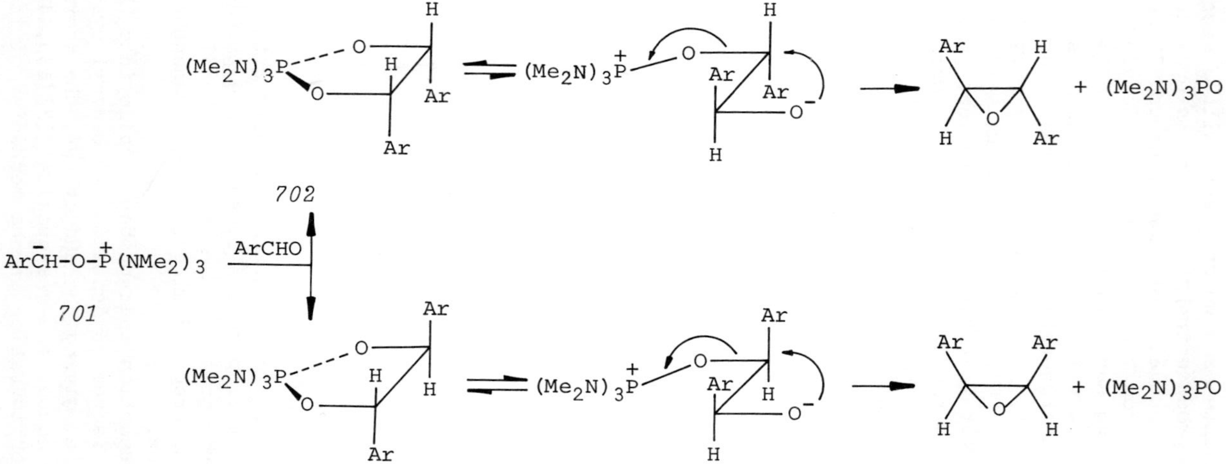

701 702 703

O_3 → *705* $(Me_2N)_3P$ → *706*

is the same as for the reaction of *703*.

1,3,2-dioxaphospholanes (*710*) analogous to *707* are also available from 1,2-diols, through conversion to the cyclic phosphites *709*, followed by oxidation with diethyl peroxide. They decompose thermally to the inverted epoxides (*711*) and triethyl phosphite; the stereospecificity of this reaction has been demonstrated by the conversion of (+)- and *meso*-2,3-butanediol, respectively, into *cis*- and *trans*-2-butene oxide (441).

$2\ PhCOCN + P(OEt)_3$ → *707* —140°C→ *708* $+ OP(OEt)_3$

—PCl_2OEt→ *709* —Et_2O_2→

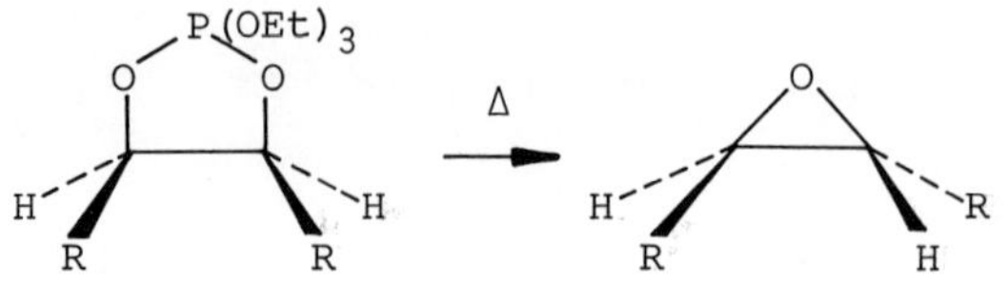

710 711

G. Epoxides from Transannular Peroxides

Ascaridole (*712*) is converted in good yield into the *cis*-diepoxide *713* on heating (442). This reaction is likely to involve a slow homolytic cleavage of the O-O bond, followed by a rapid attack by the two radical oxygen atoms on the double bonds. The treatment of *712* with triphenylphosphine produces the unsaturated epoxide *715*, probably through the zwitterionic intermediate *714* (443). The same reactions are also given by other

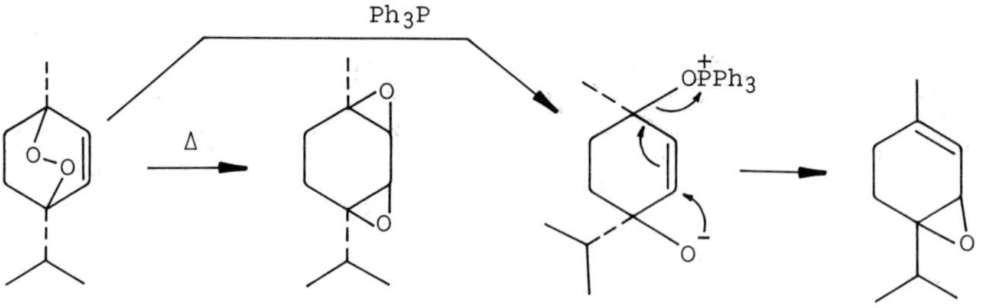

712 713 714 715

transannular peroxides that can be prepared by the dye-sensitized photooxidation of cisoid dienes, such as the levopimaric acid derivative *716*, which yields *717* and *718* (175), and the cyclopentane derivative *719*, which forms *720* (444). Also the peroxy ketone *721*, available by the oxidation of the dienone with diperoxychromium(VI) oxide etherate, is converted into the diepoxide *722* at 210°C (445).

Of particular preparative interest is the completely stereoselective conversion of cyclopentadiene endoperoxide (*723*) into 58% of the cis compound *724*, 7% of the diepoxide *725* being formed as a side product. The reaction involves a thermal fission of the intermediate, possibly through an electrocyclic four-center mechanism. Compound *724* is converted quantitatively into the trans isomer *726* by triphenylphosphine (446).

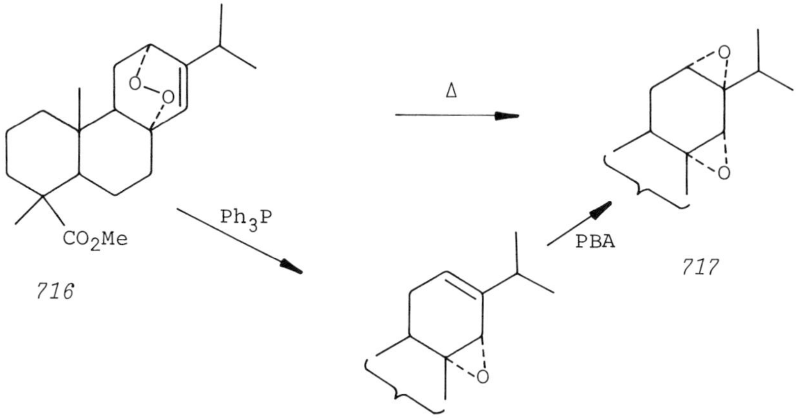

CO_2Me

716

Δ

Ph_3P

PBA

717

718

Ph Ph Me Ph

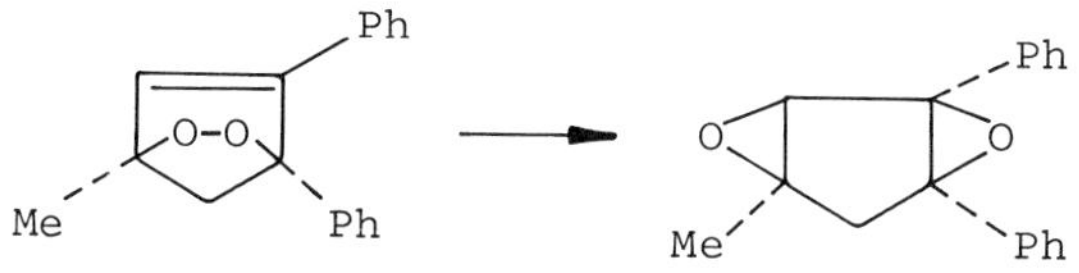

719

Ph Ph Me Ph

720

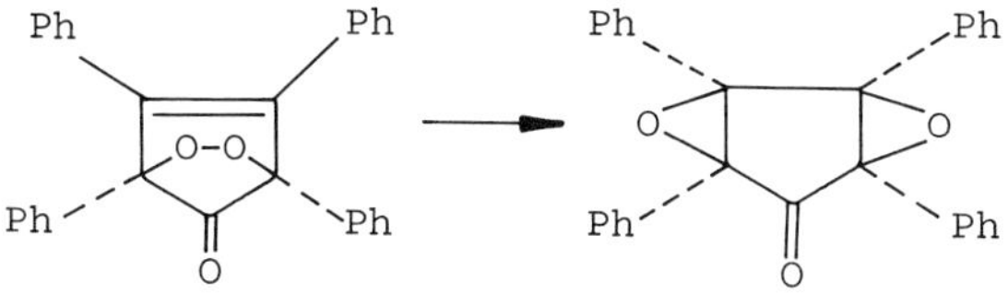

Ph Ph Ph Ph O

721

Ph Ph Ph Ph O

722

Δ CHO CHO

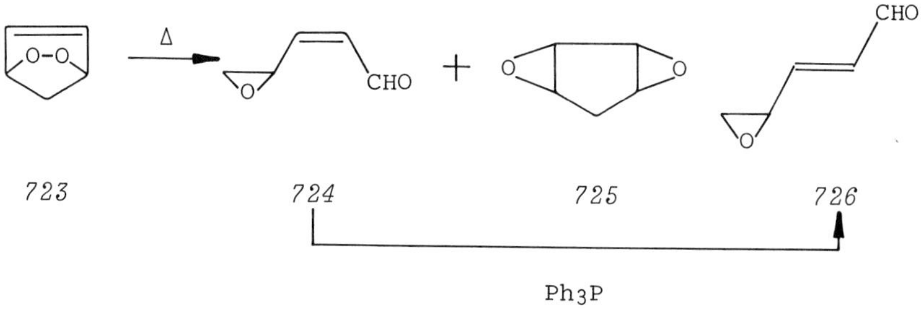

723

724

725

726

Ph_3P

III. CONCLUDING REMARKS

It is hoped that this chapter adequately stresses the vast amount of work that has been dedicated and is still being dedicated to the stereospecific synthesis of epoxides, and the many pathways that are now available for achieving this goal. The steric course of some of the reactions involved in these syntheses may sometimes still be quite unpredictable because of inadequate knowledge of mechanism, transition-state geometry, solvent effects, etc., but an accurate choice of type of reaction and conditions of operation will very often offer a safe way to obtain with high stereospecificity a particular diastereoisomer of a given epoxide, a useful intermediate for further chemical architectural work.

Whenever possible, an attempt has been made in this review to offer some explanation for the observed steric course of the reaction under discussion. It must be pointed out that many such proposals are purely speculative or rest on rather weak experimental basis since, in contrast with the large choice of data that are available on preparations and product analysis, only little quantitative mechanistic work has been carried out in epoxide syntheses. Nevertheless, as has been experienced in many of the developments of modern organic synthesis, intuitive mechanistic assumptions, no matter how rough and naive, can turn out to be quite useful as working hypotheses in planning synthetic sequences, and several of the reactions that have been discussed here exemplify this point.

REFERENCES

1. A. Wurtz, *Compt. Rend.*, *48*, 101 (1859).
2. D. Swern, *Org. Reactions*, *7*, 378 (1953).
3. R. E. Parker and N. S. Isaacs, *Chem. Rev.*, *59*, 737 (1959).
4. A. Rosowsky, *The Chemistry of Heterocyclic Compounds*, A. Weissberger, Ed., Vol. 19, Wiley-Interscience, New York, 1964, p. 1.
5. M. S. Malinovskii, *Epoxides and their Derivatives*, Israel Program for Scientific Translations, Jerusalem, 1965.
6. J. G. Buchanan and H. Z. Sable, *Stereoselective Epoxide Cleavages*, in *Selective Organic Transformations*, B. S. Thyagarajan, Ed., Vol. 2, Wiley-Interscience, New York, 1972.
7. R. C. Fahey, *Topics in Stereochemistry*, E. L. Eliel and N. L. Allinger, Eds., Vol. 3, Wiley-Interscience, New York, 1968, p. 294.
8. D. N. Kirk and M. P. Hartshorn, *Steroid Reaction Mechanisms*,

Elsevier, Amsterdam, 1968, p. 71.
9. D. Swern, *J. Amer. Chem. Soc., 69,* 1692 (1947).
10. Y. Ogata and I. Tabushi, *J. Amer. Chem. Soc., 83,* 3440, 3444 (1961).
11. B. M. Lynch and K. H. Pausacker, *J. Chem. Soc., 1955,* 1525.
12. L. P. Witnauer and D. Swern, *J. Amer. Chem. Soc., 72,* 3364 (1950).
13. D. R. Campbell, J. O. Edwards, J. Maclachlan, and K. Polgar, *J. Amer. Chem. Soc., 80,* 5308 (1958).
14. G. Berti, F. Bottari, B. Macchia, and F. Macchia, *Tetrahedron, 21,* 3277 (1965).
15. P. L. Barili, G. Bellucci, B. Macchia, F. Macchia, and G. Parmigiani, *Gazz. Chim. Ital., 101,* 300 (1971).
16. P. D. Bartlett, *Record Chem. Progr., 11,* 47 (1950).
17. P. L. Barili, G. Bellucci, G. Berti, F. Marioni, A. Marsili, and I. Morelli, *Chem. Commun., 1970,* 1437, and other unpublished work from this laboratory.
18. R. D. Bach and H. F. Henneike, *J. Amer. Chem. Soc., 92,* 5589 (1970).
19. W. H. T. Davison, *J. Chem. Soc., 1951,* 2456.
20. D. Swern, L. P. Witnauer, C. R. Eddy, and W. E. Parker, *J. Amer. Chem. Soc., 77,* 5537 (1955).
21. P. Renolen and J. Ugelstad, *J. Chim. Phys., 57,* 634 (1960).
22. R. Kavčič and B. Plesničar, *J. Org. Chem., 35,* 2033 (1970).
23. N. N. Schwartz and J. H. Blumbergs, *J. Org. Chem., 29,* 1976 (1964).
24. G. Berti and F. Bottari, *Gazz. Chim. Ital., 89,* 2380 (1959).
25. G. Berti and F. Bottari, *J. Org. Chem., 25,* 1286 (1960).
26. V. N. Sapunov and N. N. Lebedev, *Zh. Organ. Khim., 2,* 225 (1966).
27. H. Kwart and D. M. Hoffman, *J. Org. Chem., 31,* 419 (1966).
28. K. D. Bingham, G. D. Meakins, and G. H. Whitham, *Chem. Commun., 1966,* 445.
29. H. Kwart, P. S. Starcher, and S. W. Tinsley, *Chem. Commun., 1967,* 335.
30. A. Ažman, B. Borštnik, and B. Plesničar, *J. Org. Chem., 34,* 971 (1969).
31. M. L. Sassiver and J. English, *J. Amer. Chem. Soc., 82,* 4891 (1960).
32. G. Ponsinet, G. Ourisson, and G. Charles, *Bull. Soc. Chim. France, 1967,* 4453.
33. T. Sasaki, S. Eguchi, and T. Ishii, *J. Org. Chem., 35,* 249 (1970).
34. T. Sasaki, S. Eguchi, and M. Ohno, *J. Org. Chem., 33,* 676 (1968).
35. B. Rickborn and S.-Y. Lwo, *J. Org. Chem., 30,* 2212 (1965).
36. H. B. Henbest, *Proc. Chem. Soc., 1963,* 159.

37. D. Y. Curtin, *Record Chem. Progr.*, *15*, 111 (1954).
38. J. C. Richer and C. Freppel, *Can. J. Chem.*, *46*, 3709 (1968).
39. J. C. Leffingwell and E. E. Royals, *Tetrahedron Lett.*, *1965*, 3829.
40. R. M. Bowman, A. Chambers, and W. R. Jackson, *J. Chem. Soc.*, *C*, *1966*, 612.
41. W. F. Newhall, *J. Org. Chem.*, *29*, 185 (1964).
42. E. E. Royals and J. C. Leffingwell, *J. Org. Chem.*, *31*, 1937 (1966).
43. R. Wylde and J. M. Teulon, *Bull. Soc. Chim. France*, *1970*, 758.
44. N. A. LeBel and G. G. Ecke, *J. Org. Chem.*, *30*, 4316 (1965).
45. K. Marks and H. Kuczynski, *Roczniki Chem.*, *39*, 1259 (1965).
46. P. R. Jefferies and B. Milligan, *J. Chem. Soc.*, *1956*, 2363.
47. P. M. McCurry, *Tetrahedron Lett.*, *1971*, 1841.
48. H. B. Henbest and J. J. McCullough, *Proc. Chem. Soc.*, *1962*, 74.
49. C. Altona and M. Sundaralingam, *Tetrahedron*, *26*, 925 (1970).
50. F. Sweet and R. K. Brown, *Can. J. Chem.*, *46*, 707 (1968).
51. R. C. Ewins, H. B. Henbest, and M. A. McKervey, *Chem. Commun.*, *1967*, 1085.
52. A. Casadevall, E. Casadevall, and M. Lasperas, *Bull. Soc. Chim. France*, *1968*, 4506.
53. J. C. Jallageas, A. Casadevall, and E. Casadevall, *Bull. Soc. Chim. France*, *1969*, 4047.
54. A. Casadevall, E. Casadevall, and H. Mion, *Bull. Soc. Chim. France*, *1968*, 4498.
55. R. Bucourt and D. Hainaut, *Bull Soc. Chim. France*, *1965*, 1366.
56. M. Maynadier, A. Casadevall, and E. Casadevall, *Compt. Rend.*, *269*, 506 (1969).
57. Y. Kitahara, T. Kato, and S. Kanno, *J. Chem. Soc.*, *C*, *1968*, 2397.
58. G. Berti, O. Livi, and D. Segnini, *Tetrahedron Lett.*, *1970*, 1401.
59. J. E. McMurry, *Tetrahedron Lett.*, *1970*, 3731.
60. H. Inouye, T. Yoshida, S. Tobita, and M. Okigawa, *Tetrahedron*, *26*, 3905 (1970).
61. C. C. J. Culvenor, G. M. O'Donovan, and L. W. Smith, *Aust. J. Chem.*, *20*, 757 (1967).
62. J. C. Jallageas, *Compt. Rend.*, *270*, 1253 (1970).
63. H. C. Brown, W. J. Hammar, J. H. Kawakami, I. Rothberg, and D. L. V. Jagt, *J. Amer. Chem. Soc.*, *89*, 6381 (1967).
64. H. Christol and Y. Piétrasanta, *Tetrahedron Lett.*, *1966*, 6471, 6477.
65. G. Baddeley and E. K. Baylis, *J. Chem. Soc.*, *1965*, 4933.
66. G. Mehta, G. L. Chetty, U. R. Nayak, and S. Dev, *Tetrahedron*,

24, 3775 (1968).
67. H. Hikino, T. Kohama, and T. Takemoto, *Tetrahedron*, *25*, 1037 (1969).
68. B. A. Nagasampagi, L. Yankov, and S. Dev, *Tetrahedron Lett.*, *1968*, 1913.
69. C. Ehret and G. Ourisson, *Tetrahedron*, *25*, 1785 (1969).
70. B. A. Arbuzov and A. R. Vil'chinskaya, *Izv. Akad. Nauk SSSR, Ser. Khim.*, *1967*, 954.
71. P. J. Kropp, D. C. Heckert, and T. J. Flautt, *Tetrahedron*, *24*, 1385 (1968).
72. S. P. Acharya, *Tetrahedron Lett.*, *1966*, 4117.
73. K. Gollnick, S. Schröter, G. Ohloff, G. Schade, and G. O. Schenck, *Ann.*, *687*, 14 (1965).
74. K. Gollnick and G. Schade, *Ann.*, *721*, 133 (1969).
75. P. J. Kropp, *J. Amer. Chem. Soc.*, *88*, 4926 (1966).
76. A. Tahara, O. Hoshino, and T. Ohsawa, *Chem. Pharm. Bull. (Tokyo)*, *17*, 54 (1969).
77. R. C. Cambie and W. A. Denny, *Aust. J. Chem.*, *22*, 1699 (1969).
78. Ref. 8, p. 70.
79. D. Swern, *Chem. Rev.*, *45*, 1 (1949).
80. C. Djerassi, *Steroid Reactions*, Holden-Day, San Francisco, 1963, p. 597.
81. K. D. Bingham, T. M. Blaiklock, R. C. B. Coleman, and G. D. Meakins, *J. Chem. Soc.*, *C*, *1970*, 2330.
82. R. Villotti, C. Djerassi, and H. J. Ringold, *J. Amer. Chem. Soc.*, *81*, 4566 (1959).
83. J. W. Blunt, M. P. Hartshorn, and D. N. Kirk, *Tetrahedron*, *22*, 1421 (1966).
84. A. Bowers, L. Cuéllar Ibáñez, and H. J. Ringold, *Tetrahedron*, *7*, 138 (1959).
85. S. Bernstein and R. Littell, *J. Org. Chem.*, *26*, 3610 (1961).
86. J. H. Fried, G. E. Arth, and L. H. Sarett, *J. Amer. Chem. Soc.*, *81*, 1235 (1959).
87. A. Bowers, E. Denot, M. B. Sanchez, and H. J. Ringold, *Tetrahedron*, *7*, 153 (1959).
88. A. Bowers and H. J. Ringold, *J. Amer. Chem. Soc.*, *80*, 3091 (1958).
89. J. Bascoul and A. Crastes de Paulet, *Bull. Soc. Chim. France*, *1966*, 945.
90. C. Djerassi and J. Fishman, *J. Amer. Chem. Soc.*, *77*, 4291 (1955).
91. H. B. Kagan and J. Jacques, *Bull. Soc. Chim. France*, *1960*, 871.
92. P. N. Rao and L. R. Axelrod, *J. Chem. Soc.*, *1961*, 4769.
93. T. Nambara and J. Fishman, *J. Org. Chem.*, *27*, 2131 (1962).
94. F. Sondheimer, S. Bernstein, and R. Mechoulam, *J. Amer.*

Chem. Soc., 82, 3209 (1960).
95. H. Hasegawa, Y. Sato, and K. Tsuda, *Chem. Pharm. Bull. (Tokyo), 9,* 409 (1961).
96. H. Ishii, *Chem. Pharm. Bull. (Tokyo), 10,* 354 (1962).
97. T. Kubota and K. Takeda, *Tetrahedron, 10,* 1 (1960).
98. E. L. McGinnis, G. D. Meakins, J. E. Price, and M. C. Styles, *J. Chem. Soc., 1965,* 4379.
99. I. Malunowicz, *Bull. Acad. Polon. Sci. Ser. Sci. Chim., 10,* 311 (1962).
100. A. Labache-Combier, J. Levisalles, J. P. Pete, and H. Rudler, *Bull. Soc. Chim. France, 1963,* 1689.
101. M. Mousseron-Canet and F. Crouzet, *Bull. Soc. Chim. France, 1968,* 3023.
102. A. Fürst and Pl. A. Plattner, *Helv. Chim. Acta, 32,* 275 (1949).
103. P. Francois and J. Levisalles, *Bull. Soc. Chim. France, 1968,* 318.
104. G. P. Cotterrell, T. G. Halsall, and M. J. Wriglesworth, *J. Chem. Soc., C, 1970,* 1503.
105. J. Bascoul, C. Reliand, A. Guinot, and A. Crastes de Paulet, *Bull. Soc. Chim. France, 1968,* 4074.
106. R. G. Carlson and N. S. Behn, *J. Org. Chem., 32,* 1363 (1967).
107. A. V. Karmenitzky and A. A. Akhrem, *Tetrahedron, 18,* 705 (1962).
108. W. G. Dauben, G. J. Fonken, and D. S. Noyce, *J. Amer. Chem. Soc., 78,* 2579 (1956).
109. J. C. Richer, *J. Org. Chem., 30,* 324 (1965).
110. J. A. Marshall and R. D. Carroll, *J. Org. Chem., 30,* 2748 (1965).
111. M. Cherest, H. Felkin, and N. Prudent, *Tetrahedron Lett., 1968,* 2199, 2205.
112. E. L. Eliel and Y. Senda, *Tetrahedron, 26,* 2411 (1970).
113. P. Geneste, G. Lamaty, and J. P. Roque, *Tetrahedron Lett., 1970,* 5007, 5011, 5015.
114. J. J. Uebel, *Tetrahedron Lett., 1967,* 4751.
115. R. G. Carlson and N. S. Behn, *Chem. Commun., 1968,* 339.
116. E. Demole and H. Wuest, *Helv. Chim. Acta, 50,* 1314 (1967).
117. G. Büchi, M. S. Wittenau, and D. M. White, *J. Amer. Chem. Soc., 81,* 1968 (1959).
118. B. N. Blackett, J. M. Coxon, M. P. Hartshorn, B. L. J. Jackson, and C. N. Muir, *Tetrahedron, 25,* 1479 (1969).
119. J. D. Ballantine and P. J. Sykes, *J. Chem. Soc., C, 1970,* 731.
120. J. Klein, E. Dunkelblum, E. L. Eliel, and Y. Senda, *Tetrahedron Lett., 1968,* 6127.
121. J. M. Coxon, M. P. Hartshorn, and D. N. Kirk, *Tetrahedron,*

23, 3511 (1967).
122. D. D. Keane, W. I. O'Sullivan, E. M. Philbin, R. M. Simons, and P. C. Teague, *Tetrahedron*, *26*, 2533 (1970).
123. K. Watanabe, C. N. Pillai, and H. Pines, *J. Amer. Chem. Soc.*, *84*, 3934 (1962).
124. H. Kwart and T. Takeshita, *J. Org. Chem.*, *28*, 670 (1963).
125. H. C. Brown, J. H. Kawakami, and S. Ikegami, *J. Amer. Chem. Soc.*, *92*, 6914 (1970).
126. P. v. R. Schleyer, *J. Amer. Chem. Soc.*, *89*, 699 (1967).
127. P. G. Gassman and J. H. Dygos, *Tetrahedron Lett. 1970*, 4749.
128. Yu. K. Yur'ev and N. S. Zefirov, *Zh. Obshch. Khim.*, *31*, 840 (1961).
129. S. P. Acharya and H. C. Brown, *J. Org. Chem.*, *35*, 196 (1970).
130. Z. Chabudzinski and Z. Rykowski, *Bull. Acad. Polon. Sci., Ser. Sci. Chim.*, *15*, 95 (1967).
131. M. Barthélémy and Y. Bessière-Chrêtien, *Tetrahedron Lett.*, *1970*, 4265.
132. W. A. M. Davies, J. B. Jones, and A. R. Pinder, *J. Chem. Soc.*, *1960*, 3504.
133. A. Gagneux and C. A. Grob, *Helv. Chim. Acta*, *42*, 2006 (1959).
134. D. J. Goldsmith, *J. Amer. Chem. Soc.*, *84*, 3913 (1962).
135. W. Dittmann and F. Stürzenhofecker, *Ann.*, *688*, 57 (1965).
136. T. W. Craig, G. R. Harvey, and G. A. Berchtold, *J. Org. Chem.*, *32*, 3743 (1967).
137. R. W. Gleason and J. T. Snow, *J. Org. Chem.*, *34*, 1963 (1969).
138. A. C. Cope, B. S. Fisher, W. Funke, J. M. McIntosh, and M. A. McKervey, *J. Org. Chem.*, *34*, 2231 (1969).
139. R. Pauncz and D. Ginsburg, *Tetrahedron*, *9*, 40 (1960); P. J. Hendra and D. B. Powell, *Spectrochim. Acta*, *17*, 913 (1961).
140. M. Barrelle, A. Feugier, and M. Apparu, *Compt. Rend.*, *271*, 519 (1970).
141. N. Heap, G. E. Green, and G. H. Whitham, *J. Chem. Soc., C*, *1969*, 160.
142. J. K. Crandall and D. R. Paulson, *J. Org. Chem.*, *33*, 3291 (1968).
143. C. W. Shoppee, J. C. Coll, and R. E. Lack, *J. Chem. Soc., C*, *1968*, 1581.
144. H. B. Henbest and R. A. L. Wilson, *J. Chem. Soc.*, *1957*, 1958.
145. R. Albrecht and Ch. Tamm, *Helv. Chim. Acta*, *40*, 2216 (1957).
146. P. Chamberlain, M. L. Roberts, and G. H. Whitham, *J. Chem.*

Soc., B, 1970, 1374.
147. M. Narayanaswamy, V. M. Sathe, and A. S. Rao, *Chem. Ind. (London), 1969,* 921.
148. H. Hikino, N. Suzuki, and T. Takemoto, *Chem. Pharm. Bull. (Tokyo), 15,* 1395 (1967).
149. H. Z. Sable, K. A. Powell, H. Katchian, C. B. Niewoehner, and S. B. Kadlec, *Tetrahedron, 26,* 1509 (1970).
150. E. Glotter, S. Greenfield, and D. Lavie, *Tetrahedron Lett., 1967,* 5261.
151. M. Lj. Milhailović, J. Foršek, Lj. Lorenc, Z. Maksimović, H. Fuhrer, and J. Kalvoda, *Helv. Chim. Acta, 52,* 459 (1969).
152. R. Steyn and H. Z. Sable, *Tetrahedron, 25,* 3579 (1969).
153. H. Z. Sable, T. Anderson, B. Tolbert, and Th. Posternak, *Helv. Chim. Acta, 46,* 1157 (1963).
154. E. J. Langstaff, R. Y. Moir, R. A. B. Bannard, and A. A. Casselman, *Can. J. Chem., 46,* 3649 (1968).
155. Z. Rykowski and Z. Chabudzinski, *Roczniki Chem., 43,* 1427 (1969).
156. J. M. Coxon, E. Dansted, M. P. Hartshorn, and K. E. Richards, *Tetrahedron, 24,* 1193 (1968).
157. J. Katsuhara, *Bull. Chem. Soc. Japan, 42,* 2593 (1969).
158. A. C. Cope, J. K. Heeren, and V. Seeman, *J. Org. Chem., 28,* 516 (1963).
159. A. C. Cope, A. H. Keough, P. E. Peterson, H. E. Simmons, and G. W. Wood, *J. Amer. Chem. Soc., 79,* 3900 (1957).
160. J. L. Pierre, P. Chautemps, and P. Arnaud, *Bull. Soc. Chim. France, 1969,* 1317.
161. C. J. Cheer and C. R. Johnson, *J. Amer. Chem. Soc., 90,* 178 (1968).
162. H. E. Simmons and R. D. Smith, *J. Amer. Chem. Soc., 80,* 5323 (1958).
163. S. Winstein and J. Sonnenberg, *J. Amer. Chem. Soc., 83,* 3235 (1961).
164. J. H. H. Chan and B. Rickborn, *J. Amer. Chem. Soc., 90,* 6406 (1968).
165. C. D. Poulter, E. C. Friedrich, and S. Winstein, *J. Amer. Chem. Soc., 91,* 6892 (1969).
166. B. Rickborn and J. H. H. Chan, *J. Org. Chem., 32,* 3576 (1967).
167. S. Greenfield, E. Glotter, D. Lavie, and Y. Kashman, *J. Chem. Soc., C, 1967,* 1460.
168. T. H. Campion and G. A. Morrison, *Tetrahedron Lett., 1968,* 1.
169. J. Joska, J. Fajkoś, and F. Śorm, *Coll. Czech. Chem. Comm., 31,* 298 (1966).
170. W. G. Dauben, G. A. Boswell, W. Templeton, J. W. McFarland, and G. H. Berezin, *J. Amer. Chem. Soc., 85,* 1672 (1963).

171. H. Wehrli, C. Lehmann, P. Keller, J. J. Bonet, K. Schaffner, and O. Jeger, *Helv. Chim. Acta, 49,* 2218 (1966).
172. H. B. Henbest and B. Nicholls, *J. Chem. Soc., 1957,* 4608.
173. R. Zurflüh, E. N. Wall, J. B. Siddall, and J. A. Edwards, *J. Amer. Chem. Soc., 90,* 6224 (1968).
174. P. Garside, T. G. Halsall, and G. M. Hornby, *J. Chem. Soc., C, 1969,* 716.
175. W. Herz, R. C. Ligon, H. Kanno, W. H. Schuller, and R. V. Lawrence, *J. Org. Chem., 35,* 3338 (1970).
176. H. Christol, D. Duval, and G. Solladie, *Bull. Soc. Chim. France, 1968,* 689.
177. M. Mousseron, M. Mousseron-Canet, and G. Philippe, *Compt. Rend., 258,* 3705 (1964).
178. A. G. Armour, G. Büchi, A. Eschenmoser, and A. Storni, *Helv. Chim. Acta, 42,* 2233 (1959).
179. A. C. Darby, H. B. Henbest, and I. McClenaghan, *Chem. Ind. (London), 1962,* 462.
180. B. Tolbert, R. Steyn, J. A. Franks, and H. Z. Sable, *Carbohydr. Res., 5,* 62 (1967).
181. M. Nakajima and N. Kurihara, *Chem. Ber., 94,* 515 (1961).
182. A. Hasegawa and H. Z. Sable, *J. Org. Chem., 31,* 4149 (1966).
183. G. Wolczunowicz, F. G. Cocu, and Th. Posternak, *Helv. Chim. Acta, 53,* 2275 (1970).
184. M. Mousseron-Canet and J. C. Guilleux, *Bull. Soc. Chim. France, 1966,* 3853.
185. J. M. Coxon, M. P. Hartshorn, and C. N. Muir, *Tetrahedron, 25,* 3925 (1969).
186. G. M. L. Cragg and G. D. Meakins, *J. Chem. Soc., 1965,* 2054.
187. K. D. Bingham, G. D. Meakins, and J. Wicha, *J. Chem. Soc., C, 1969,* 510.
188. M. Mousseron-Canet, B. Labeeuw, and J. C. Lanet, *Bull. Soc. Chim. France, 1968,* 2125.
189. J. C. Lanet and M. Mousseron-Canet, *Bull. Soc. Chim. France, 1969,* 1751.
190. H. H. Appel, C. J. W. Brooks, and K. H. Overton, *J. Chem. Soc., 1959,* 3322.
191. H. B. Henbest and B. Nicholls, *J. Chem. Soc., 1959,* 221.
192. M. Mousseron-Canet and B. Labeeuw, *Bull. Soc. Chim. France, 1968,* 4165, 4171.
193. M. Mousseron-Canet and J. C. Guilleux, *Bull. Soc. Chim. France, 1966,* 3858.
194. J. Perez Ruelas, J. Iriarte, F. Kincl, and C. Djerassi, *J. Org. Chem., 23,* 1744 (1958).
195. A. D. Cross, E. Denot, R. Acevedo, R. Urquiza, and A. Bowers, *J. Org. Chem., 29,* 2195 (1964).

196. S. G. Levine, N. H. Eudy, and C. F. Leffler, *J. Org. Chem.*, *31*, 3995 (1966).
197. L. Nédélec, *Bull. Soc. Chim. France*, *1970*, 2548.
198. J. C. Guilleux and M. Mousseron-Canet, *Bull. Soc. Chim. France*, *1967*, 24.
199. C. H. Heathcock and Y. Amano, *Tetrahedron*, *24*, 4917 (1968).
200. J. A. Marshall and A. R. Hochstetler, *J. Org. Chem.*, *31*, 1020 (1966).
201. R. M. Bowman, A. Chambers, and W. R. Jackson, *J. Chem. Soc.*, *C*, *1966*, 1296.
202. K. Piatkowski and H. Kuczynski, *Roczniki Chem.*, *35*, 239, 1579 (1961).
203. J. Wolinsky, R. O. Hutchins, and J. H. Thorstenson, *Tetrahedron*, *27*, 753 (1971).
204. G. V. Pigulevskii and G. V. Markina, *Dokl. Akad. Nauk SSSR*, *63*, 677 (1948).
205. L. Goodman, S. Winstein, and R. Boschan, *J. Amer. Chem. Soc.*, *80*, 4312 (1958).
206. A. Hasegawa and H. Z. Sable, *J. Org. Chem.*, *31*, 4154 (1966).
207. G. Lukacs and D. K. Fukushima, *J. Org. Chem.*, *34*, 2707 (1969).
208. K. Ponsold and W. Preibisch, *J. Prakt. Chem.*, *25*, 26 (1964).
209. G. Drefahl and K. Ponsold, *Chem. Ber.*, *91*, 271 (1958).
210. D. K. Fukushima, M. Smulowitz, J. S. Liang, and G. Lukacs, *J. Org. Chem.*, *34*, 2702 (1969).
211. M. Lasperas, A. Casadevall, and E. Casadevall, *Bull. Soc. Chim. France*, *1970*, 2580.
212. N. S. Crossley, A. C. Darby, H. B. Henbest, J. J. McCullough, B. Nicholls, and M. F. Stewart, *Tetrahedron Lett.*, *1961*, 398.
213. G. Bellucci, F. Marioni, and A. Marsili, *Tetrahedron*, *28*, 3393 (1972).
214. D. B. Roll and A. C. Huitric, *J. Pharm. Sci.*, *55*, 942 (1966).
215. M. Nakajima, A. Hasegawa, and N. Kurihara, *Chem. Ber.*, *95*, 2708 (1962).
216. J. Meinwald and L. Hendry, *Tetrahedron Lett.*, *1969*, 1657.
217. E. J. Corey and R. Noyori, *Tetrahedron Lett.*, *1970*, 311.
218. A. P. Gray, D. E. Heitmeier, and H. Kraus, *J. Amer. Chem. Soc.*, *84*, 89 (1962).
219. J. A. Berson and S. Suzuki, *J. Amer. Chem. Soc.*, *80*, 4341 (1958).
220. A. P. Gray and D. E. Heitmeier, *J. Org. Chem.*, *30*, 1226 (1965).
221. G. I. Poos and J. D. Rosenau, *J. Org. Chem.*, *28*, 665 (1963).

222. R. B. Woodward, F. E. Bader, H. Bickel, A. J. Frey, and R. W. Kierstead, *Tetrahedron 2,* 1 (1958).
223. G. I. Fray, R. J. Hilton, and J. M. Teire, *J. Chem. Soc., C, 1966,* 592.
224. N. Langlois and B. Gastambide, *Bull. Soc. Chim. France, 1965,* 2966.
225. G. Eggart, P. Keller, C. Lehmann, and H. Wehrli, *Helv. Chim. Acta, 51,* 940 (1968).
226. V. F. Kucherov, G. M. Segal, and I. N. Nazarov, *Izv. Akad. Nauk SSSR, Otd. Khim. Nauk, 1959,* 673, 682, 1253.
227. V. F. Kucherov, G. M. Segal, and I. N. Nazarov, *Zh. Obshch. Khim., 29,* 804 (1959).
228. D. R. Boyd and M. A. McKervey, *Quart. Rev. (London), 22,* 111 (1968).
229. R. M. Bowman and M. F. Grundon, *J. Chem. Soc., C, 1967,* 2368.
230. F. Montanari, I. Moretti, and G. Torre, *Chem. Commun., 1969,* 135.
231. D. R. Boyd, D. M. Jerina, and J. W. Daly, *J. Org. Chem., 35,* 3170 (1970).
232. G. B. Payne, P. H. Deming, and P. H. Williams, *J. Org. Chem., 26,* 659 (1961).
233. G. B. Payne, *Tetrahedron, 18,* 763 (1962).
234. Y. Ogata and Y. Sawaki, *Tetrahedron, 20,* 2065 (1964).
235. G. Farges and A. Kergomard, *Bull. Soc. Chim. France, 1969,* 4476.
236. T. Mori, K. H. Yang, K. Kimoto, and H. Nozaki, *Tetrahedron Lett., 1970,* 2419.
237. R. J. Ferrier and N. Prasad, *J. Chem. Soc., C, 1969,* 575.
238. E. Weitz and A. Scheffer, *Chem. Ber., 54,* 2327 (1921).
239. Ref. 4, p. 57.
240. Ref. 80, p. 595.
241. R. D. Temple, *J. Org. Chem., 35,* 1275 (1970).
242. C. A. Bunton and G. J. Minkoff, *J. Chem. Soc., 1949,* 665.
243. H. O. House and R. S. Ro, *J. Amer. Chem. Soc., 80,* 2428 (1958).
244. N. H. Cromwell and R. A. Setterquist, *J. Amer. Chem. Soc., 76,* 5752 (1954).
245. H. B. Henbest and W. R. Jackson, *J. Chem. Soc., C, 1967,* 2459.
246. H. E. Zimmerman, L. Singer, and B. S. Thyagarajan, *J. Amer. Chem. Soc., 81,* 108 (1959).
247. H. H. Wasserman and N. E. Aubrey, *J. Amer. Chem. Soc., 77,* 590 (1955).
248. A. Padwa and N. C. Das, *J. Org. Chem., 34,* 816 (1969).
249. H. O. House and D. J. Reif, *J. Amer. Chem. Soc., 79,* 6491 (1957).

250. G. Ohloff and G. Uhde, *Helv. Chim. Acta, 53,* 531 (1970).
251. G. B. Payne, *J. Org. Chem., 24,* 2048 (1959).
252. J. V. Murray and J. B. Cloke, *J. Amer. Chem. Soc., 56,* 2749 (1934).
253. G. B. Payne and P. H. Williams, *J. Org. Chem., 26,* 651 (1961); G. B. Payne, *ibid.,* 663.
254. A. Robert and A. Foucaud, *Bull. Soc. Chim. France, 1969,* 4528.
255. M. Igarashi, M. Akano, A. Fujimoto, and H. Midorikawa, *Bull. Chem. Soc. Japan, 43,* 2138 (1970).
256. H. Newman and R. B. Angier, *Tetrahedron, 26,* 825 (1970).
257. B. Zwanenburg and J. ter Wiel, *Tetrahedron Lett., 1970,* 935.
258. M. Igarashi and M. Midorikawa, *Bull. Chem. Soc. Japan, 40,* 2624 (1967).
259. E. Klein and G. Ohloff, *Tetrahedron, 19,* 1091 (1963).
260. J. Katsuhara, H. Yamasaki, and N. Yamamoto, *Bull. Chem. Soc. Japan, 43,* 1584 (1970).
261. J. Katsuhara, *Bull. Chem. Soc. Japan, 42,* 2391 (1969).
262. W. R. Jackson and A. Zurqiyah, *J. Chem. Soc., B, 1966,* 49.
263. R. T. Gray and H. E. Smith, *Tetrahedron, 23,* 4229 (1967).
264. H. Newman, *J. Org. Chem., 35,* 3990 (1970).
265. J. Katsuhara, *Bull. Chem. Soc. Japan, 41,* 2700 (1968).
266. B. A. Brady, M. M. Healey, J. A. Kennedy, W. I. O'Sullivan, and E. M. Philbin, *Chem. Commun., 1970,* 1434.
267. C. W. Shoppee, S. K. Roy, and B. S. Goodrich, *J. Chem. Soc., 1961,* 1583.
268. F. Khuong-Huu, D. Herlem, and M. Beneche, *Bull. Soc. Chim. France, 1970,* 2702.
269. B. Löken, S. Kaufmann, G. Rosenkranz, and E. Sondheimer, *J. Amer. Chem. Soc., 78,* 1738 (1956).
270. J. R. Bull, *Tetrahedron Lett., 1968,* 5959.
271. D. J. Collins, *J. Chem. Soc., 1959,* 3919.
272. M. Tomoeda, T. Furuta, and T. Koga, *Chem. Pharm. Bull. (Tokyo), 13,* 1078 (1965).
273. M. E. Kuehne and J. A. Nelson, *J. Org. Chem., 35,* 161 (1970).
274. Ref. 8, p. 201.
275. H. Wehrli, C. Lehmann, T. Iizuka, K. Schaffner, and O. Jeger, *Helv. Chim. Acta, 50,* 2403 (1967).
276. H. B. Henbest, W. R. Jackson, and I. Malunowicz, *J. Chem. Soc., C, 1967,* 2469.
277. J. A. Saboz, T. Iizuka, H. Wehrli, K. Schaffner, and O. Jeger, *Helv. Chim. Acta, 51,* 1362 (1968).
278. G. Rio, B. Muller, and F. Laréze, *Compt. Rend., 268,* 1157 (1969).
279. G. B. Payne and P. H. Williams, *J. Org. Chem., 24,* 54

(1959).
280. G. G. Allan and A. N. Neogi, *J. Catal.*, *16*, 197 (1970).
281. M. Igarashi and H. Midorikawa, *J. Org. Chem.*, *32*, 3399 (1967).
282. B. G. Christensen, W. J. Leanza, T. R. Beattie, A. A. Patchett, B. H. Arison, R. E. Ormond, F. A. Kuehl, G. Albers-Schonberg, and O. Jardetzky, *Science*, *166*, 123 (1969).
283. E. J. Glamkowski, G. Gal, R. Purick, A. J. Davidson, and M. Sletzinger, *J. Org. Chem.*, *35*, 3510 (1970).
284. N. C. Yang and R. A. Finnegan, *J. Amer. Chem. Soc.*, *80*, 5845 (1958).
285. H. W. Moore, *J. Org. Chem.*, *32*, 1996 (1967).
286. C. E. Griffin and S. K. Kundu, *J. Org. Chem.*, *34*, 1532 (1969).
287. G. B. Payne, *J. Org. Chem.*, *25*, 275 (1960).
288. K. Maruyama, R. Goto, and S. Kitamura, *Nippon Kagaku Zasshi*, *81*, 1780 (1960); *Chem. Abstr.*, *56*, 2399i (1962).
289. B. Muckenstrum, *Tetrahedron Lett.*, *1968*, 6139.
290. P. S. Bailey, *Chem. Rev.*, *58*, 945 (1958).
291. P. S. Bailey and A. G. Lane, *J. Amer. Chem. Soc.*, *89*, 4473 (1967).
292. R. W. Murray, R. D. Youssefyeh, and P. R. Story, *J. Amer. Chem. Soc.*, *89*, 2429 (1967).
293. W. J. Wechter and G. Slomp, *J. Org. Chem.*, *27*, 2549 (1962).
294. Ref. 4, p. 86.
295. K. B. Wiberg, *Oxidation in Organic Chemistry*, Part A, Academic Press, New York, 1965, p. 125.
296. W. J. Hickinbottom, D. Peters, and D. G. M. Wood, *J. Chem. Soc.*, *1955*, 1360.
297. G. E. M. Moussa and N. F. Eweiss, *J. Appl. Chem.*, *19*, 313 (1969).
298. A. Behr, *Chem. Ber.*, *5*, 277 (1872).
299. J. Roček and J. C. Drozd, *J. Amer. Chem. Soc.*, *92*, 6668 (1970).
300. A. K. Awasthy and J. Roček, *J. Amer. Chem. Soc.*, *91*, 991 (1969).
301. W. A. Mosher, F. W. Steffgen, and P. T. Lansbury, *J. Org. Chem.*, *26*, 670 (1961).
302. J. S. Littler, *Tetrahedron*, *27*, 81 (1971).
303. I. G. Guest, J. G. L. Jones, B. A. Marples, and M. J. Harrington, *J. Chem. Soc.*, *C*, *1969*, 2360.
304. E. Glotter, S. Greenfield, and D. Lavie, *J. Chem. Soc.*, *C*, *1968*, 1646.
305. P. S. Kalsi, K. S. Kumar, and M. S. Wadia, *Chem. Ind. (London)*, *1971*, 31.

306. J. M. Constantin and L. H. Sarett, *J. Amer. Chem. Soc.*, *74*, 3908 (1952).
307. S. Marmor, *J. Org. Chem.*, *28*, 250 (1963).
308. D. H. Rosenblatt and G. H. Broome, *J. Org. Chem.*, *28*, 1290 (1963).
309. A. Robert and A. Foucaud, *Bull. Soc. Chim. France*, *1969*, 2531.
310. A. Nickon and W. L. Mendelson, *J. Amer. Chem. Soc.*, *87*, 3921 (1965).
311. S. Winstein and R. B. Henderson, in *Heterocyclic Compounds*, R. C. Elderfield, Ed., Vol. 1, John Wiley, New York, 1950, p. 8.
312. Ref. 4, p. 94.
313. J. Fried and E. F. Sabo, *J. Amer. Chem. Soc.*, *75*, 2273 (1953).
314. E. E. van Tamelen and K. B. Sharpless, *Tetrahedron Lett.*, *1967*, 2655.
315. J. W. Cornforth and D. T. Green, *J. Chem. Soc.*, *C*, *1970*, 846.
316. W. Fickett, H. K. Garner, and H. J. Lucas, *J. Amer. Chem. Soc.*, *73*, 5063 (1951).
317. C. C. Tung and A. J. Speziale, *J. Org. Chem.*, *28*, 2009 (1963).
318. P. Baret, J.-L. Pierre, and R. Heilmann, *Bull. Soc. Chim. France*, *1967*, 4735.
319. G. Bert, F. Bottari, B. Macchia, and V. Nuti, *Ann. Chim. (Rome)*, *54*, 1253 (1964).
320. G. Berti, F. Bottari, G. Lippi, and B. Macchia, *Tetrahedron*, *24*, 1959 (1968).
321. C. A. Grob and R. A. Wohl, *Helv. Chim. Acta*, *49*, 2175 (1966).
322. G. Berti, F. Bottari, P. L. Ferrarini, and B. Macchia, *J. Org. Chem.*, *30*, 4091 (1965).
323. R. E. Lutz, R. L. Wayland, Jr., and H. G. France, *J. Amer. Chem. Soc.*, *72*, 5511 (1950).
324. J. W. Cornforth, R. H. Cornforth, and K. K. Mathew, *J. Chem. Soc.*, *1959*, 112.
325. J. W. Cornforth, R. H. Cornforth, and K. K. Mathew, *J. Chem. Soc.*, *1959*, 2539.
326. E. P. Kohler and M. Tishler, *J. Amer. Chem. Soc.*, *57*, 217 (1935).
327. M. S. Kharash and O. Reinmuth, *Grignard Reactions of Nonmetallic Substances*, Prentice-Hall, New York, 1954, p. 181.
328. Ref. 4, p. 119, 132.
329. E. E. van Tamelen and T. J. Curphey, *Tetrahedron Lett.*, *1962*, 121.
330. E. E. van Tamelen, *Acc. Chem. Res.*, *1*, 111 (1968).

331. E. L. Eliel and D. W. Delmonte, *J. Org. Chem.*, *21*, 596 (1956).
332. S. Mitsui and S. Imaizumi, *Nippon Kagaku Zasshi*, *86*, 219 (1965); *Chem. Abstr.*, *63*, 4133f (1965).
333. J. C. Pommelet, N. Manisse, and J. Chuche, *Compt. Rend.*, *270*, 1894 (1970).
334. M. L. Lewbart, *J. Org. Chem.*, *33*, 1695 (1968).
335. N. A. Le Bel and R. F. Czaja, *J. Org. Chem.*, *26*, 4768 (1961).
336. R. M. Moriarty and T. Adams, *Tetrahedron Lett.*, *1969*, 3715.
337. J. M. Coxon, M. P. Hartshorn, and A. J. Lewis, *Aust. J. Chem.*, *24*, 1009 (1971).
338. D. C. Kleinfelter and J. H. Long, *Tetrahedron Lett.*, *1969*, 347.
339. C. Anselmi, P. L. Barili, G. Berti, B. Macchia, F. Macchia, and L. Monti, *Tetrahedron Lett.*, *1970*, 1743.
340. W. Cocker and D. H. Grayson, *Tetrahedron Lett.*, *1969*, 4451.
341. R. G. Carlson and R. Ardon, *J. Org. Chem.*, *36*, 216 (1971).
342. M. Kočor, P. Lenkowski, A. Mironowicz, and L. Nowak, *Bull. Acad. Polon. Sci., Ser. Sci, Chim.*, *14*, 79 (1966).
343. T. Nambara, K. Shimada, and S. Goya, *Chem. Pharm. Bull. (Tokyo)*, *18*, 453 (1970).
344. J. W. Blunt, M. P. Hartshorn, and D. N. Kirk, *Tetrahedron*, *21*, 559 (1965).
345. B. O. Lindgren and C. M. Svahn, *Acta Chem. Scand.*, *24*, 2699 (1970).
346. D. G. Hey, G. D. Meakins, and M. W. Pemberton, *J. Chem. Soc., C*, *1966*, 1331.
347. J. Klinot, K. Waisser, L. Streinz, and A. Vystrcil, *Coll. Czech. Chem. Commun.*, *35*, 3610 (1970).
348. H. Kuczynski and M. Walkowicz, *Roczniki Chem.*, *37*, 955 (1963).
349. B. Rickborn and J. Quartucci, *J. Org. Chem.*, *29*, 2476 (1964).
350. H. Kuczynski and Z. Chabudzinski, *Roczniki Chem.*, *34*, 177 (1960).
351. Z. Chabudzinski and H. Kuczynski, *Bull. Acad. Polon. Sci., Ser. Sci. Chim.*, *9*, 519 (1961); *Roczniki Chem.*, *36*, 1173 (1962).
352. G. Berti, B. Macchia, and F. Macchia, *Tetrahedron*, *24*, 1755 (1968).
353. J. C. Leffingwell and R. E. Shackelford, Fr. Pat. 2,002,595; *Chem. Abstr.*, *72*, 90673k (1970).
354. F. H. Newth, *Quart, Rev.*, *13*, 30 (1959).
355. Ref. 4, p. 150.

356. M. Davis and V. Petrow, *J. Chem. Soc.*, *1949*, 2536.
357. A. S. Hallsworth and H. B. Henbest, *J. Chem. Soc.*, *1957*, 4604.
358. A. T. Rowland and H. R. Nace, *J. Amer. Chem. Soc.*, *82*, 2833 (1960).
359. G. Berti, F. Bottari, and A. Marsili, *Tetrahedron*, *25*, 2939 (1969).
360. R. J. W. Cremlyn, D. L. Garmaise, and C. W. Shoppee, *J. Chem. Soc.*, *1953*, 1847.
361. R. Wylde and F. Forissier, *Bull. Soc. Chim. France*, *1969*, 4508.
362. M. Chérest, H. Felkin, J. Sicher, F. Sipos, and M. Tichy, *J. Chem. Soc.*, *1965*, 2513.
363. A. C. Cope and E. R. Trumbull, *Org. Reactions*, *11*, 352, 389 (1960).
364. B. Witkop and C. M. Foltz, *J. Amer. Chem. Soc.*, *79*, 197 (1957).
365. J. Read and I. G. M. Campbell, *J. Chem. Soc.*, *1930*, 2377.
366. M. Svoboda and J. Sicher, *Coll. Czech. Chem. Commun.*, *23*, 1540 (1958).
367. W. H. Puterbaugh and C. R. Hauser, *J. Amer. Chem. Soc.*, *86*, 1394 (1964).
368. P. Moreau, A. Casadevall, and E. Casadevall, *Bull. Soc. Chim. France*, *1969*, 2013.
369. P. Moreau and E. Casadevall, *Compt. Rend.*, *268*, 1909 (1969).
370. G. Berti, B. Macchia, F. Macchia, and L. Monti, *J. Org. Chem.*, *33*, 4045 (1968).
371. Y. Pocker and B. P. Ronald, *J. Org. Chem.*, *35*, 3362 (1970); *J. Amer. Chem. Soc.*, *92*, 3385 (1970).
372. M. Fetizon and P. Foy, *Compt. Rend.*, *263*, 821 (1966); *Coll. Czech. Chem. Commun.*, *35*, 440 (1970).
373. J. M. Coxon, M. P. Hartshorn, and D. N. Kirk, *Tetrahedron*, *21*, 2489 (1965).
374. G. Snatzke, *Ann.*, *686*, 167 (1965).
375. B. Ellis, V. Petrow, and B. Waterhouse, *J. Chem. Soc.*, *1960*, 2596,
376. E. R. Kaplan, K. Naidu, and D. E. A. Rivett, *J. Chem. Soc.*, *C*, *1970*, 1656.
377. H. Wynberg, E. Boelema, J. H. Wieringa, and J. Strating, *Tetrahedron Lett.*, *1970*, 3613.
378. H. B. Henbest and T. I. Wrigley, *J. Chem. Soc.*, *1957*, 4596.
379. Ref. 4, p. 137.
380. J. J. Riehl and L. Thil, *Tetrahedron Lett.*, *1969*, 1913.
381. C. L. Stevens and T. H. Coffield, *J. Amer. Chem. Soc.*, *80*, 1919 (1958).

382. D. Ricard and J. Cantacuzene, *Bull. Soc. Chim. France, 1969*, 628.
383. J. Cantacuzene and R. Jantzen, *Tetrahedron, 26*, 2429 (1970).
384. J. Cantacuzene, M. Atlani, and J. Anibié, *Tetrahedron Lett., 1968*, 2335.
385. J. Cantacuzene and M. Atlani, *Tetrahedron, 26*, 2447 (1970).
386. M. S. Newman and B. J. Magerlein, *Org. Reactions, 5*, 413 (1949).
387. M. Ballester, *Chem. Rev., 55*, 283 (1955).
388. Ref. 4, p. 106.
389. H. E. Zimmerman and L. Ahramjian, *J. Amer. Chem. Soc., 82*, 5459 (1960).
390. V. R. Valente and J. L. Wolfhagen, *J. Org. Chem., 31*, 2509 (1966).
391. C. C. Tung, A. J. Speziale, and H. W. Frazier, *J. Org. Chem, 28*, 1514 (1963).
392. J. Seyden-Penne, M. C. Roux-Schmitt, and A. Roux, *Tetrahedron, 26*, 2649 (1970).
393. J. Seyden-Penne and A. Roux, *Compt. Rend., 267*, 1067 (1968).
394. B. Deschamps and J. Seyden-Penne, *Compt. Rend., 271*, 1097 (1970).
395. J. Seyden-Penne, C. Gibert, and D. Bernard, *Compt. Rend., 263*, 895 (1966).
396. J. D. White, J. B. Bremner, M. J. Dimsdale, and R. L. Garcea, *J. Amer. Chem. Soc., 93*, 281 (1971).
397. F. W. Bachelor and R. K. Bansal, *J. Org. Chem., 34*, 3600 (1969).
398. J. Villieras and J. C. Combret, *Compt. Rend., 272*, 236 (1971).
399. J. Villieras, B. Castro, and N. N. Ferracutti, *Bull. Soc. Chim. France, 1970*, 1450.
400. P. F. Vogt and D. F. Tavares, *Can. J. Chem., 47*, 2875 (1969).
401. F. Bohlmann and G. Haffer, *Chem. Ber., 102*, 4017 (1969).
402. D. F. Tavares, R. E. Estep, and M. Blezard, *Tetrahedron Lett., 1970*, 2373.
403. C. D. Gutsche, *Org. Reactions, 8*, 364 (1954).
404. C. D. Gutsche and D. Redmore, *Carbocyclic Ring Expansion Reactions*, Academic Press, New York, 1968, p. 81.
405. G. W. Cowell and A. Ledwith, *Quart. Rev., 24*, 119 (1970).
406. Ref. 4, p. 158.
407. J. N. Bradley, G. W. Cowell, and A. Ledwith, *J. Chem. Soc., 1964*, 4334.
408. C. D. Gutsche and J. E. Bowers, *J. Org. Chem., 32*, 1203 (1967).

409. R. E. Bowman, A. Campbell, and W. R. N. Williamson, *J. Chem. Soc.*, *1964*, 3846.
410. G. Facchinetti, F. Pietra, and A. Marsili, *Tetrahedron Lett.*, *1971*, 393.
411. D. A. Prins and T. Reichstein, *Helv. Chim. Acta*, *24*, 945 (1941).
412. J. A. Marshall and J. J. Partridge, *J. Org. Chem.*, *33*, 4090 (1968).
413. R. G. Carlson and N. S. Behn, *J. Org. Chem.*, *33*, 2069 (1968).
414. P. A. Hart and R. A. Sandmann, *Tetrahedron Lett.*, *1969*, 305.
415. R. S. Bly, F. B. Culp, and R. K. Bly, *J. Org. Chem.*, *35*, 2235 (1970).
416. A. W. Johnson and R. B. LaCount, *J. Amer. Chem. Soc.*, *83*, 417 (1961).
417. E. J. Corey and M. Chaykovsky, *J. Amer. Chem. Soc.*, *84*, 3782 (1962).
418. V. Franzen and H. E. Driessen, *Tetrahedron Lett.*, *1962*, 661.
419. E. J. Corey and M. Chaykovsky, *J. Amer. Chem. Soc.*, *84*, 867 (1962).
420. E. J. Corey and M. Chaykovsky, *J. Amer. Chem. Soc.*, *87*, 1353 (1965).
421. E. J. Corey and M. Chaykovsky, *Org. Syn.*, *49*, 78 (1969).
422. A. W. Johnson, *Ylid Chemistry*, Academic Press, New York, 1966, p. 328.
423. A. W. Johnson and J. O. Martin, *Chem. Ind. (London)*, *1965*, 1726.
424. C. E. Cook. R. C. Corley, and M. E. Wall, *J. Org. Chem.*, *33*, 2789 (1968).
425. J. B. Bryan Jones and R. Grayshan, *Chem. Commun.*, *1970*, 741.
426. R. S. Bly, C. M. Du Bose, and G. B. Konizer, *J. Org. Chem.*, *33*, 2188 (1968).
427. A. W. Johnson, V. J. Hruby, and J. L. Williams, *J. Amer. Chem. Soc.*, *86*, 918 (1964).
428. E. J. Corey and W. Oppolzer, *J. Amer. Chem. Soc.*, *86*, 1899 (1964).
429. E. J. Corey, M. Jautelat, and W. Oppolzer, *Tetrahedron Lett.*, *1967*, 2325.
430. J. Adams, L. Hoffman, and B. M. Trost, *J. Org. Chem.*, *35*, 1600 (1970).
431. R. W. LaRochelle, B. M. Trost, and L. Krepski, *J. Org. Chem.*, *36*, 1126 (1971).
432. M. J. Hatch, *J. Org. Chem.*, *34*, 2133 (1969).
433. C. R. Johnson, M. Haake, and C. W. Schroeck, *J. Amer. Chem.*

Soc., *92*, 6594 (1970).
434. C. R. Johnson and C. W. Schroeck, *J. Amer. Chem. Soc.*, *90*, 6852 (1968).
435. C. R. Johnson and G. F. Katekar, *J. Amer. Chem. Soc.*, *92*, 5753 (1970).
436. V. Mark, *J. Amer. Chem. Soc.*, *85*, 1884 (1963).
437. F. Ramirez, A. S. Gulati, and C. P. Smith, *J. Org. Chem.*, *33*, 13 (1968).
438. M. S. Newman and S. Blum, *J. Amer. Chem. Soc.*, *86*, 5598 (1964).
439. T. Mukaiyaka, I. Kuwajima, and K. Ono, *Bull. Chem. Soc. Japan*, *38*, 1954 (1965).
440. J. H. Boyer and R. Selvarajan, *J. Org. Chem.*, *35*, 1229 (1970).
441. D. B. Denney and D. H. Jones, *J. Amer. Chem. Soc.*, *91*, 5821 (1969).
442. J. Boche and O. Runquist, *J. Org. Chem.*, *33*, 4285 (1968).
443. G. O. Pierson and O. A. Runquist, *J. Org. Chem.*, *34*, 3654 (1969).
444. G. Rio and M. Charifi, *Compt. Rend.*, *268*, 1960 (1969).
445. H. W. S. Chan, *Chem. Commun.*, *1970*, 1550.
446. K. H. Schulte-Elte, B. Willhalm, and G. Ohloff, *Angew. Chem.*, *81*, 1045 (1969); *Angew. Chem. Intern. Ed. Engl.*, *8*, 985 (1969).

THE ELECTRONIC STRUCTURE AND STEREOCHEMISTRY OF SIMPLE CARBONIUM IONS

VOLKER BUSS,* PAUL VON R. SCHLEYER, AND LELAND C. ALLEN

Department of Chemistry, Princeton University
Princeton, New Jersey

*Present address: Max-Planck Institut für biophysikalische Chemie Göttingen-Nikolausberg, Germany.

I. INTRODUCTION

Elucidation of the structures of reaction intermediates poses challenging problems to experimental and theoretical chemistry because these short-lived molecules cannot be isolated and subjected to the conventional methods of structure determination. Of all organic intermediates, carbonium ions have been investigated most extensively (1,2). Because of their unstable and transient character, little experimental structural data is available, except for the X-ray analyses of a few aryl-substituted cations (1,3a). However, the lifetimes of carbonium ions can be greatly increased if they are generated in strong acid media (4), and in this way, even simple tertiary and secondary carbonium ions are formed as relatively long-lived species. Our present knowledge of carbonium ion geometries derives largely from spectroscopic investigations in these acid media. For example, the proton resonance (pmr) and infrared (ir) spectra of certain allyl derivatives can be rationalized only by assuming a symmetrical, planar, completely delocalized structure (*1a*) (5). Similarly, pmr, carbon nuclear magnetic resonance (cmr), and more recently ESCA (core electron) spectra have been employed in an investigation of the structure of the nonclassical 2-norbornyl cation (*2*) (6,7). The preferred bisected conforma-

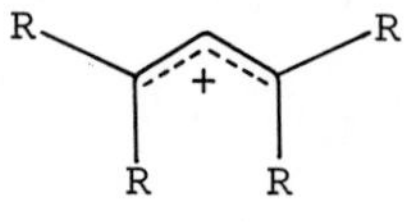

1a 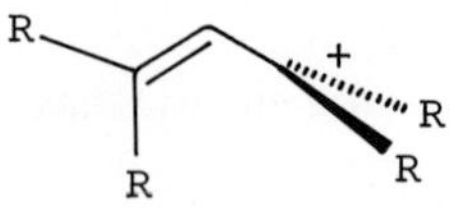 *1b* 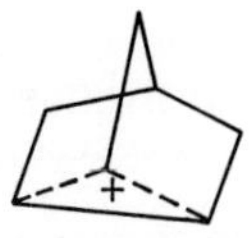 *2*

3a *3b*

tion of the cyclopropylcarbinyl cation *3a* has been ascertained from its pmr spectra and other experimental evidence (8,9). The barrier to rotation of the $C^+(CH_3)_2$ group (*3a* → *3b*) has been determined by nuclear magnetic double resonance to be 13.7 kcal/mole (10). Much symmetry information also can be gained from ir and Raman studies. For example, laser-generated Raman and low-energy ir spectra (11) of simple tertiary alkyl cations are consistent (12) with a planar (13) or near-planar structure.

Preferred planarity of carbonium ions is suggested by analysis of bridgehead-substituted polycyclic systems where the intermediate cationic species is constrained to a nonplanar arrangement because of angle strain (3). The rates of solvolysis of bridgehead derivatives, which are taken to be a measure of the ease of formation of the corresponding ions, are generally strongly suppressed. Vinyl cations, *4*, are examples of highly strained cations in a "two-membered ring." Their behavior should give indications of the effect of angle strain on ground-state conformations. The rate deceleration for generation of cyclic vinyl cations as the ring size decreases (14) suggests a linear structure; thus, planar structures may be inferred for distorted species such as the cyclopropyl cation (15). The unfavorable perpendicular geometries of the cyclopropylcarbinyl (*3b*) and the

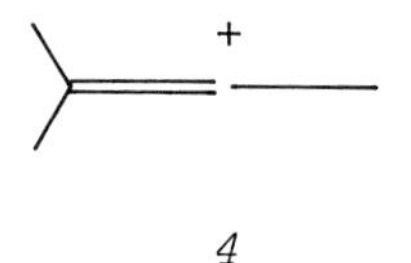

4

allyl (*1b*) cations were studied in conformationally locked allyl and cyclopropylcarbinyl adamantane derivatives (16).

Product analyses shed further light on carbonium ion structures. For example, the generally observed racemization of optically active systems in reactions involving carbioniumlike ions is strong evidence for planar reaction intermediates (17). The occurrence of systems with rearranged structures is indicative of the possible intervention of nonclassical, bridged systems as intermediates or at least as low-lying transition states.

Although many insights can be gained through experimental investigation as seen by the examples given above, the possibilities for conformational analysis of carbonium ions using purely experimental techniques are limited. "Molecular tweezers" (3) that adjust bond and dihedral angles, as well as bonding and nonbonding distances, are hard to find and are realized only in polycyclic strained cage molecules. Theoretical calculations, on the other hand, have the advantage that questions may be answered that are hard to put into experimental form. They also serve a useful purpose in the interpretation of data to which new meanings may be given. In order to reduce the computational problem to manageable size, approximations are necessary, and one of the principal objectives of this chapter is to discuss and give examples of those currently in use. Semiempirical schemes have been by far the most popular, and many such calculations have been reported in the literature. Unfortunately, the ability of these schemes to make *a priori* predictions is regarded as low by most chemists. On the other hand, the recent use of *ab initio* molecular orbital methods for simple carbonium ions has greatly increased the confidence with which meaningful *a priori* statements can be made. Probably the best strategy for treating larger systems is to calibrate and reparameterize semiempirical molecular orbital schemes by the most analogous simple ion for which an *ab initio* molecular orbital calculation can be carried out.

II. COMPUTATIONAL METHODS

Introduced more than forty years ago, quantum mechanics has become a well-established discipline; according to its basic postulates, the geometry, energy, and all other observable properties can be calculated exactly by solving Schrödinger's equation for the appropriate system of nuclei and electrons. However, even with the availability of high-speed digital computers, the complete solution of Schrödinger's equation can be carried out only for systems with three or four electrons. In order to handle systems sufficiently large to interest organic chemists, there have been two general types of approximation that are practical and useful enough to have received widespread attention (18). Both of these start from the Hartree-Fock approximation, i.e., the assumption of statistically independent particles which represent the true many-electron wave function as a product (appropriately antisymmetrized for identical electrons) of optimized one-electron functions (molecular orbitals) distributed over the nuclei. From a historical viewpoint, it is surprising that the molecular orbital representation has won out over the valence bond representation, because even today most of the qualitative treatments of bonding in organic textbooks adopt a conceptual framework based on the valence bond approach. From a quantitative standpoint, molecular orbitals are superior because they are mathematically and physically simpler and better defined. But the descriptive format provided by molecular orbital theory is quite different from valence bond ideas, and this partly accounts for the long delay in an appreciable impact of physical theory on organic chemistry. It has also been true that few of those interested in quantitative physical theory have had an adequate knowledge of organic chemistry, and vice versa.

The first approach based on the Hartree-Fock scheme is *ab initio*; i.e., the kinetic energy of all of the nonrelativistic interactions between electrons and between electrons and nuclei are calculated exactly with a trial set of molecular orbitals that are as close as possible to a true Hartree-Fock solution. The possibility of using *ab initio* methods for the solution of problems in organic chemistry has come about in the last few years through great advances in digital computer technology and economics, and through the cleverness of quantum chemists in devising new mathematical approximations and sophisticated computer programs. The underlying reason for confidence in the *ab initio* approach is that it is a mathematically well-defined approximation to the Schrödinger equation and preserves the correct balance between the various kinetic and potential energy terms. Physically, the single important omission in the Hartree-Fock approximation is instantaneous electron-electron interactions (correlation energy), which at large internuclear separations give rise to the well-known London dispersion forces. This is an area of continuing research, but tabulation of heats of hydrogenation derived from wave functions close to the Hartree-Fock limit show that relative energies for different isomers or different species are obtained with good accuracy (19,20). This strongly suggests that one is justified in believing that quantitatively useful results can be obtained on carbonium ions where direct experimental evidence is weak. The art in using *ab initio* techniques enters with the choice of the atomic orbital basis set used in the construction of the molecular orbitals. The computer time required for a given molecule is directly dependent on the size and quality of the basis set employed, and there is, therefore, strong pressure to use as small and simple a basis as possible. At present there are essentially two quality levels of basis sets that are close enough to the Hartree-Fock limit to give meaningful results. These are as follows: (*a*) the single exponential per atomic orbital, implemented either directly or through Pople's STO-3G and STO-4G

Gaussian representations (20). The majority of results discussed here employ this set. (*b*) The Hartree-Fock atomic orbital: basis sets for this level often employ variationally determined functions so that part of the form of the atomic orbitals is allowed to adjust itself to the molecular environment. One such basis set is Whitten's three-Gaussian *s*-group plus Huzinaga's five-Gaussian *p*-orbital representation for carbon, combined with Whitten's five *s* Gaussians grouped four and one (21,22). Where no basis is specified below, this set is to be understood (23). A similar basis set at this level is Pople's 4-31G (20). Of course, if computer time is available, one always prefers to use the sets at level *b*. The output of an *ab initio* molecular orbital calculation is a set of coefficients that weight the various atomic orbitals in each molecular orbital along with the one-electron energy associated with that molecular orbital, the total energy, and any other expectation values, such as dipole moment, that may have been specified for evaluation.

The second approach deriving from the Hartree-Fock approximation is the collection of semiempirical molecular orbital schemes. All these share with the *ab initio* approach an output that gives atomic orbital weightings and one-electron energies for the molecular orbitals. Their claim for attention stems from much shorter digital computer computation times (100-1000 times faster). The schemes that have been most employed in carbonium ion research are CNDO/2 (24), INDO (24), NDDO (24), and MINDO (25). These all have a common parentage and their main simplification is neglect of differential overlap. This drastic approximation does not have a straightforward rationalization, but it eliminates the vast majority of the time-consuming multicenter, two-electron integrals, which are the principal computation time bottleneck of the *ab initio* approach. An effective electronegativity term, which is evaluated by use of experimental data on atomic spectra, replaces the kinetic and one-electron potential energy terms. Beyond this general parameterization, Dewar (25) has modified INDO to specifically achieve a better predictive capability for bond lengths and heats of formation. For carbonium ions a troublesome inherent fault—a direct consequence of differential overlap neglect—is the indiscriminate bias toward more highly connected structures (e.g., bridged over open ions). This obviously eliminates its use for many important problems. Extended Hückel theory is an even simpler and faster scheme than those already noted, but it yields even less accurate results, and the saving in computer time has been sufficiently small so that it has not been a determining factor. This scheme is therefore of only limited interest to the subject matter in this article.

III. THE SIMPLEST CATIONS

A. The Methyl Cation

Though there are smaller systems that warrent experimental and theoretical interest, such as CH^+ and CH_2^+ (26), the methyl cation, CH_3^+, is the simplest tricoordinate carbonium ion, and as such its analysis forms the basis for the discussion of the larger alkylcarbonium ions. There is no direct experimental evidence for the geometry of the methyl cation, a high-energy species that has probably never been observed outside a mass spectrometer, but there seems to be little doubt as to the planar structure of this molecule. This conforms with the Nyholm-Gillespie predictions for AX_3 compounds (27) as well as with the structures of isoelectronic three-coordinate boron compounds like $B(CH_3)_3$. The dimerization of BH_3 which makes it rather elusive probably does not occur in the case of CH_3^+ because of the increased repulsion between the two positively charged nuclei (28). Numerous MO calculations of the semiempirical (29, 30) as well as the *ab initio* type (26,30b,31,15,18c) have been performed on CH_3^+. In no case was a geometry other than planar D_{3h} reported, and the earlier predictions by Walsh (32) are confirmed in this case. There is, however, one surprising aspect of the geometry of the methyl cation: the C-H bond distance is rather large, around 1.09 Å, longer than in the methyl radical and in the same range as CH_4. A decrease in the coordination number of carbon is usually accompanied by a decrease of its bonding radius (33), an effect that may be rationalized either by the change of the hybridization state of the central atom (34,35) or by the decreased repulsion between the ligands, i.e., their reduced space requirement (36). One might expect, then, the C-H bond in CH_3^+ to be significantly shorter than in methane, a conclusion that is not borne out by the results of the calculations. Recent *ab initio* (26) and semiempirical calculations (37) indicate that the opposite result may be expected. The reason for this may be seen more clearly if one considers the formation of the carbonium ion from its neutral hydrocarbon precursor in two steps: removal of a hydrogen atom to form the radical and loss of an electron to give the cation. Minimization

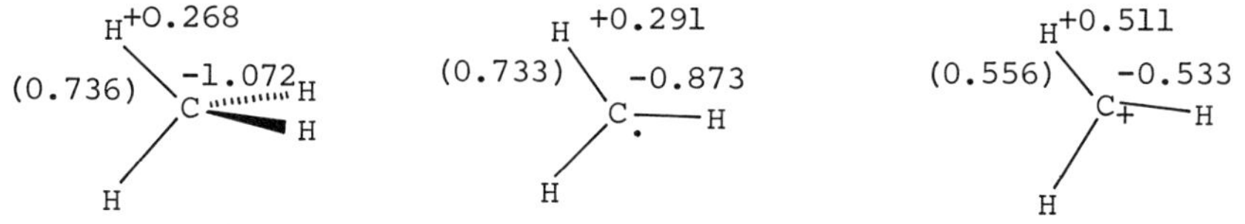

H +0.268 (0.736) C −1.072 H H H

H +0.291 (0.733) C −0.873 H H

H +0.511 (0.556) C+ −0.533 H H

of the energy of the planar methyl radical gives a geometry with a C-H distance of 1.076 Å (38), in agreement with other *ab initio* calculations (26,30). Obviously, this is a case where hybridization or repulsion arguments hold. In the methyl cation, the minimized C-H distance is 1.091 Å (39,40), a significant increase over the radical and about the same as in methane. Looking at the population analyses given above (counting the odd electron of the radical which is localized in the π-type *p*-orbital of carbon), it turns out that removal of this electron only decreases the negative charge on the carbon atom from -0.873 to -0.533. The electrons are shifted toward the carbon nucleus from the hydrogens, whose nuclear potential cannot effectively compete with the much lower-lying carbon orbitals, and from the bonding region between carbon and the hydrogens. As the population analysis reveals, both the hydrogen and C-H overlap populkations decrease significantly upon formation of the positively charged species. The increased repulsion of the more positive hydrogens as well as the shift of electrons out of the bonding region result in the increase of the C-H bond distance. In a slightly different way, these arguments have been applied earlier to explain the increased flattening of the pyramidal ground state structure along the isoelectronic series CH_3^-, NH_3, OH_3^+ (41). The effect of angle strain on the preferred geometry of the methyl cation (as well as the radical and the anion) has been investigated recently (15), since it is not immediately obvious what should be expected. Angle strain is known experimentally to induce nonplanarity in radicals and anions, and the possibility has been suggested that the 7-norbornyl cation might have a nonplanar arrangement around the carionic center due to the rather small (∿95°) C-C^+-C angle (42). A nonplanar carbonium ion might

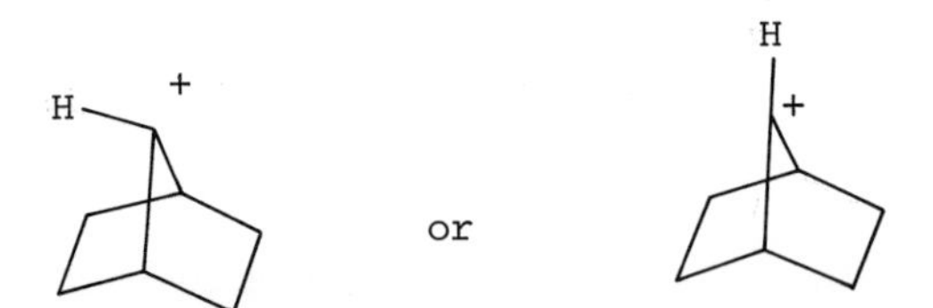

account for the observation (42) that solvolysis of a deuterium-tagged 7-norbornyl derivative leads to products with predominant retention of configuration. To test this hypothesis, *ab initio* calculations were performed on the methyl cation (15) with one of the H-C^+-H angles, θ, confined first to 120° and then to 90° to simulate angle strain. For each value of θ the energy of the planar species was computed as well as the energies that resulted when the third hydrogen was bent out of the plane formed by the remaining atoms by different degrees of ϕ. The results are shown in Figure 1. It turns out that both the free and the

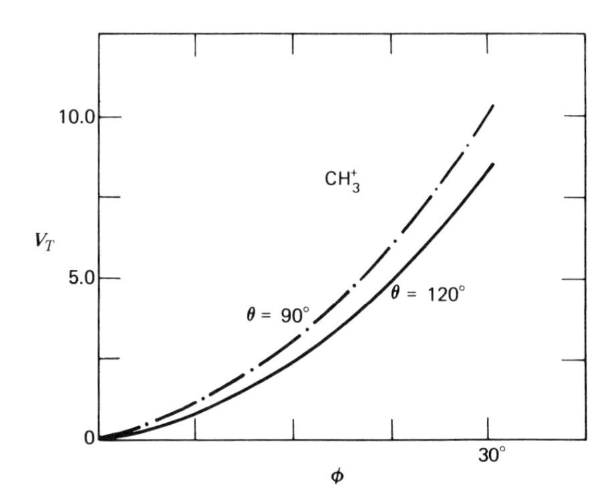

Fig. 1. Effect of angle strain on H_2C^+-H out-of plane bending.

angle-strained methyl cations have a planar equilibrium geometry. However, the barrier to out-of-plane deformation increases upon deformation of the system; i.e., the angle-strained methyl cation prefers planarity more than the free (θ = 120°) cation. Even conceding that the leaving group and the solvent (which are not included in the calculations) may have a considerable effect on the structures of these charged species, we are reasonably sure that the qualitative trend of Figure 1 is correct. Vinyl cations, which can be considered extreme models of angle-strained cations (in a two-membered ring), probably are linear (if they have C_{2v} symmetry), as is indicated by both solvolysis studies (14) and MO calculations (29,26).

The concept of angle strain was originally introduced to rationalize the behavior of three-coordinate systems under angle deformation, and it was argued (43) that the planar transition state of an angle-strained system is subjected to more angle strain relative to the pyramidal ground state than the unstrained compound resulting in an increased barrier to inversion (Fig. 2). This argument accounts qualitatively for the experimentally observed behavior of angle-strained carbanions and amines. It is based on the assumption that a bond angle of 120° is the best "unstrained" value for AX_3 species, and consequently, if one

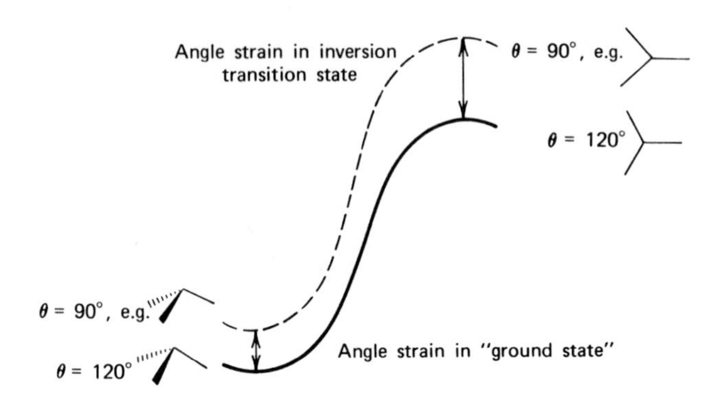

Fig. 2. Effect of angle strain on inversion barrier.

angle X-A-X is reduced to a value less than 120°, the third atom X will move out of plane to make the remaining X-A-X angles approach 120° as closely as possible. This argument does not differentiate between a carbanion and a carbonium ion. It predicts for the angle-strained methyl cation either a pyramidal ground state or a planar ground state with a *reduced* out-of-plane bending force constant as compared to the free ion. One can illustrate this conclusion by application of the Westheimer strain model (44). In this scheme an unstrained value of 120° is assumed for three-coordinate compounds, and as expected, an angle-strained carbonium ion is computed to have a pyramidal ground-state geometry (45). However, the idea that deformed ($\theta < 120°$) carbonium ions are nonplanar is incorrect. The origin of the planar structure arises chiefly from the change of the nuclear repulsion during out-of-plane deformation. Not only is the total energy change (i.e., the destabilization during out-of-plane deformation) more than 90% accounted for by the repulsion between the nuclei, but also the increased barrier to pyramidal deformation of the angle-strained system is the result of an equally increased nuclear repulsion term. Minimization of ligand repulsion accounts for the planar D_{3h} symmetry of the unstrained ion. When two ligands are held at a small angle, the remaining ligand occupies a position as far as possible from the others; i.e., a planar structure is still preferred. Moreover, during out-of-plane deformation, the repulsion between the ligands increases more rapidly the smaller the angle θ. Conversely, the larger the value of θ, the less the nuclear repulsion changes for this deformation. For $\theta = 180°$ there would be no change in repulsion at all. The different behavior of the methyl anion is due to the increased relief of electronic repulsion of the strained system in going from the planar transition state to the pyramidal ground state.

Though the potential energy surface of the methyl cation is now well established, extrapolation of these results to alkyl-substituted methyl cations should be done with caution for two reasons:

1. The energy to deform the planar cation center to one of locally tetrahedral symmetry is very similar for the methyl and ethyl cation (46), but the bending process is more complex in the latter. A large increase in nuclear repulsion is moderated by a large increase in coulombic attraction between the nuclei and the electrons. This is presumably due to the ability of the carbon-carbon bond, through use of the *p*-orbitals, to adjust more efficiently to the new nuclear configuration than the C-H bonds. Qualitatively, however, the above reached conclusions remain valid.

2. In certain conformations, some cations do not possess a plane of symmetry passing through the four nuclei of the cation center. A typical example is the "staggered" form of the ethyl cation, *5a*. Because the interaction of the methylene protons

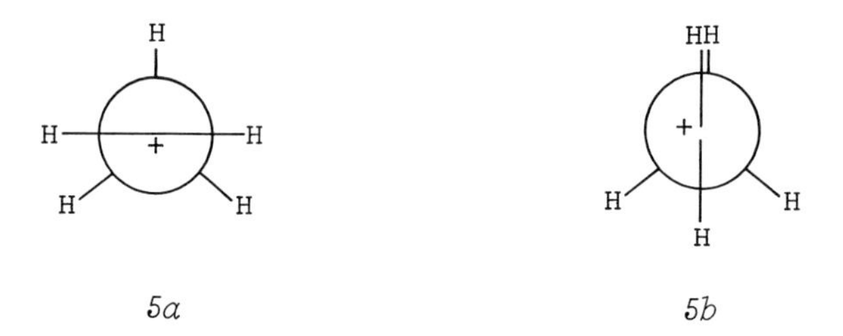

with the β-hydrogens is different on the two sides of the CH_2^+ group (one hydrogen above vs. two hydrogens below), a highly improbably cancelling of interactions would have to occur to render the planar configuration the stable equilibrium geometry. Indeed, one can argue that because of the missing symmetry plane or C_2 axis, the cation theoretically cannot be planar (47). Upon complete minimization of the geometry of *5a* (the only restriction being the retention of a single reflection plane perpendicular to the CH_2^+ group) it was found (26,46) that the plane containing the C^+H_2 nuclei is displaced 2.9° upward from the C-C bond corresponding to a small distortion toward a staggered ethane-type geometry. It should be noted that the symmetry argument presented above is a *necessary* but not a *sufficient* condition for a molecule to exhibit a certain geometry. Cation

5a cannot be planar because it lacks the proper symmetry element. Cation *5b*, though possessing this reflection plane, does not necessarily have to be planar. Molecules often do not adopt the geometry of highest symmetry. However, the potential energy surface for out-of-plane deformation in *5b* is symmetric with respect to the planar conformation, and because of the very small energy differences involved, the structure could be planar on a time-averaged basis (even the average structure of *5a* would be nonplanar). Similarly, the solvent or counterion shell may render carbonium ions nonplanar. Such a species, e.g., an ion pair, would effectively destroy the symmetry of the system. This in no way touches upon the conclusion we reached in the case of the methyl cation: carbonium ions themselves tend strongly to prefer a planar structure. Whether they will achieve it or not depends on the symmetry and preferred electronic distribution of the carbon species itself and its environment.

B. The Methane Radical Cation (Methanium Ion)

The CH_4^+ system is of particular interest both from an experimental and a theoretical point of view. It is the simplest member of the class of alkane radical cations which are formed as primary reaction products in the radiolysis of alkanes by energetic particles. As such, they are involved in photoelectron spectroscopy (48) and mass spectrometry (49). In the latter, they represent the well-known parent molecular ions of alkane decompositions. The theoretical interest in CH_4^+ stems from the high tetrahedral symmetry of the parent molecule. In methane, the highest occupied molecular orbital is triply degenerate, and the radical cation, lacking an electron, is therefore expected to undergo Jahn-Teller distortion (50). Numerous calculations have been performed on the potential energy surface of this system (26,51). The two obvious deformations from tetrahedral symmetry are elongation along a C_3 axis leading to a species with C_{3v} symmetry (trigonal distortion), and tetragonal distrotion along a C_2 axis leading to systems of the D_{2d} and finally D_{4h} point group (Figs. 3 and 4). From Figures 3 and 4 it becomes clear why Jahn-Teller distortions play such a dominant role in the structure of CH_4^+. During both a tetragonal (Fig. 3) and trigonal distrotion (Fig. 4), two of the orbitals of the triply degenerate t_2 group vis. t_{2x} and t_{2y}, are stabilized owing to better overlap between the carbon and hydrogen atomic orbitals. Only the t_{2z} orbital in the tetragonal distortion suffers a destabilizing effect because the hydrogens move toward the nodal plane of the carbon $2p_z$ orbital. In the trigonal case, overlap between the carbon $2p_z$ orbital and the apical hydrogen is diminished. A distorted minimum energy

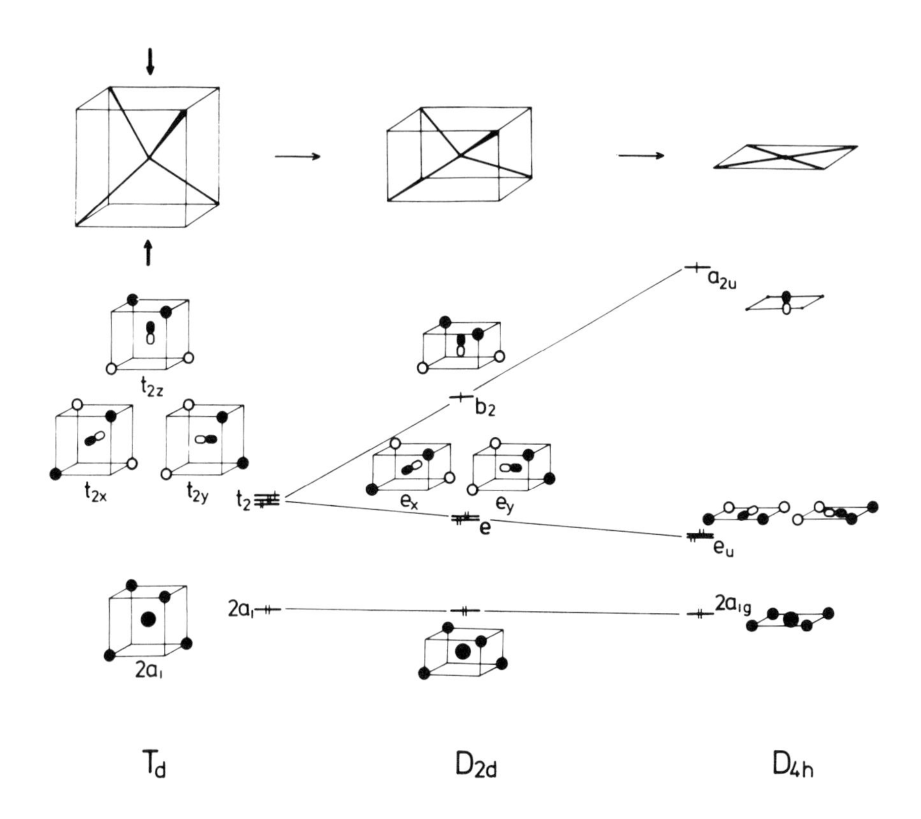

Fig. 3. Effect of tetragonal distortion on the energy levels of CH_4^+: schematic. Lowering of energy niveau indicates stabilization of the particular molecular orbital. For each m.o., the group theoretical symbol is given, together with the carbon 2s or 2p orbital and the corresponding hydrogen s orbital combination.

geometry results from the compromise between the energy lowering through occupancy of the doubly degenerate orbitals against energy raising through occupation of the t_{2z} orbital. D_{2d} structure, *6a*, has a lower total energy than the trigonal pyramid *6b* (26,51a-d). Another interesting structure has been proposed (51c) which is planar, with two hydrogens bonded to carbon by sp^2 hybrids ($\angle$H-C-H $\simeq$ 120°), and the other two by a three-center, two-electron bond (*6c*). This structure turns out to have an energy slightly lower than the trigonal pyramid but considerably

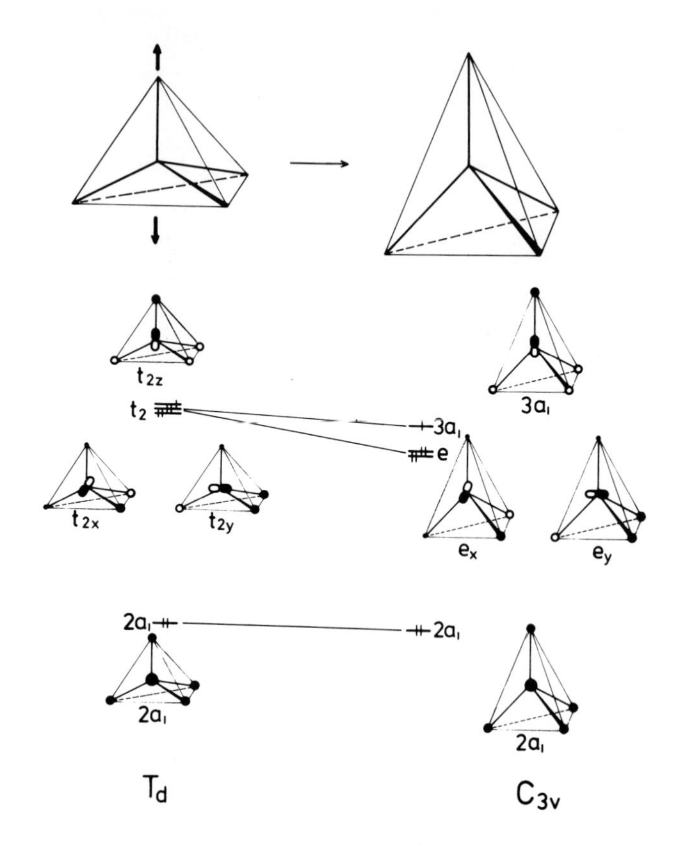

Fig. 4. Effect of trigonal distortion on the energy levels of CH_4: schematic. For explanation, see Fig. 3. Note that local C_{3v} symmetry was assumed for the tetrahedral species.

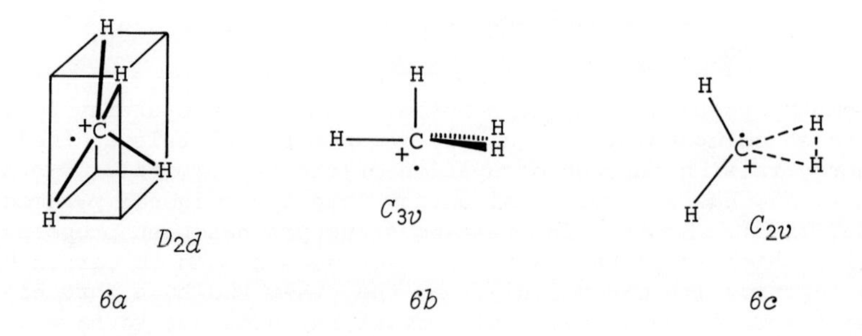

higher than *6a* (51b).

C. Protonated Methane

Numerous semiempirical and *ab initio* calculations (26,52, 53) reveal the current interest in this species. It became particularly relevant to organic chemistry when Olah (52) showed its probable intermediacy in the reactions of methane in superacid solutions. There are a number of different symmetries to be discussed for CH_5^+, the trigonal bipyramid, D_{3h} (*7a*), the square pyramid, C_{4v} (*7b*), and species containing only reflection planes (*7c* or *7d*) of point group C_s. Early studies indicated a

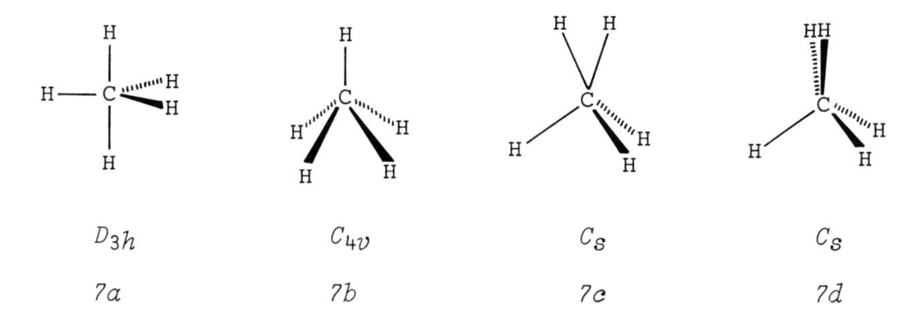

preference for the more symmetric structures *7a* and *7b*, but the work of Kutzelnigg et al. (53h) seems to have settled the question in favor of the C_s configuration, in agreement with other investigations (26,53d-i).

The C_s structure containing a three-center bond looks somewhat like a hydrogen molecule attached to CH_3^+ (53f) and suggests a generally preferred front side attack of single bonds in the interactions with electrophiles (51e). Note the similarity to the proposed structure *6c* for CH_4^+ and to certain structures of the $C_3H_7^+$ class, such as corner-protonated cyclopropane. Like other sixfold barriers (46,54), the barrier to rotation of the "H_2" group (*7c* $\rightleftarrows$ *7d*) is negligible (53h). As in *6c*, there are two types of bonds, three two-center, two-electron bonds and a three-center, two-electron bond between H_2 and the CH_3^+ moiety. An analogous structure has been calculated for the C-H protonated form of $C_2H_7^+$, but the C-C protonated form is even more stable (26).

IV. ALKYL CARBONIUM IONS

A. The Ethyl Cation

The ethyl cation is the simplest cation to exhibit a torsional barrier and a hyperconjugative stabilization effect. Also it is the smallest system in which the problem of classical vs. nonclassical (bridged) structures arises. It is not surprising, then, that a number of calculations, both *ab initio* and semiempirical, have appeared in the literature dealing with the $C_2H_5^+$ system. The results do not always concur, owing to the previously noted deficiencies in semiempirical methods. The structure of the ethyl cation is best discussed in terms of its electronic distribution shown below (38). Unfortunately, separation of the charge density into atomic and overlap populations according to the Mulliken population alaysis (55) is often quite sensitive to the type of basis set used. Therefore, a Mulliken

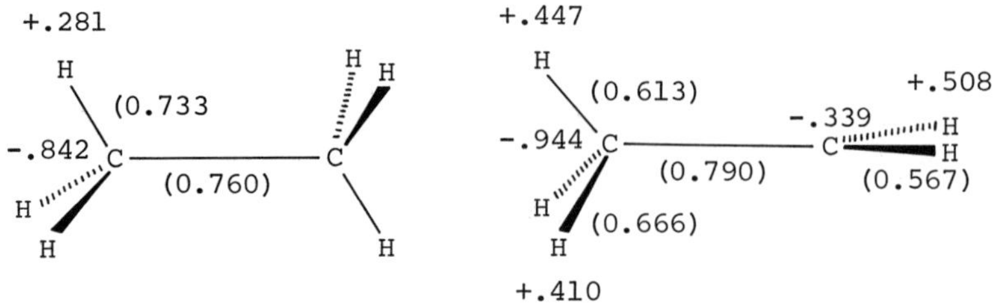

+.281 +.447
(0.733 (0.613) +.508
-.842 -.944 -.339
(0.760) (0.790) (0.567)
(0.666)
+.410

analysis of ethane and the ethyl cation calculated with an identical basis set is given above. According to the atomic populations, all hydrogens of the ethyl cation carry a roughly equal share of the positive charge, higher than that in ethane. The carbon atoms in the ethyl cation are both negatively charged, but although the carbinyl carbon has become more positive upon formation of the cation, the negative charge on the methyl carbon has *increased* during this process. Even when compared to the neutral hydrocarbon, the analysis does not support the contention that the carbinyl carbon is the positive center of the molecule, since the positive charge is quite effectively delocalized.

The charge on the β-hydrogens is a sensitive function of their orientation with respect to the empty *p*-orbital. The highest number of electrons is transferred from the hydrogen that forms the smallest dihedral angle with that orbital. The overlap population of the methyl C-H bonds changes similarly, but 90° out of phase. The electronic distribution in the methyl group is best rationalized by assuming a hyperconjugative interaction of the empty *p*-orbital with the proper linear combination of the three β-hydrogen orbitals (34). Indeed, by suppressing hyperconjugation (by omitting the empty *p*-orbital from the calculation) its population is forced to be zero by preventing its interaction with the other basis functions, and the discrimination of these hydrogens with respect to their charges all but vanishes (38). The accumulation of negative charge on the carbon atom next to the positive center is at first sight surprising, but this is a characteristic way in which positively charged electron-deficient species like the ethyl cation stabilize themselves. Positively charged species, like carbonium ions, are not simply stabilized by dispersal of their excess charge, but by optimal utilization of their electrons for filling the regions of low coulombic potential. As we have seen above, removing H^- from ethane to form the ethyl cation *increases* the negative charge on the methyl carbon, and also the cationic carbon is far from developing its formal positive charge. The same principle—less bonding population, more atomic population—applied to the methyl cation was the reason for its surprisingly large C-H bond distance.

The stabilization of cationic centers by alkyl substitution is a well-established experimental fact (1). Generally, the order of stability of carbonium ions is tertiary > secondary > primary. Likewise, the formation of the ethyl cation from ethane is a more favorable process than formation of the methyl cation from methane. Quantitatively, the experimental value for the free enthalpy of the reaction is 39 kcal/mole at room temperature

$$C_2H_5^+ + CH_4 \longrightarrow CH_3^+ + C_2H_6$$

(56); i.e., the extra stabilization of the ethyl over the methyl cation due to the methyl group substitution is close to 40 kcal/mole. *Ab initio* values for this reaction range from 18 (38) to 30.9 kcal/mole (26), modified CNDO gives 30 kcal/mole (37), and MINDO/2 gives 50.7 kcal/mole (25e). Because of neglect of correlation effects, *ab initio* molecular orbital theory does not yield an accurate *absolute* value for binding energies (may be up to 50% in error), but an analysis of the effect of alkyl substitution on carbonium ions in terms of *ab initio* wave functions is highly informative.

It has long been assumed that methyl groups are more electron releasing than hydrogen (57). For example, this assumption has been invoked to account qualitatively for the stabilizing effect at positively charged sp^2 and sp centers. Recent experimental evidence and analysis of semiempirical (30) and *ab initio* wave functions (38) show that this explanation is superficial (58); the conventional view of the electronic effects of alkyl groups requires alteration. A comparison of the population analyses for the ethyl and methyl cations reveals that in the former the methyl group has donated 0.323 electrons to the positive center, whereas in the methyl cation each hydrogen has transferred 0.515 electrons, indicating a larger electron-releasing character of hydrogen than of methyl. However, two points should be kept in mind. First, the difference, 0.192 electrons, is smaller than the corresponding difference, 0.268 electrons, in the neutral precursors, ethane and methane. Since hyperconjugation and induction are more important in carbonium ions than in alkanes, the methyl group is a better electron donor in the former. Secondly, population analyses depend strongly on the basis set used. According to our analysis (38) the stabilization of carbonium ions is due to an increased utilization, in the cation, of the most stable available valence orbitals, the carbon 2*s* orbitals, for both atomic and overlap population. This analysis in terms of very strong bonds to the CH_2^+ center of the alkyl group is supported by Kollmar's semi-empirical analysis of the problem (59).

Hyperconjugation, as depicted below, has considerable effects on both the energy and geometry of the ethyl cation. The minimum energy C-C bond length is calculated to be 1.47 (37) to 1.48 Å (26), shorter than in ethane. Suppression of hyperconjugation as described above results in an increase of the C-C bond distance to 1.558 Å, as expected if the right-hand form depicted above does not contribute to the electronic configuration (60). Concomitantly, the total energy of the system is raised by 11.8 kcal/mole (*ab initio*) in surprisingly good agreement with a recent MINDO result (61).

As has been discussed at length already, the minimum energy conformation of the ethyl cation, *5a*, does not contain a planar cationic center but rather corresponds to a partially bridged species with a nonplanar trigonal center leading to what may be called an approximately "staggered" arrangement. Nonplanarity is not limited to the ethyl cation but is also found in the 1-propyl and *t*-butyl cations. Another class of intermediates, the tricoordinate carbon free radicals, show out-of-plane bending force constants considerably smaller than in carbonium ions (15);

i.e., radicals should resist pyramidal deformation less than do carbonium ions. If the molecular dissymmetry is the reason for the nonplanarity in the ethyl cation, one might reasonably expect an increased influence of molecular structure on the preferred conformations of such free radicals; this is indeed what is found theoretically. In the most stable conformation of the ethyl radical (corresponding to the minimum energy geometry of the ethyl cation) the plane of the CH_2^+ atoms is tilted not 2.9°, but 22.9° (9), toward the C-C bond (26), but part of this difference is due to the use of the STO-3G basis set which gives a nonplanar methyl radical (26). Nonplanarity is likewise to be expected in other alkyl-substituted radicals in their minimum energy configurations, if the methyl or other groups are arranged such as to exclude the symmetry plane (or C_2 axis) of the molecule from containing the radical carbon and its three nearest neighbors. Available spectroscopic evidence, especially α-hydrogen (62) and ^{13}C hyperfine splitting constants (63), suggests that the odd electron in the ethyl radical occupies an orbital with slightly more *s*-character than in the methyl radical. The long-standing controversy (64) as to the planar (63) or nonplanar (65) structure of the cyclohexyl radical may be resolved at once with this symmetry argument. In none of its conformations except that with the ring planar (C_{2v} symmetry), which is certainly prohibitively high in energy, does this radical have local C_s symmetry containing the HC(—C)—C_2 nuclei. Thus it *cannot* be planar in its most stable conformation.

1. *The Potential Energy Surface of* $C_2H_5^+$

If locally tetrahedral symmetry around the methyl group and a planar CH_2^+ group are assumed, both *ab initio* (29,46,66a-c) and semiempirical schemes (29,30a,67) predict a zero rotational barrier (*5a*-*5b*) for the ethyl cation. There are good reasons why this barrier should be zero:

1. The total nuclear repulsions between the hydrogens at opposite ends of the C-C bond are independent of the angle of rotation. There are two moderately bad interactions at dihedral angles of 30° each in *5a*, and one "bad" eclipsing and one "good" staggering in *5b*.

2. The hyperconjugative stabilization is also independent of the rotation angle. At every angle, linear combinations of the $p\pi$-orbital on the methyl carbon and of the three hydrogen 1*s* orbitals can be formed of proper symmetry to mix with the empty C^+ orbital. Localization of both the staggered and

eclipsed ethyl cations gives identical hyperconjugation energies.

If all geometrical parameters in *5a* and *5b* are allowed to vary (the only restriction being the retention of the vertical reflection plane in each case), the bond angles distort from the canonical values and the energy of structure *5a* is lowered by 1.25, that of structure *5b* by 1.03 kcal/mole (46). Because of the greater ability of the staggered conformation *5a* to take advantage of stabilizing deformations, the rotational barrier is no longer zero but 0.22 kcal/mole. The qualitative agreement between semiempirical and *ab initio* calculations is lost when the question of bridged vs. classical structures of carbonium ions is considered. The classical ethyl cation can undergo a degenerate rearrangement (1,2 hydride shift) whose product is a classical ethyl cation with the methyl and methylene carbons interchanged. The symmetrical intermediate or transition state in this reaction is the nonclassical or bridged ethyl cation *8*, sometimes called protonated ethylene. The little

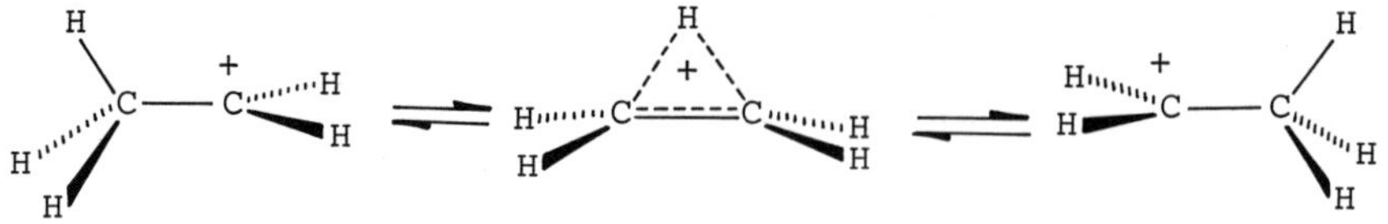

5 *8* *5*

available experimental evidence indicates that an appreciable barrier to rearrangement of labeled ethyl derivatives exists (68). This barrier is obtained by several *ab initio* calculations (29,46,66a,c,d) which place the bridged system *8* 3 to 12 kcal/mole higher in energy than the open ion *5*. In addition, it has been shown that *8* is not even an intermediate, but a true transition state. Quite the opposite result obtains with semiempirical methods. According to NDDO and CNDO the bridged species should be considerably more stable than the open system, 33 (29) and 10 kcal/mole (59), respectively. There are several more cases which show the preference of semiempirical methods for bridged structures over their classical, open isomers (69a). For example, the cyclopropyl-allyl cation rearrangement has been sutdied quite extensively by several semiempirical methods (30c, 37,69,70). One of the early investigations by CNDO showed the cyclopropyl cation to be more stable than the open-chain allyl cation by the ridiculous value of 300 kcal/mole (30c), and a difference of 81 kcal/mole has been reported recently (69a). Even the reparameterized CNDO method of Kollmar and Fischer, claimed to yield reasonable energies and geometries for hydrocarbons and carbonium ions, still favors the cyclic cation by 25 kcal/mole after a full geometry search (37). The same problem exists for MINDO/2 (25d,e). Experimentally, the spontaneous ring opening of the cyclopropyl to the allyl cation is observed, and the energy difference is estimated to be at least 18 kcal/mole in favor of allyl (71). *Ab initio* calculations are in decidedly better agreement with experiment. STO-3G yields a 27.8 kcal/mole lower total energy for the allyl cation than for its cyclic counterpart, and with the extended 4-31G basis set, the energy difference increases to 46.5 kcal/mole (which seems somewhat large) (72). Other cases where semiempirical schemes favor cyclic over open structures may be found in the $C_2H_3^+$ class (vinyl vs. protonated ethylene) (29), the propyl cations, and the butyl cations. Semiempirical schemes neglect many of the two-electron integrals which are included in an *ab initio* treatment, and since these integrals represent electron repulsion and exchange interactions, their neglect gives too much weight to the attractive one-electron terms (29). For this reason semiempirical methods generally favor the closed, more condensed (cyclic) structure over the open forms.

B. The Propyl Cations

1. *The Energy Surface of $C_3H_7^+$*

The potential energy surface of carbonium ions of composition $C_3H_7^+$ is expected to include the 1- and 2-propyl cations (*9* and *10*) and the corner- (*11*), edge- (*12*), and face- (*13*) protonated cyclopropanes. There are numerous rearrangement paths interconnecting these different species. Thus the 1-propyl

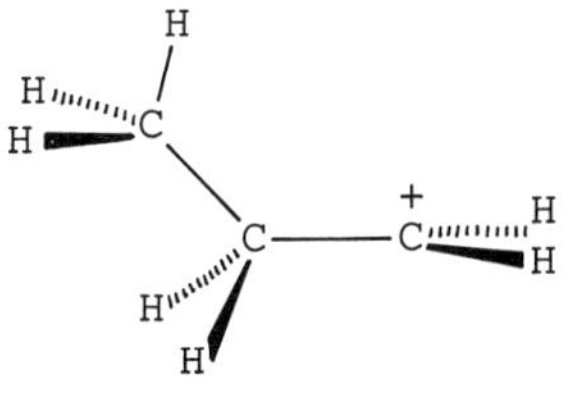

9

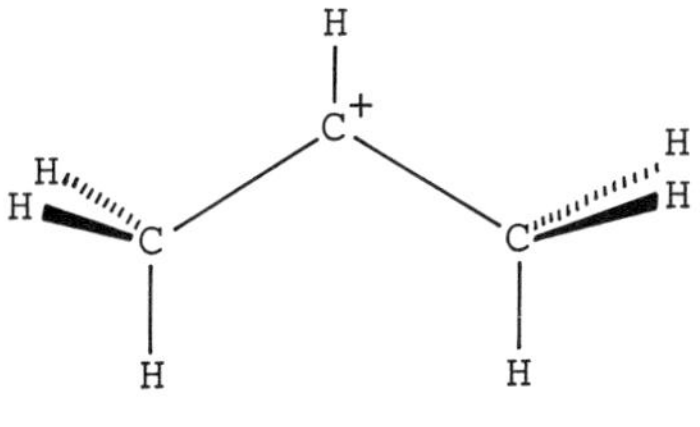

10

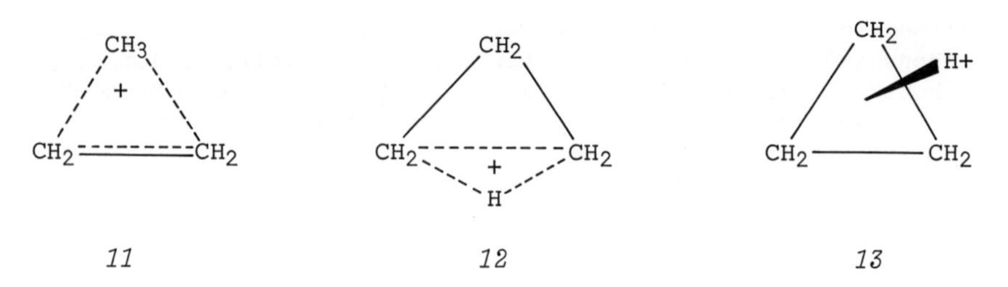

cation has the choice to rearrange degenerately by way of a 1,2 methyl shift via *11*, by two consecutive 1,2 hydride shifts with the isopropyl cation *10* as intermediate, or by a 1,3 hydride shift involving edge-protonated cyclopropane, *12*. In view of the numerous experimental data relating to $C_3H_7^+$ rearrangements (73), there was early theoretical interest in them. Hoffmann computed protonated cyclopropane structures with extended Hückel theory (30a). Relatively complete energy surfaces have been computed with CNDO (74), NDDO (29), a modified CNDO scheme (75), MINDO/2 (25e), and *ab initio* (76). Not surprisingly, all the semiempirical calculations predict one of the cyclic isomers, edge-protonated cyclopropane *12*, to be the most stable species—even more stable than the isopropyl cation!

Experimentally it is found (77) that isopropyl cation *10* is stable in SO_2ClF-SbF_5 solution below 0°C, with scrambling of protons being observed between 0 and 40°C—possibly due to a reversible rearrangement via the 1-propyl cation and protonated cyclopropane. The 2-propyl cation has recently been determined to be more stable than the primary isomer *9* by 16 kcal/mole in the gas phase (78). This value is satisfactorily reproduced by the only full-scale *ab initio* investigation to date including complete geometry optimization (76). Geometries were determined by an STO-3G search, and for the energy minima so obtained an extended 4-31G basis calculation was then carried out (Table 1).

Particular points of interest are as follows:

1. Of all $C_3H_7^+$ systems, the classical secondary 2-propyl cation *10* is the lowest energy form. Its stability is calculated to be 17 kcal/mole higher than the primary cation *9*.

2. Contrary to semiempirical results, edge-protonated cyclopropane *12* is found to have a higher energy than either the corner-protonated form, *11*, or the 1-propyl cation, and the face-protonated ring structure, *13*, in agreement with all previous calculations, is so high in energy that a major role in the interconversion of these ions is prohibited.

3. The 1-propyl cation, *9*, and the corner-protonated

Table 1 Relative Energies of $C_3H_7^+$ Cations (kcal/mole)

	Cation	Symmetry	STO-3G	4-31G
10	2-Propyl	C_{2v}	0	0
9	1-Propyl	C_s	20.5	16.9
11	Corner-protonated cyclopropane	C_s	22.8	17.3
12	Edge-protonated cyclopropane	C_{2v}	27.1	27.1
13	Face-protonated cyclopropane	C_{3v}	161.0	139.6

cyclopropane, *11*, have similar energies, but only *9* corresponds, besides *10*, to a potential minimum in the complete $C_3H_7^+$ surface. Of the three possible rearrangement paths of the 1-propyl cation, the one involving *11* has the lowest activation energy of 0.4 kcal/mole, whereas the 1,2 and 1,3 hydride shifts require 16 and 10 kcal/mole, respectively.

Apart from their relative energies, the structures of some of these species present some highly unusual features.

2. *The 1-Propyl Cation*

What is the minimum energy conformation of this ion? Is there a considerable barrier to rotation of either the CH_2^+ or the CH_3 groups? What is the relative importance of C-H vs. C-C hyperconjugation? For a long time it has been tacitly assumed

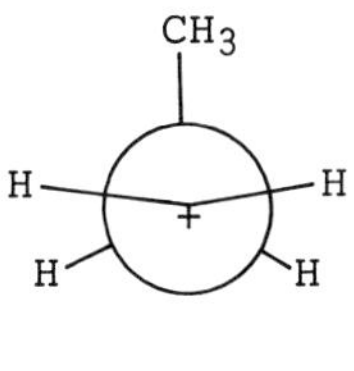

9a 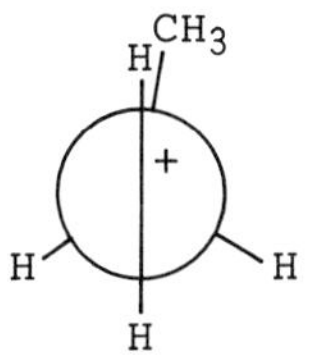 *9b* 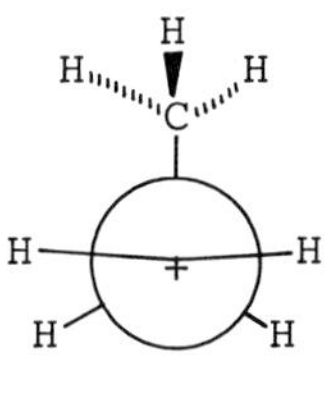 *9c* 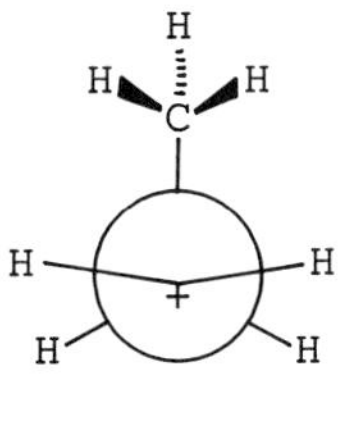 *9d*

that the torsional barriers in carbonium ions, like that of the ethyl cation, were not appreciable (79) and differences in the dihedral arrangement of the substituent at the α-carbon atom were therefore neglected. Two *ab initio* calculations (using a basis set of sufficient quality to reproduce the rotational barrier of ethane in quantitative agreement with experiment) show that there exists a definite preference for a certain conformation in the 1-propyl cation (76). The geometry that eclipses the empty *p*-orbital of the cation center with the β-C-C bond, *9a*, is predicted to be more stable than its 90° rotamer, *9b*, by 2.3 to 2.5 kcal/mole (76). This difference should be largest in a primary system and smaller in the more stable secondary and tertiary ions. It should be noted, however, that even in these latter cases the assumption of free rotation of the cation center in alkylcarbonium ions is incorrect unless the system happens to have a sixfold barrier.

Isomers *9a* and *9b* differ most in the degree by which they can be stabilized by C-H and C-C hyperconjugation. In *9b*, the β-C-C bond is orthogonal to the empty *p*-orbital and thus cannot mix into hyperconjugative resonance with this orbital. Localization of this ion in the manner described for the ethyl cation gives a destabilization of 11.8 kcal/mole, identical with the value found for the ethyl cation (60). This was to be expected since the interaction with the two β-hydrogens is identical in *9b* and *5b*. In the stable configuration, *9a*, the dihedral angle between the C-C bond and the empty orbital is zero. C-C hyperconjugation should be most effective in this geometry. The destabilization due to the localization process in this ion amounts to 12.3 kcal/mole, or 0.5 kcal more than in the C-H hyperconjugative form, *9b*. Although this difference is not enough to account for the total barrier, it certainly does *not* support the contention long held by chemists that C-H bonds are more effective in stabilizing a positive charge through nonbonding resonance than C-C bonds (57).

A striking feature of the conformational analysis of the 1-propyl cation is the considerable variation of the barrier height with substitution at the β-carbon atom. Table 2 shows the dependence of the barrier on five different substituents, the values varying between 3.73 for methyl and 0.87 kcal/mole for CN. This is in contrast to the behavior of similarly substituted propanes which show almost no variation in the calculated barriers to rotation of the terminal methyl group (76). Electron shifts are severely limited by symmetry restrictions. In II, the empty *p*-orbital is perpendicular to the C-X bond; stabilization of the positive charge through overlap of the $2p_y(C^+)$ orbital with either the $2p_x(C_3)$ or the $2p_y(C_2)$ orbitals is impossible. This is reflected in the population of the $2p_x(C^+)$ orbital which is essentially independent of the substituent X. In the perpendicular form I, this "empty" orbital overlaps strongly. An electron-releasing substituent X (e.g., CH_3) lowers the energy of I relative to II thereby increasing the barrier height, whereas an electron-withdrawing substituent

Table 2 Rotational Barriers of β-Carbon-Substituted 1-Propyl Cations and Propanes

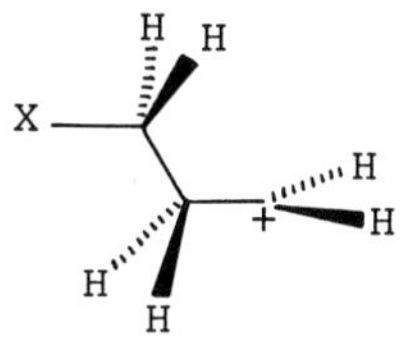

I 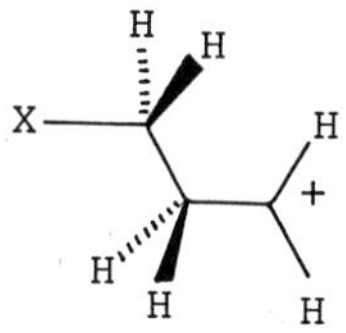 II 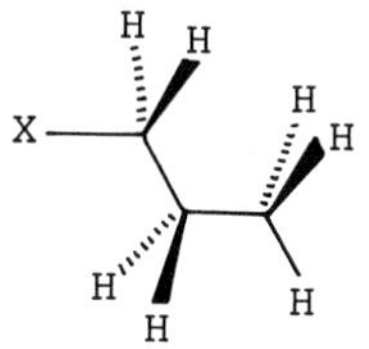 III 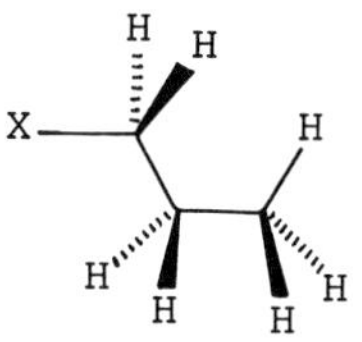 IV

X	Barrier I → II	Barrier III → IV
CH_3	3.73	3.63
H	2.52	3.69
F	2.11	3.46
OH	0.91	3.49
CN	0.87	3.64

(e.g., X = CN) destabilizes I and leads to a lower barrier.

When compared with the corresponding conformation, *5a*, of the ethyl cation, the difference of the interactions above and below the CH_2^+ group in the propyl cation conformation, *9a*, is seen to be more pronounced. Therefore it is not surprising that the deviation of the cationic center from planarity is increased in the latter: The CH_2^+ bending angle toward the C^+-C bond is 5.4° in *9c* as compared to 2.9° in the ethyl cation. As shown below, the minimum energy conformation of the 1-propyl cation corresponds to a highly distorted, very weakly bridged structure with an 83° C-C-C angle.

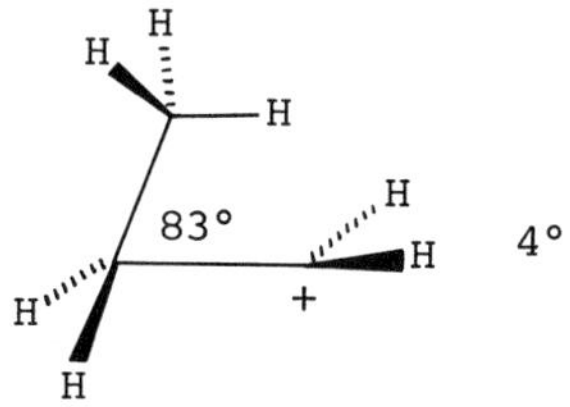

83° 4°

9e

In addition to the large deviation of the bond angles in this 1-propyl cation from tetrahedral and trigonal values, the long β-C-C bond length (1.632 Å) is indicative of a bridging deformation; in the bridged ethyl cation, the distance between the carbon atoms is 1.403 Å. Similarly lengthened bond distances are exhibited by the bridge bonds of the protonated cyclopropane structures (discussed below). The increased bond distances in these nonclassical species may be thought of as three-center bonds which are occupied by two electrons and are typical of bridged structures; the increase in nuclear repulsion due to the reduced screening effect of the electrons tends to keep the nuclei apart. The geometry of diborane may be used as a model; the bridging hydrogens have a B-H distance of 1.34 Å, considerably larger than the B-H distance of the terminal hydrogens of 1.19 Å (80). It is also interesting to note that the methyl-methylene-eclipsed form (*9c* and *9e*) is more stable by 0.5 kcal/mole than the methyl-staggered form *9d*, despite the very small C-C-C angle of 83.4° and the eclipsed conformation. This situation reflects the tendency of the pivotal hydrogen atom to bridge with the empty *p*-orbital on C^+. "The structure of the 1-propyl cation illustrates the dangers inherent in the arbitrary distinction between 'classical' and 'non-classical' carbonium ions. Even 'classical' carbonium ions may have structures which differ appreciably from those normally assumed" (76).

3. The 2-Propyl Cation

The calculations on the ethyl cation have already shown that there exists no preferred conformation between a planar cation center and a methyl group bonded to it. Thus one would expect the isopropyl cation to adopt a geometry that minimizes the steric interactions between the two methyl ligands and the hydrogen atom. Of the three relevant structures, *10a* to *c*, only *10c* is suspect on this basis if the trigonal angle is held at

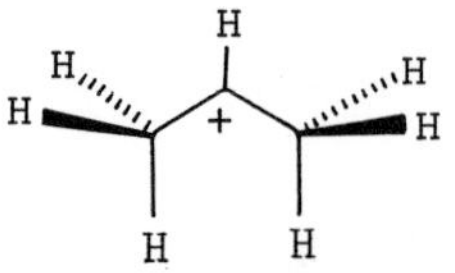

10a 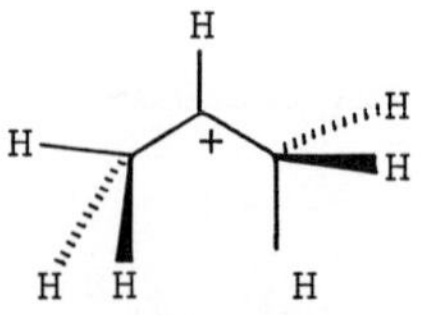 *10b* 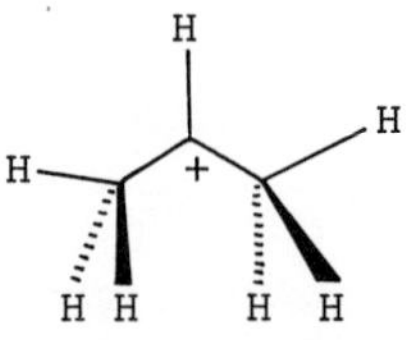 *10c*

120°. When free variation of this angle is allowed, *10a* turns out to be the preferred geometry with the C-C-C angle widened to 125.3°, whereas it is 124° in *10b* and 123.9° in *10c*.

The protonated cyclopropane structures provide additional examples for the effect of bridging on bond distances. In the corner-protonated form *11*, the bridging C-C bond length is 1.803 Å (vs. 1.399 Å for the other C-C bond), and in edge-protonated cyclopropane *12* the corresponding values are 1.849 and 1.516 Å. As in other sixfold rotational barriers (*7c* → *7d*, *5a* → *5b*), rotation of the methyl group in corner-protonated cyclopropane is practically free.

C. Higher Alkylated Carbonium Ions

The data accumulated on smaller systems allow one to predict with considerable confidence the effect of alkyl group substitution on carbonium ions.

1. *Rotational Barriers of β,β-Dialkyl-Substituted 1-Propyl Cations*

The barrier to rotation of the methylene group in the isobutyl cation, *14a* → *14b*, provides a check for internal consistency of theoretical calculations. This is useful in view of the paucity of experimental data on carbonium ion conformations. The isobutyl cation, *14b*, differs from the *n*-propyl cation, *9a*, in that each β-hydrogen is exchanged by methyl and vice versa.

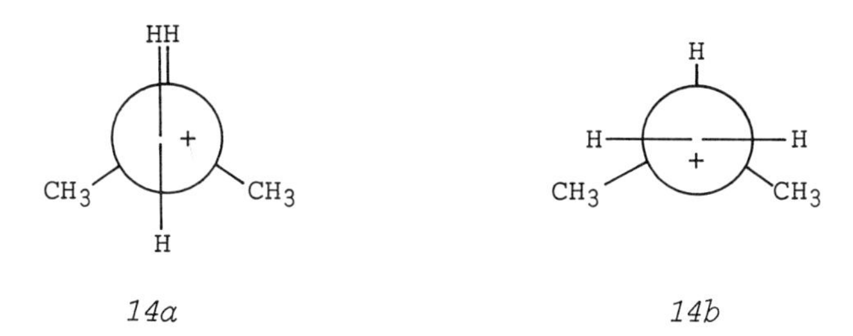

14a *14b*

If the preference for the C-CH_3 bond to be coplanar with the empty *p*-orbital in the *n*-propyl cation follows a simple two-fold cosine potential function, it is easy to show that the same function applied to the isobutyl cation predicts the conformation *14a* (which has both C-CH_3 bonds forming dihedral angles of 30° with the empty *p*-orbital) to have the lowest energy and the barrier to be the same as in the *n*-propyl cation. STO-3G results are in agreement with this prediction. The "bisected" form, *14a*, is preferred over *14b* by 2.7 kcal/mole (76). For the process *14a* → *14b*, CNDO yields a barrier height of 3.7 kcal/mole (81), though a value for the propyl cation barrier was not quoted. Incorporation of the two methyl groups into an increasingly smaller ring system leads to an increased preference of the cation for the "bisected" conformation, the energy difference using STO-3G being 4.1 kcal/mole in the cyclobutyl- (*15a* → *15b*) and 17.5 kcal/mole in the cyclopropylcarbinyl system (*16a* → *16b*) (76). The corresponding CNDO values (81) are

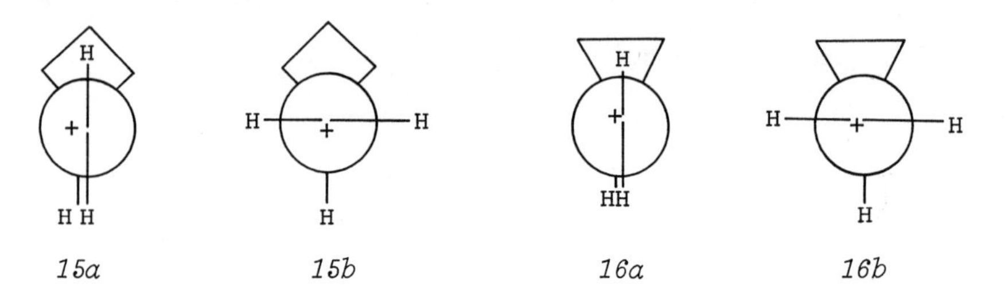

15a *15b* *16a* *16b*

7.4 and 25.1 kcal/mole. The well-documented energy difference between the cyclopropylcarbinyl cation conformations, *16a* and *16b*, is thus revealed to be not a special case but only one example of a spectrum ranging from the isobutyl cation low barrier to the high barrier in the two-membered ring system of the allyl cation.

2. t-*Butyl Conformations*

Techniques have been developed recently which allow the Raman spectra of stable cations to be determined, and the vibrational analysis of the *t*-butyl cation and of several alkylated derivatives in strong acid solution has been reported (12). According to Raman and low-energy ir data, the *t*-butyl cation has a planar or near-planar carbon skeleton, and overall C_{3v} symmetry is indicated from analysis of the selection rules. The structure formed (*17a*) has one hydrogen of each CH_3 group above the plane of the four carbons and the other two symmetrically to the left and to the right below this plane. For the

C_{3v}

17a

C_{3h}

17b

planar geometry, this is certainly not the geometry of minimum repulsion among the methyl groups. Recent *ab initio* results (82) indicate this C_{3v} structure to be 1.5 kcal/mole higher in energy than the C_{3h} symmetry species, *17b* (in which the in-plane hydrogens of the methyl groups "snuggle" between the hydrogens of the neighboring hydrogen ligand). The preferred structure for the isoelectronic trimethyl boron is probably C_{3h} (*17b*) in the gas phase. However, in the liquid phase the *t*-butyl cation might prefer structure *17a* due to the presence of the gegenion.

V. UNSATURATED CARBONIUM IONS

Two types of unsaturated carbonium ions can be distinguished depending on the position of the unsaturation center with respect to the positive carbon. In the vinyl cations, e.g., *18*, the double bond is directly attached to the C^+ carbon, reducing its

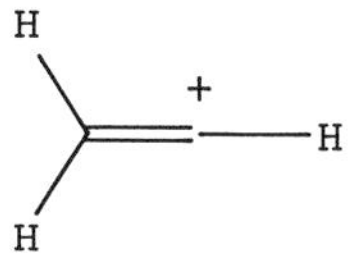

18 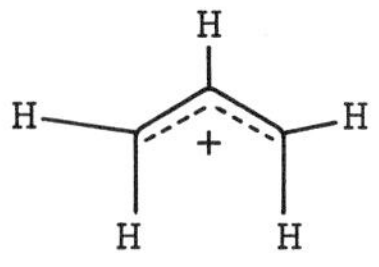 *19* 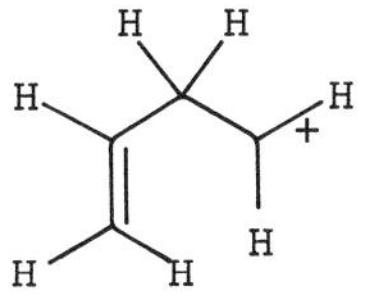 *20*

coordination number to two. The stability of vinyl cations is reduced, and thus only suitably substituted vinyl cations may be observed as intermediates (14). If a double (or triple) bond is separated by one or more atoms from the cation center, a stabilizing interaction through donation of π-electron density onto the electron-deficient carbon results. Simple examples are allyl (*19*) and the homoallyl cation, *20*.

A. The Vinyl Cation

Recently there has been growing interest in the chemistry of vinyl cations because of the development of "super" leaving groups. These permit simple alkyl-substituted systems to be studied solvolytically. The decreasing reactivity of cyclic vinyl trifluoromethane sulfonates with smaller ring size suggests (14,83) a preferred linear structure for the vinyl cation intermediates. This preferred geometry is in accord with the results of several theoretical investigations (26,29,84-87). If a double bond can be taken to represent the extreme case of angle strain (in a "two-membered ring" system), the linear vinyl cation geometry has its analogy in the presumably planar structure of the cyclopropyl cation and of angle-strained carbonium ions.

The energy of the vinyl cation relative to its nonclassical hydrogen-bridged isomer, protonated acetylene, *21*, is of interest. Calculations by Hoffmann (EHT) (30a) and Yonezawa (88)

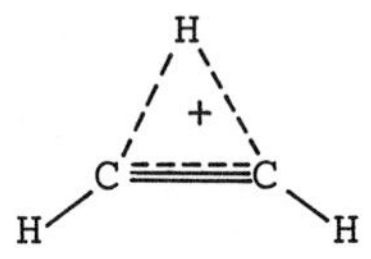

21

indicated the open ion *18* to be more stable, but the semi-empirical NDDO method favored the cyclic structure by 32 kcal/mole (29). Two recent *ab initio* calculations indicate a preference for the classical ion, *18*, by 25 (29) and 18 kcal/mole (84).

B. The Allyl Cation

Delocalization of the π-electrons of a double bond into the empty *p*-orbital of a cation center leads to a highly geometry-dependent stabilization of the electronic configuration. For the allyl cation, *19*, a planar ground state has been observed experimentally (5,6) in accord with theoretical calculations (82,88). The barrier to rotation of a terminal methylene group, *19* → *19a*, has been the subject of two *ab initio* studies which report values of 45 (88) and 35 kcal/mole (82). By an analysis of the substituent effects on experimental barriers of substituted allyl cations, a range of 38-43 kcal/mole has been estimated for the rotational barrier for the allyl cation itself (89).

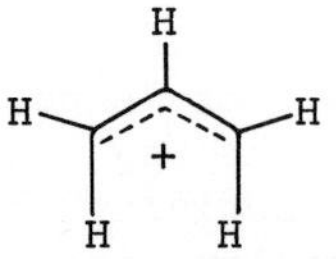

19 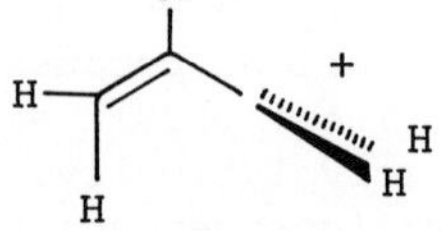 *19a*

Allyl products are generally the only species observed from solvolysis of a cyclopropyl derivative. This highly exothermic process is an example of a concerted reaction controlled by the conservation of orbital symmetry and thus predictable on the basis of the Woodward-Hoffmann rules (90). Localized orbitals have been employed in the analysis of $C_4H_7^+$ cations (91, 92). Here the lack of proper symmetry elements makes application of the Woodward-Hoffmann rules difficult.

In general, the virtue of molecular orbital calculations of concerted processes lies in the fact that the reaction paths and the nodal properties of the molecular orbitals are largely controlled by symmetry and are often reproduced by even the crudest schemes. A case in point is the cyclopropyl to allyl interconversion. The ring opening of the planar cyclopropyl cation, *22*, can occur either conrotatory or disrotatory with respect to the vertical reflection plane of the molecule:

Disrotatory Conrotatory

22 *19* *22*

According to the principle of conservation of orbital symmetry, the disrotatory process is allowed in the ground state (thermal process), whereas conrotatory ring opening is forbidden. This discrimination of possible isomerization paths is correctly reproduced by EHT (93), CNDO (94), MINDO (25d), and *ab initio* calculations (82,95), though the relative energies of the two cations, *19* and *22*, differ by as much as 400 kcal/mole. Although the calculations predict the planar arrangement around the positive center in *22* to be the most stable geometry for the free cation (25d,82,93-95), the presence of the leaving group must be taken into account when discussing a solvolysis process. For the cyclopropyl cation *in statu nascendi* (93), two modes of disrotatory ring opening, Dis 1 and Dis 2 can be differentiated (96), both symmetry allowed. In Dis 1 the cyclopropyl bond opens trans to the leaving group X with the substituents trans to X moving outward, but in Dis 2 those groups move inward and the C-C bond of the cyclopropane ring breaks cis to the C-X bond. Experimental results indicate a highly stereospecific ring-opening process according to the Dis 1 path (97) in agreement with all calculations reported to date (25d,60,93-95).

VI. NONPLANAR CARBONIUM IONS

Symmetry of the molecule and of the solvent (gegenion) shell permitting, carbonium ions undoubtedly prefer a planar structure. Because of geometrical restrictions, e.g., at the bridgehead positions of polycyclic systems, many carbonium ions are prevented from achieving this favored geometry and therefore show a decreased reactivity (transition-state destabilization) of many orders of magnitude compared to unrestricted carbonium ions. The decreased stability of the pyramidally deformed methyl cation has been discussed at length (15). Calculations of preferred conformations of alkyl substituted carbonium ions with a nonplanar cation center offer the opportunity to assess the importance of electronic effects vs. strain effects in bridgehead reactivities. The rotational barriers of the ethyl and 1-propyl cations, with the C^+H_2 group rigidly fixed in a tetrahedral geometry, have been calculated at the *ab initio* level (76). The results are summarized in Table 3.

From a rotational barrier close to zero in the planar ethyl cation, the barrier height rises to 2.8 kcal/mole in the tetrahedral ethyl cation, *23a* → *23b*, approximately the value for ethane. Similarly, the propyl cation barrier increases from 2.5 (planar C^+H_2 group) to 6.04 kcal/mole (tetrahedral cation center, *24a* → *24b*). The stable geometries (*23a* and *24a*) have the "empty" sp^3 lobe trans to the hydrogen and methyl group, respectively. Although the nature of this stabilizing "trans" effect remains to be fully clarified, an experimental verification of this unexpected phenomenon seems to have been found. Solvolysis rate constants of bridgehead-substituted polycyclic derivatives vary over 19 powers of ten (98). By use of a computer conformational analysis scheme (molecular mechanics), it has been possible to correlate these solvolysis rates with the calculated strain energy differences in the parent hydrocarbon and the corresponding carbonium ion (taken to be a model for the transition state of the solvolysis reaction) (98). The correlation works as well as it does apparently because errors in the strain energy calculation of the hydrocarbon and of the ion tend to cancel when their difference is taken. Figure 5

Table 3 Relative Energies of Tetrahedral Alkyl Cations (kcal/mole)

23a	0.0
23b	+2.8
24a	0.0
24b	+6.0

shows several compounds of the series which have been calculated.

Also included in this figure is the 10-perhydrotriquinacyl system, which shows a large deviation. The acetolysis of *25* is nearly 10^9 times slower than that expected on the basis of conformational analysis calculations (98,99). The reason for this discrepancy probably lies in the torsional arrangement of the β-C-C and C-C bonds in the carbonium ion formed from *25* since it is different from the other compounds considered in Figure 5. The Newman projections of two of these cations, the

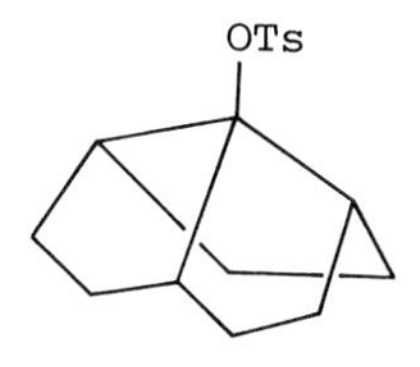

25

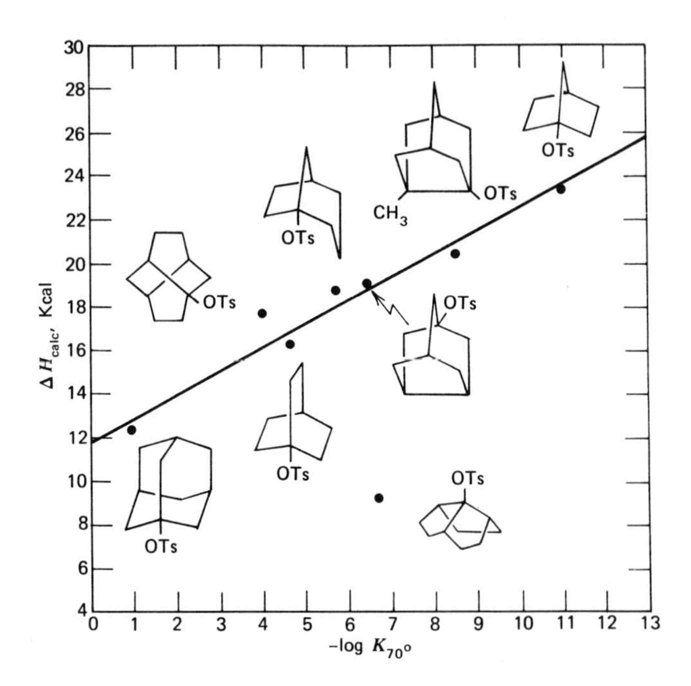

Fig. 5. Correlation of solvolysis rates with calculated hydrocarbon-cation strain differences.

1-adamantyl, *26*, and the 1-bicyclo[2.2.2]octyl, *27*, reveal that in both of these cases, a carbon-carbon bond is eclipsed trans to the empty sp^x orbital, corresponding to the most stable geometry of the tetrahedral 1-propyl cation, *24a*. On the other hand, the 10-perhydrotriquinacyl geometry, *28*, lacks this favorable arrangement; all bonds are more or less eclipsed depending on the degree of nonplanarity of the tricoordinate carbon in the transition state. Although the conformational analysis calculation does take into account the "torsional" strain exhibited by *28* (and also by the parent hydrocarbon corresponding to *25*), it does not reflect the electronic destabilization of *28* due to the lack of a C-C or C-H bond trans to the empty orbital on the electron-deficient carbon. According to STO-3G calculations (76), *28* (with the tricoordinate carbon center in a planar geometry) corresponds to the least stable conformation, *29*, of the isobutyl cation, and this is 2.9 kcal/mole higher in energy

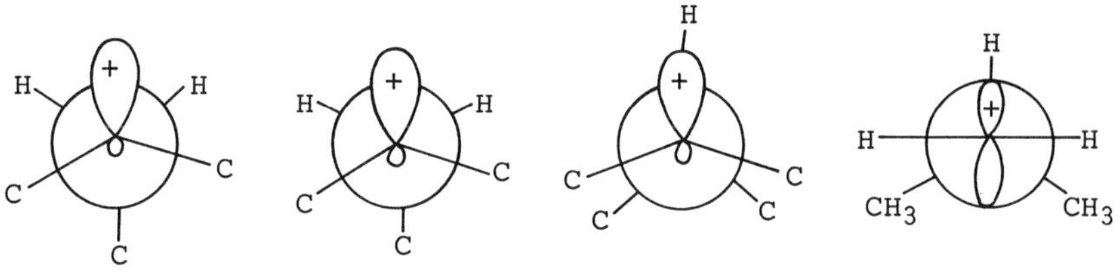

26 *27* *28* *29*

than the bisected form of the ion, *14a*. A quantitative evaluation of this electronic effect and an analysis of the energy difference of 14 kcal/mole between the predicted and observed stability of the transition state of the solvolysis reaction of *25*, is still lacking. The tetrahedral isobutyl cation barrier should be calculated in the same manner as that carried out for the barrier *24a* → *24b*. Comparison of the energies for corresponding rotameric forms of propane and the 1-propyl cation and of isobutane and the isobutyl cation, i.e., ΔA and ΔB, would yield an estimate of the magnitude of the electronic destabilization in *28* relative to *24a*. Tertiary cationic systems also should be investigated in a similar way.

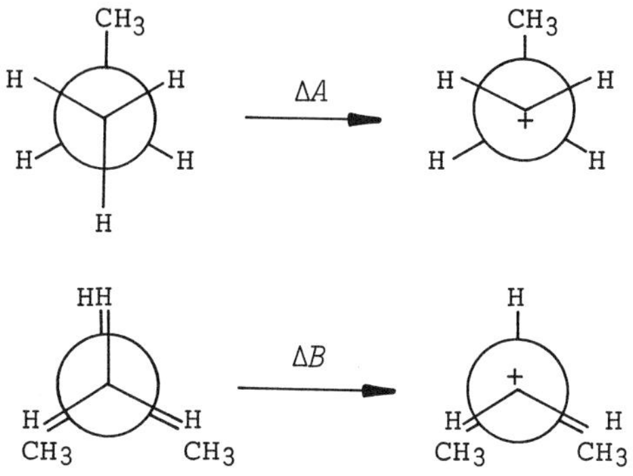

REFERENCES

1. D. Bethell and V. Gold, *Carbonium Ions. An Introduction*, Academic Press, New York, 1967.
2. *Carbonium Ions*, G. Olah and P. v. R. Schleyer, Eds., Wiley-Interscience, New York, Vol. I, 1968; Vol. II, 1970; Vol. III, 1972.

3. (a) R. C. Fort and P. v. R. Schleyer, *Advan. Alicyclic Chem.*, *1*, 283 (1966); (b) R. C. Fort, Jr., in ref. 2, Vol. III, Chap. 32.
4. G. A. Olah and J. A. Olah in ref. 2, Vol. II, Chap. 17.
5. N. C. Deno in ref. 2, Vol. II, p. 783; N. C. Deno, *Chem. Eng. News*, *42* (40), 88 (1964).
6. G. D. Sargent, review in ref. 2, Vol. III, Chap. 24.
7. G. A. Olah, A. M. White, J. R. DeMember, A. Commeyras, and C. Y. Lui, *J. Amer. Chem. Soc.*, *92*, 4627 (1970); G. A. Olah, private communication. See G. Klopman, *J. Amer. Chem. Soc.*, *91*, 89 (1969) for semiempirical calculations of these structures.
8. G. A. Olah, D. P. Kelly, C. L. Jeuell, and R. D. Porter, *J. Amer. Chem. Soc.*, *92*, 2544 (1970); G. A. Olah, C. L. Jeuell, D. P. Kelly, and R. D. Porter, *J. Amer. Chem. Soc.*, *94*, 146 (1972).
9. C. U. Pittman and G. A. Olah, *J. Amer. Chem. Soc.*, *87*, 2998 (1965); N. C. Deno, J. S. Lui, J. G. Turner, D. N. Lincoln, and R. E. Fruit, *J. Amer. Chem. Soc.*, *87*, 3000 (1965); cf. C. D. Poulter and S. Winstein, *J. Amer. Chem. Soc.*, *91*, 3650 (1969). Reviews: H. G. Richey, Jr., in ref. 2, Vol. III, Chap. 25; K. B. Wiberg, B. A. Hess, Jr., and A. J. Ashe, III, in ref. 2, Vol. III, Chap. 26.
10. D. S. Kabakoff and E. Namanworth, *J. Amer. Chem. Soc.*, *92*, 3234 (1970).
11. For a general account of laser Raman spectroscopy, see M. R. Booth and R. J. Gillespie, *Endeavour*, *29*, 89 (1970); D. A. Long, *Chem. Britain*, *7*, 108 (1971).
12. G. A. Olah, J. R. DeMember, A. Commeyras, and J. L. Bribes, *J. Amer. Chem. Soc.*, *93*, 459 (1971).
13. Here and in the following *planar* refers only to the geometry of the cationic carbon and the three atoms directly bonded to it.
14. W. D. Pfeifer, C. A. Bahn, P. v. R. Schleyer, C. Bocher, C. E. Harding, K. Hummel, M. Hanack, and P. J. Stang, *J. Amer. Chem. Soc.*, *93*, 1513 (1971); M. Hanack, *Acc. Chem. Res.*, *3*, 209 (1970). Review: H. G. Richey and J. M. Richey in ref. 2, Vol. II, Chap. 21, p. 899.
15. V. Buss, P. v. R. Schleyer, and L. C. Allen, *J. Amer. Chem. Soc.*, in press.
16. P. v. R. Schleyer and V. Buss, *J. Amer. Chem. Soc.*, *91*, 5880 (1969); J. C. Martin and B. Ree, *J. Amer. Chem. Soc.*, *91*, 5882 (1969); B. Ree and J. C. Martin, *J. Amer. Chem. Soc.*, *92*, 1660 (1970); V. Buss, R. Gleiter, and P. v. R. Schleyer, *J. Amer. Chem. Soc.*, *93*, 3927 (1971).
17. A. Streitwieser, Jr., *Solvolytic Displacement Reactions*, McGraw-Hill, New York, 1962.

18. (a) A review of the present status of quantum chemistry has been given by L. C. Allen, *Ann. Rev. Phys. Chem.*, *20*, 315 (1969). Also see (b) W. G. Richards and J. H. Horsley, *ab Initio Molecular Orbital Calculations for Chemists*, Oxford University Press, London, 1970; (c) W. G. Richards, T. E. H. Walker, and R. K. Hinkley, *A Bibliography of ab initio Molecular Wave Functions*, Oxford University Press, London, 1971.
19. L. C. Snyder, *J. Chem. Phys.*, *46*, 3602 (1967); L. C. Snyder and H. Basch, *J. Amer. Chem. Soc.*, *91*, 2189 (1969); W. J. Hehre, R. Ditchfield, L. Radom, and J. A. Pople, *J. Amer. Chem. Soc.*, *92*, 4796 (1970), and other papers in the same series.
20. J. A. Pople, *Acc. Chem. Res.*, *3*, 217 (1970).
21. J. L. Whitten, *J. Chem. Phys.*, *39*, 349 (1963); *ibid.*, *44*, 359 (1966).
22. S. Huzinaga, *J. Chem. Phys.*, *42*, 1293 (1965).
23. The digital computer program used to implement the molecular orbital method is described by S. Rothenberg, P. Kollman, M. E. Schwartz, E. F. Hayes, and L. C. Allen, *Intern. J. Quant. Chem.*, *3*, 715 (1970).
24. J. A. Pople and D. L. Beveridge, *Approximate Molecular Orbital Theory*, McGraw-Hill, New York, 1970; J. A. Pople, D. P. Santry, and G. A. Segal, *J. Chem. Phys.*, *43*, S129 (1965); J. A. Pople and G. A. Segal, *J. Chem. Phys.*, *43*, S136 (1965); *44*, 3289 (1966); D. P. Santry and G. A. Segal, *J. Chem. Phys.*, *47*, 158 (1967); J. A. Pople, D. L. Beveridge, and P. A. Dobosh, *J. Chem. Phys.*, *47*, 2026 (1967).
25. (a) M. J. S. Dewar, *The Molecular Orbital Theory of Organic Chemistry*, McGraw-Hill, New York, 1969; (b) N. C. Baird and M. J. S. Dewar, *J. Chem. Phys.*, *50*, 1262 (1969); (c) M. J. S. Dewar and E. Haselbach, *J. Amer. Chem. Soc.*, *92*, 590 (1970); (d) M. J. S. Dewar and S. Kirschner, *J. Amer. Chem. Soc.*, *93*, 4290-4292 (1971); (e) N. Bodor and M. J. S. Dewar, *ibid.*, *93*, 6675 (1971); *94*, 5303 (1972).
26, (a) W. A. Lathan, W. J. Hehre, and J. A. Pople, *J. Amer. Chem. Soc.*, *93*, 808 (1971); (b) W. A. Lathan, W. J. Hehre, L. A. Curtin, and J. A. Pople, *J. Amer. Chem. Soc.*, *93*, 6377 (1971).
27. R. J. Gillespie, *J. Chem. Educ.*, *47*, 18 (1970).
28. For a discussion of diborane as a model system for carbonium ions, see R. E. Davis and A. S. N. Murthy, *Tetrahedron*, *24*, 4595 (1968); also W. Kutzelnigg et al., *Chem. Phys. Lett.*, *7*, 503 (1970).
29. R. Sustmann, J. E. Williams, Jr., M. J. S. Dewar, L. C. Allen, and P. v. R. Schleyer, *J. Amer. Chem. Soc.*, *91*, 5350 (1969).
30. R. Hoffmann, *J. Chem. Phys.*, *40*, 2480 (1964); (b) J. E.

Williams, R. Sustmann, L. C. Allen, and P. v. R. Schleyer, *J. Amer. Chem. Soc.*, *91*, 1037 (1969); (c) N. S. Isaacs, *Tetrahedron*, *25*, 3555 (1969).
31. (a) R. E. Kari and I. G. Csizmadia, *J. Chem. Phys.*, *46*, 1817 (1967); (b) G. v. Buenau, G. Diercksen, and H. Preuss, *Intern. J. Quant. Chem.*, *1*, 645 (1967); (c) S. D. Peyerimhoff, R. J. Buenker, and L. C. Allen, *J. Chem. Phys.*, *45*, 734 (1966); (d) B. Joshi, *J. Chem. Phys.*, *46*, 875 (1967).
32. A. D. Walsh, *J. Chem. Soc.*, *1953*, 2296.
33. Compare the decreasing C-H bond lengths in the series ethane, ethylene, acetylene.
34. C. A. Coulson, *Valence*, Oxford University Press, 1961, Chap. 8.
35. K. Mislow, *Introduction to Stereochemistry*, Benjamin, New York, 1965, Chap. 1.
36. L. S. Bartell, *J. Chem. Educ.*, *45*, 754 (1968).
37. H. Kollmar and H. O. Smith, *Theoret. Chim. Acta*, *20*, 65 (1971).
38. J. E. Williams, V. Buss, and L. C. Allen, *J. Amer. Chem. Soc.*, *93*, 6867 (1971).
39. (a) K. Morokuma, L. Pedersen, and M. Karplus, *J. Chem. Phys.*, *48*, 4801 (1968); (b) P. Millie and G. Berthier, *Intern. J. Quant. Chem.*, *2*, 67 (1968).
40. STO-3G gives $r_{C-H}(CH_4)$ = 1.083 Å and $r_{C-H}(CH_3^+)$ = 1.120 Å; the corresponding values with the extended 4-31G basis set are 1.081 and 1.076 Å (26).
41. See ref. 3a, pp. 287-289. Also see L. C. Allen in *Quantum Theory of Atoms, Molecules, and the Solid State*, P.-O. Löwdin, Ed., Academic Press, New York, 1966, p. 60.
42. (a) P. G. Gassman, J. M. Hornback, and J. L. Marshall, *J. Amer. Chem. Soc.*, *90*, 6238 (1968); (b) F. B. Miles, *J. Amer. Chem. Soc.*, *90*, 1265 (1968).
43. D. J. Cram, *Fundamentals of Carbanion Chemistry*, Academic Press, New York, 1965.
44. J. E. Williams, P. J. Stang, and P. v. R. Schleyer, *Ann. Rev. Phys. Chem.*, *19*, 531 (1968).
45. R. C. Bingham, private communication.
46. J. E. Williams, V. Buss, L. C. Allen, P. v. R. Schleyer, W. A. Lathan, W. J. Hehre, and J. A. Pople, *J. Amer. Chem. Soc.*, *92*, 2141 (1970).
47. This argument is due to K. Mislow.
48. A. D. Baker, C. Baker, C. R. Brundle, and D. W. Turner, *Intern. J. Mass Spectry. Ion Phys.*, *1*, 285 (1968).
49. For recent reviews, see (a) F. Cacace, *Advances in Physical Organic Chemistry*, V. Gold, Ed., Vol. 8, Academic Press, New York, 1970, p. 79; (b) T. W. Bentley and R. A. W.

Johnstone, *ibid.*, p. 150.

50. (a) H. A. Jahn and E. Teller, *Proc. Roy. Soc. (London) Ser. A.*, *161*, 220 (1937); (b) H. A. Jahn, *Proc. Roy. Soc. (London) Ser. A.*, *164*, 117 (1938).
51. (a) F. A. Brimm and J. Godoy, *Chem. Phys. Lett.*, *6*, 336 (1970); (b) J. Arents and L. C. Allen, *J. Chem. Phys.*, *53*, 73 (1970); (c) R. N. Dixon, *Mol. Phys.*, *20*, 113 (1971); (d) G. S. Handler and H. W. Joy, *Intern. J. Quant. Chem.*, *5*, 529 (1970); (e) G. A. Olah and G. Klopman, private communication.
52. G. A. Olah, G. Klopman, and R. H. Schlosberg, *J. Amer. Chem. Soc.*, *91*, 3261 (1969).
53. (a) W. H. Fink, J. L. Whitten, and L. C. Allen, unpublished calculations; (b) A. Gamba, G. Morosi, and M. Simonetta, *Chem. Phys. Lett.*, *3*, 20 (1969); (c) J. L. Gole, *Chem. Phys. Lett.*, *3*, 577 (1969); (d) S. Ehrenson, *Chem. Phys. Lett.*, *3*, 585 (1969); (e) W. Th. A. M. van der Lugt and P. Ros, *Chem. Phys. Lett.*, *4*, 389 (1969); (f) H. Kollmar and H. O. Smith, *Chem. Phys. Lett.*, *5*, 7 (1970); (g) J. J. C. Mulder and J. S. Wright, *Chem. Phys. Lett.*, *5*, 445 (1970); (h) V. Dyczmons, V. Staemmler, and W. Kutzelnigg, *Chem. Phys. Lett.*, *5*, 361 (1970); (i) W. A. Lathan, W. J. Hehre, and J. A. Pople, *Tetrahedron Lett.*, 2699 (1970).
54. (a) J. Dale, *Tetrahedron*, *22*, 3373 (1966); (b) J. P. Lowe, *Progr. Phys. Org. Chem.*, *6*, 31 (1968).
55. R. S. Mulliken, *J. Chem. Phys.*, *23*, 1833, 1841 (1955).
56. Calculated using the compiled data of J. L. Franklin, J. G. Dillard, H. M. Rosenstock, J. T. Herron, and K. Draxl, "Ionization Potentials, Appearance Potentials, and Heats of Formation of Gaseous Positive Ions," *NSRDS-NBS 26*, National Bureau of Standards, Washington, D.C.
57. M. J. S. Dewar, *Hyperconjugation*, Ronald Press, New York, 1962.
58. For a recent review see J. F. Sebastian, *J. Chem. Educ.*, *48*, 97 (1971).
59. H. Kollmar and H. O. Smith, *Angew. Chem.*, *82*, 444 (1970).
60. V. Buss, unpublished STO-3G calculations.
61. N. C. Baird, *Theoret. Chim. Acta*, *16*, 239 (1970).
62. (a) R. W. Fessenden and R. H. Shuler, *J. Chem. Phys.*, *39*, 2147 (1963); (b) J. K. Kochi and P. J. Krusic, *J. Amer. Chem. Soc.*, *91*, 3940 (1969); (c) L. Altmann and B. Nelson, *J. Amer. Chem. Soc.*, *91*, 5163 (1969).
63. R. W. Fessenden, *J. Phys. Chem.*, *71*, 74 (1967).
64. (a) W. A. Pryor, *Free Radicals*, McGraw-Hill, New York, 1966; (b) O. Simamura, in *Topics in Stereochemistry*, E. L. Eliel and N. L. Allinger, Eds., Vol. 4, Wiley-Interscience, New York, 1969, p. 1.

65. A. Hudson, H. A. Hussain, and J. N. Murrel, *J. Chem. Soc., A, 1968,* 2336.
66. (a) G. V. Pfeiffer and J. G. Jewett, *J. Amer. Chem. Soc., 92,* 2141 (1970); (b) L. J. Massa, S. Ehrenson, and M. Wolfsberg, *Intern. J. Quant. Chem., 4,* 625 (1970); (c) F. Fratev, R. Janoschek, and H. Preuss, *Intern. J. Quant. Chem., 3,* 873 (1969); (d) D. T. Clark and D. M. J. Lilley, *Chem. Commun., 1970,* 549.
67. J. J. Dannenberg, T. D. Berke, *Abstracts of Papers, 158th National Meeting of the American Chemical Society, New York, Sept., 1969,* Abstract PHYS 163.
68. (a) P. C. Myhre and E. Evans, *J. Amer. Chem. Soc., 91,* 5641 (1969); (b) P. C. Myhre and K. S. Brown, *J. Amer. Chem. Soc., 91,* 5639 (1969).
69. E. I. Snyder, *J. Amer. Chem. Soc., 92,* 7529 (1970).
70. D. T. Clark and G. Smale, *Tetrahedron Lett., 1968,* 3673.
71. G. G. Meisels, J. Y. Park, and B. G. Giessner, *J. Amer. Chem. Soc., 92,* 254 (1970).
72. L. Radom, J. A. Pople, and P. v. R. Schleyer, unpublished calculations.
73. For recent reviews, see (a) C. J. Collins, *Chem. Rev., 69,* 541 (1969); (b) C. C. Lee, *Progr. Phys. Org. Chem., 7,* 129 (1970); (c) J. L. Fry and G. J. Karabatsos in ref. 2, Vol. II, Chap. 14.
74. H. Fischer, H. Kollmar, H. O. Smith, and K. Miller, *Tetrahedron Lett., 1968,* 582.
75. H. Kollmar and H. O. Smith, *Tetrahedron Lett., 1970,* 1833.
76. L. Radom, J. A. Pople, V. Buss, and P. v. R. Schleyer, *J. Amer. Chem. Soc., 94,* 311 (1972); aspects of $C_3H_7^+$ structures have been theoretically considered earlier by *ab initio* methods; J. D. Petke and J. L. Whitten, *J. Amer. Chem. Soc., 90,* 3338 (1968); L. Radom, J. A. Pople, V. Buss, and P. v. R. Schleyer, *J. Amer. Chem. Soc., 92,* 6380, 6987 (1970); *93,* 1813 (1971).
77. M. Saunders and E. L. Hagen, *J. Amer. Chem. Soc., 90,* 6881 (1968).
78. F. P. Lossing and G. P. Semeluk, *Can. J. Chem., 48,* 955 (1970).
79. P. v. R. Schleyer, *J. Amer. Chem. Soc., 86,* 1854, 1856 (1964).
80. L. S. Bartell and B. J. Carrol, *J. Chem. Phys., 42,* 1135 (1965).
81. K. B. Wiberg, *Tetrahedron, 24,* 1083 (1968); for other semi-empirical barrier calculations, see refs. 30 and 88.
82. L. Radom and J. A. Pople, private communication.
83. W. D. Pfeifer, C. A. Bahn, P. v. R. Schleyer, S. Bocher, C. E. Harding, K. Hummel, M. Hanack, and P. J. Stang, *J.*

Amer. Chem. Soc., *93*, 1513 (1971).

84. A. C. Hopkinson, K. Yates, and I. G. Csizmadia, *J. Chem. Phys.* *55*, 3835 (1971).
85. G. Morosi, *Rend. Ist. Lombardo Sci. Lettere, A*, *103*, 761 (1969).
86. H. Fischer, K. Hummel, and M. Hanack, *Tetrahedron Lett.*, *1969*, 2169.
87. L. Burnell, *Tetrahedron*, *20*, 2403 (1964).
88. S. D. Peyerimhoff and R. J. Buenker, *J. Chem. Phys.*, *51*, 2528 (1969).
89. V. Buss. R. Gleiter, and P. v. R. Schleyer, *J. Amer. Chem. Soc.*, *93*, 3927 (1971).
90. R. B. Woodward and R. Hoffmann, *Angew. Chem.*, *81*, 797 (1969); *Angew. Chem. Intern. Ed. Engl.*, *8*, 781 (1969); *The Conservation of Orbital Symmetry*, Academic Press, New York, 1969.
91. K. B. Wiberg, B. A. Hess, Jr., and A. J. Ashe, III. To be published.
92. G. Trindle and O. Sinanoglu, *J. Amer. Chem. Soc.*, *91*, 4054 (1969).
93. W. Kutzelnigg, *Tetrahedron Lett.*, *1967*, 4965.
94. D. T. Clark and G. Smale, *Tetrahedron*, *25*, 13 (1969).
95. D. T. Clark and D. R. Armstrong, *Theoret. Chim. Acta*, *13*, 365 (1969).
96. C. H. DePuy, *Acc. Chem. Res.*, *1*, 33 (1968).
97. P. v. R. Schleyer, T. M. Su, M. Saunders, and J. C. Rosenfield, *J. Amer. Chem. Soc.*, *91*, 5174 (1969) and references cited therein.
98. R. C. Bingham and P. v. R. Schleyer, *J. Amer. Chem. Soc.*, *93*, 3189 (1971).
99. R. C. Bingham and P. v. R. Schleyer, *Tetrahedron Lett.*, *1971*, 23.

FAST ISOMERIZATIONS ABOUT DOUBLE BONDS

HANS-OTTO KALINOWSKI

Institut für Organische Chemie der Universität, Frankfurt am Main, Germany

HORST KESSLER

Institut für Organische Chemie der Universität, Frankfurt am Main, Germany

I. INTRODUCTION

It is well known that restricted rotation about $p_\pi p_\pi$ double bonds* leads to geometric isomers (1) (*Z,E* isomers (2)). Al-

*Only $p_\pi p_\pi$ double bonds are discussed here. Very little is known about the stereochemical behavior of the $p_\pi d_\pi$ double bond (3). Rotation about such a $p_\pi d_\pi$ bond should not be strongly hindered electronically (3a) but there might be hindrance in an asymmetric case such as a sulfamoyl chloride (3b). On the other hand, the observed process in the latter compound might be caused by steric hindrance of N=S rotation (fast nitrogen inversion).

though this phenomenon was discovered and explained about 100 years ago, it is only during recent decades that the development of modern physical methods for measurement of isomerization rates has enabled us to investigate the detailed mechanisms of interconversion. The isomerization rates of double bonds as well as of partial double bonds can be very fast or very slow depending on the nature of substituents and the type of double bond (C=C, C=N, C=O, etc.).

If several possible pathways exist for the *Z,E* isomerization of a molecule, the conditions under which the measurements were made (e.g., irradiation, heating, or catalysis) determine which mechanism corresponds to the observed process, because only the fastest process is found kinetically. This article is solely concerned with thermal, uncatalyzed *Z,E* isomerizations, and includes the photochemical (4) and catalytic (4c) isomerizations only insofar as they are competitive with the thermal pathway.

Earlier reviews on thermal isomerizations of double bonds are listed in ref. 1.

II. STEREOCHEMICAL BACKGROUND

We may look at a trigonal planar molecule of type *1* from two different sides. If the ligands of A, a, b, and c, ordered

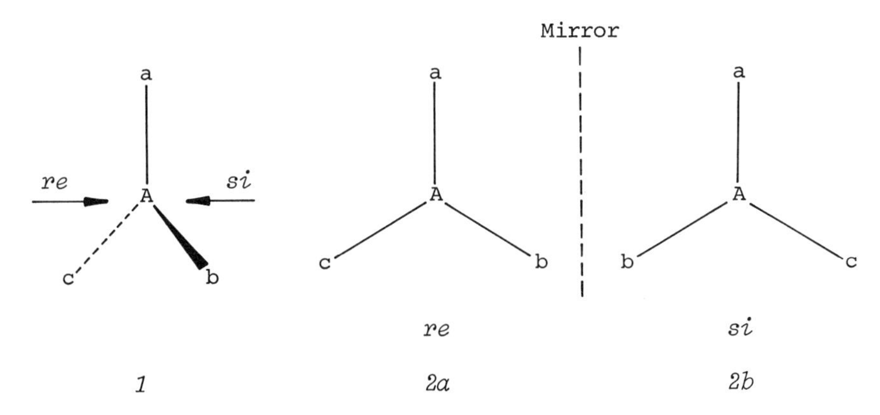

in accordance with the Cahn-Ingold-Prelog-Sequence rules (5), are arranged clockwise, we are looking at the molecule from the *re* side; if they are arranged counterclockwise, we are viewing it from the *si* side (6). The *re* and the *si* arrangements (*2a* and *2b*) are chiral in two-dimensional space. The interconversion of these two-dimensional enantiomers is possible by taking the molecules out of the plane; in three-dimensional space the enantiomerism disappears. If, however, we put together two planar chiral species of type *2* at position c, we can construct two planar diastereomeric molecules *3* and *4*.

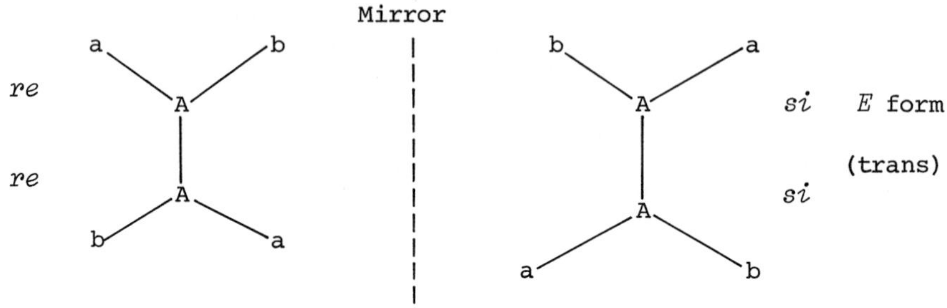

Mirror

re *re* | *si* *si* *E* form (trans)

"Racemic form"

3

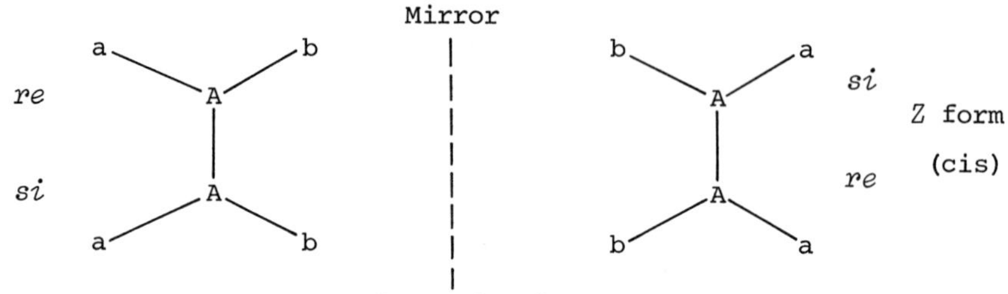

Mirror

re *si* | *si* *re* *Z* form (cis)

"Meso form"

4

The substituents a and b in each molecule are held in the plane of the paper by the maximum overlap of the *p*-orbitals at the A atoms.* The resulting *Z* and *E* molecules remain diastereomers even in three-dimensional space. Interconversion between them requires changing the chirality (*re*→*si*; *si*→*re*) of one half of the molecule by an exchange of a and b at one A. This exchange may proceed via several different pathways, e.g., rotation around the A-A bond, bond cleavage and recombination (7) or inversion if a consists of an odd electron or an electron pair. These possibilities will be discussed later on.

*These and the following conclusions are based on the sp^2-p hybridization mode. In principle the sp^3-hybridization concept with banana bonds will also fit the stereochemical arrangements at the double bond in the ground state.

Finally let us consider the question of whether *Z*,*E* isomers are configurational isomers or conformational isomers. We will see later on that there is no exact delineation between single bonds and double bonds. Formal double bonds very often have rather high barriers of *Z*,*E* isomerization. For this reason double-bond isomers have in the past been called configurational isomers; "bond cleavage" involving high energy is required for isomerization.

Recent studies have shown (see below) that isomerization of double bonds may be fast under certain conditions. On the other hand, rotation about formal single bonds may be hindered electronically or sterically to such an extent that stable isomers result. Therefore it is quite arbitrary to consider "single-bond isomers" as conformers and "double-bond isomers" as configuration isomers. We can avoid this difficulty by defining conformers as stereoisomers that can be interconverted by intramolecular rotations.* All other stereoisomers that are not conformers are configurational isomers. By this definition *Z*,*E* isomers represent conformers (8).

For "degenerate isomers," which differ only in the arrangement of designated but otherwise identical ligands, the term "topomer" has been proposed (9). The "degenerate *Z*,*E* isomeriza-

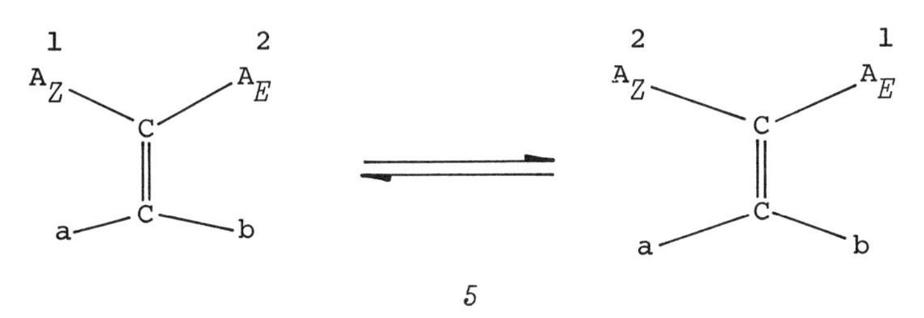

$A_Z = A_E$. The subscripts indicate location. The superscripts serve to distinguish the two equal ligands.

tion" *5* is therefore called *Z*,*E* topomerization and represents a case of diastereotopomerization.

*One must consider that the real interconversion of conformers may occur by another process (e.g., inversion of the imines); the definition only implies the *possibility* of a mutual exchange by rotation.

III. METHODS OF DETERMINATION OF FAST ISOMERIZATION RATES AT DOUBLE BONDS

Isomers can be separated if their mean lifetime is of the order of several hours, which is equivalent to a rate constant for interconversion of $k < 10^{-4}$ sec^{-1}. For separability at room temperature (T = 298°K) it thus follows from the Eyring equation[10]

$$k = \kappa \frac{RT}{hN} \exp\left(-\frac{\Delta G^{\ddagger}}{RT}\right) \quad [1]$$

κ = transmission coefficient (taken as unity)
R = gas constant
T = absolute temperature
h = Planck's constant
N = Avogadro's number

that the free energy of activation $\Delta G^{\ddagger}$ must be larger than about 23 kcal/mole. This is further illustrated in Figure 1.

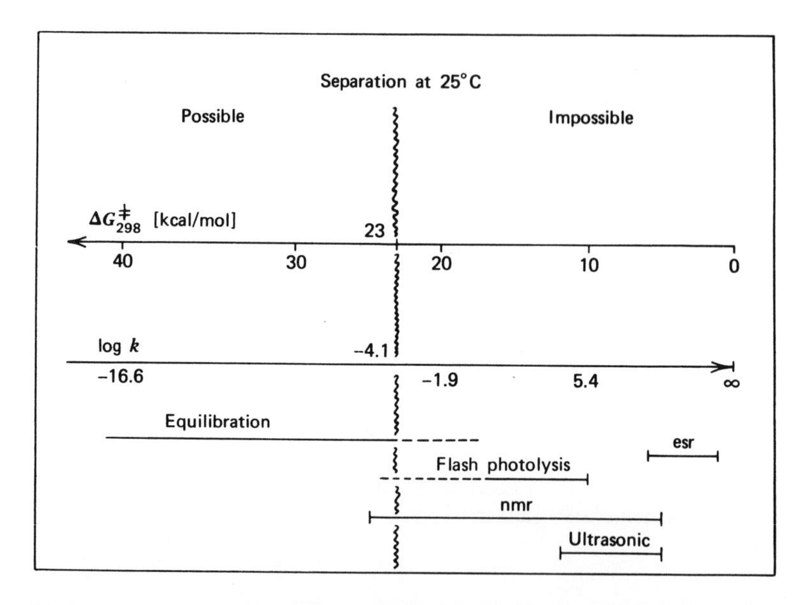

Fig. 1. Kinetic limitations of experimental methods for determination of isomerization rates.

Several methods have been used for determination of rates of double-bond isomerizations. These methods are limited both in the range of k values (see Fig. 1) and in specific applicability of each method. Direct equilibration (10,11) is possible only if we succeed in separation or enrichment of one isomer. It is therefore impossible to study topomerization (9) rates by equilibration. For dynamic nmr measurements (dnmr) (12,13), the difference in the free enthalpy between the isomers (ΔG^0) should be small or zero. Flash photolysis (14) requires $\Delta G^0 \neq 0$ and photochemical convertibility of the isomers, whereas ultrasonic absorption measurements (15) require a difference in the ground-state enthalpy ΔG^0 between Z,E isomers.

The great advantage of the newer physical methods (dnmr spectroscopy, flash photolysis, ultrasonic absorption) is that with these methods it is now possible to study fast rate processes down to barriers of about 5 kcal/mol, whereas previously, because of the necessity of separation, only "stable" isomers could be investigated.

Besides these empirical methods, there exist calculations of Z,E isomerization barriers, both *ab initio* calculations and semiempirical treatments (16-18). Generally, theoretical calculations have the disadvantage of being applicable only to small molecules and simplified model compounds. On the other hand, they give a better insight into the mechanisms of the interconversions and they are suitable for determining the whole potential energy surface.

From the temperature dependence of the k values it is possible to derive the thermodynamic parameters such as the entropy of activation $\Delta S^{\ddagger}$, the enthalpy of activation $\Delta H^{\ddagger}$, as well as the Arrhenius activation energy E_a by the Arrhenius equation (19):

$$k = A \exp\left(-\frac{E_a}{RT}\right) \qquad [2]$$

with (20)

$$A = \frac{k_B T_m}{h} \exp\left(\frac{\Delta S^{\ddagger}}{R}\right)$$

where T_m = mean absolute temperature for the range over which rate measurements are made, and

$$E_a = \Delta H^{\ddagger} + RT_m$$

The real thresholds V (which are obtained from theoretical calculations) differ in principle from the zero point activation energy E_0, but normally E_0 should be nearly equal to V (see Fig. 2). For comparison of experimental and theoretical data one

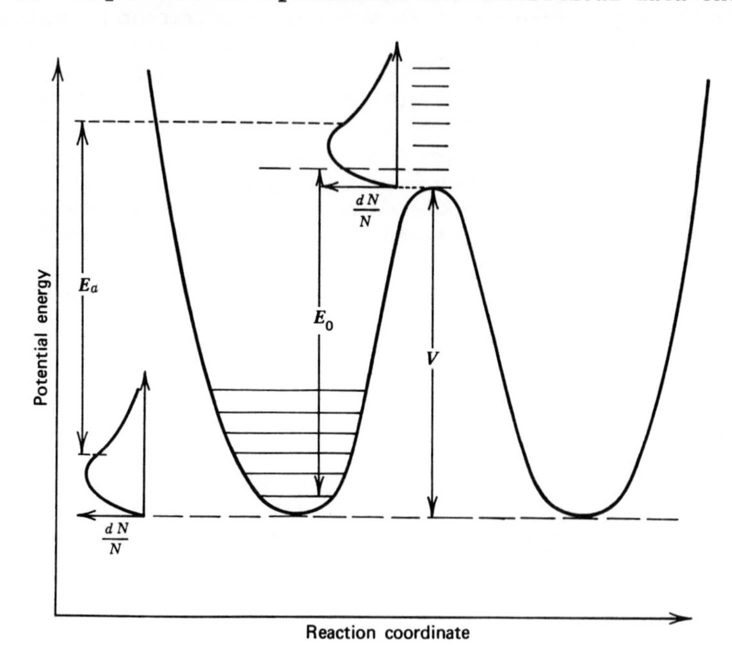

Fig. 2. Potential shape and activation energies (dN/N = population of the energy levels).

must note that the E_a or the $\Delta H^{\ddagger}$ values include also higher vibrational levels in the ground state as well as in the transition state (20,21). Very often only $\Delta G^{\ddagger}$ values at one temperature are available. The free energy of activation $\Delta G^{\ddagger}$ is temperature dependent by the $\Delta S^{\ddagger}$ part, and caution is required when comparing $\Delta G^{\ddagger}$ values. For further discussions of these problems are refs. 13 and 22.

For thermal unimolecular Z,E isomerizations one should expect low entropies of activation ("normal" preexponential factors of the Arrhenius equation, $\log A \simeq 12\text{-}13$) but it is possible that differences in solvation in the ground state and the transition state may lead to unusual $\Delta S^{\ddagger}$ values. Exact data for experimental support of this statement are not yet available.

IV. THE CARBON-CARBON DOUBLE BOND

A. Unsubstituted and Simple Ethylenes

1. *Mechanism and Barrier of Z,E Isomerization*

The olefinic C=C double bonds are the classical examples for *Z*,*E* isomerism:

Z ⇌ *E*

The calculated angular dependence of the potential energy of the electronic ground state *N*, the lowest triplet state *T*, and the lowest excited singlet state *V* are shown in Figure 3 (4,26).

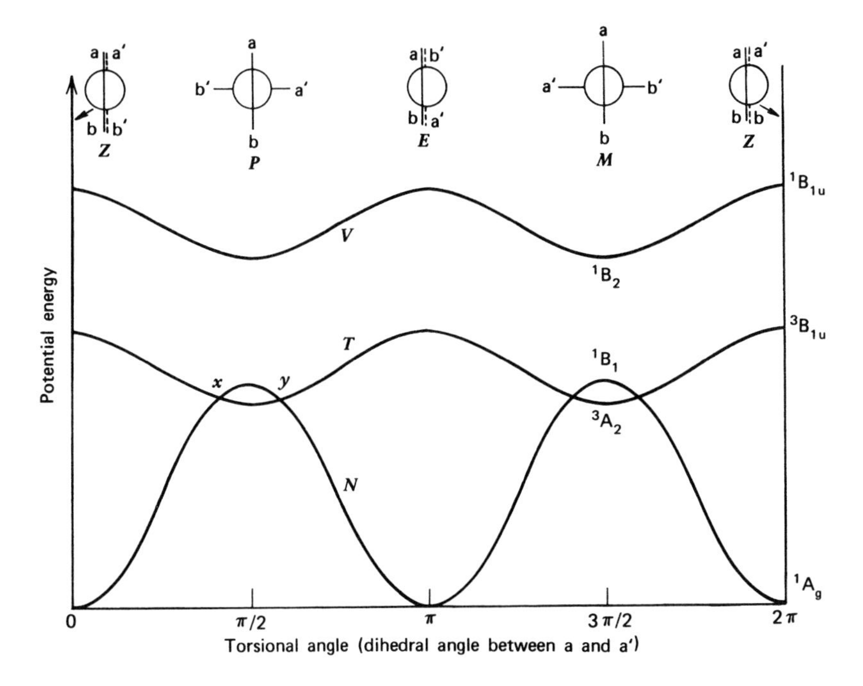

Fig. 3. Angular dependence of the *N*, *V*, and *T* states of ethylene.

Without change of the multiplicity, the only way for the isomerization $E \rightleftharpoons Z$ to occur in the ground state follows the N-state energy profile. A transition to the triplet state T at x or y (see Fig. 3) is very unlikely and would cause a low transmission coefficient κ in the Eyring equation (eq. [1]). The energy barrier of E-1,2-dideuterioethylene was found to be E_a = 65 kcal/mol with a "normal" log A value of 13 (23). The latter is consistent with the assumption that the T state is not involved in thermal isomerization of simple ethylenes (24,25). For discussions of the diradical behavior of the twisted ethylene see ref. 213.

Theoretical calculations of the ethylene barrier by the *ab initio* treatment (82.1 kcal/mol (26)), extended Hückel calculation (69 kcal/mol (27), 101.7 kcal/mol (28)), and the MINDO/2 method (54.1 kcal/mol (17)) gave differing quantitative results, but all of them show a strong fixation of the molecule in the planar form and a high stability of simple Z,E isomers as had been found experimentally long ago.

An interesting result of both the MINDO/2 and the *ab initio* SCF calculations is the nonsinusoidal behavior of the potential function for twisting in ethylene (17). In accordance with results of crystallographic measurements (29) a flatter shape for small angles of twist than the normal sine function is calculated. However, more extensive (configuration interaction) calculations showed (26b) this is a spurious result, due to the inadequacy of the SCF wave functions for nonplanar structures.

2. *Alternative Pathways for Isomerization*

Besides the uncatalyzed thermal unimolecular Z,E isomerization considered above, there are several other mechanistic possibilities for interconverting Z,E isomers. These will not be discussed in detail, although the possible routes must be recognized in order that they may be excluded as the rate-determining step.

The *photochemical* cis/trans isomerization of ethylene involves the V and/or the T state. The mechanistic pathway of the photoisomerization is sometimes quite complex (4). Normally one would assume an ($N \rightarrow V$) excitation in accordance with the Franck-Condon principle, followed by a singlet-triplet transition to the T state in the planar or twisted ethylene molecule from which the ground state Z or E (see Fig. 3) is reached. But there are several other excited states of ethylene (i.e., the Z-state and the Rydberg states) in which the rotational barrier is quite low and isomerization is possible. During the photoisomerization of ethylene, formation of hydrogen and acetylene also occurs (30). It is interesting to note that the elimination of terminal hydrogens predominates over 1,2-elimination.

$$H_2C{=}CH_2 \xrightarrow[-H_2]{h\nu} H_2C{=}C: \longrightarrow HC{\equiv}CH$$

Stepwise dissociation and recombination according to the following scheme may therefore also account for the photochemical *Z,E* isomerization of ethylene-d_2:

$$(E)\text{-}HDC{=}CDH \xrightarrow[-H\cdot]{h\nu} HDC{=}\dot{C}D \rightleftharpoons HDC{=}\dot{C}D \xrightarrow{+H\cdot} (Z)\text{-}HDC{=}CDH$$

The vinyl radical is known to have an angular structure and a low inversion barrier (about 2 kcal/mol, see below).

One might also imagine a direct thermal or photochemical dissoziation of ethylene into two carbenes followed by recombination:

$$H_2C{=}CH_2 \longrightarrow :CH_2 + :CH_2 \longrightarrow H_2C{=}CH_2$$

This process is unlikely during thermal isomerization of ethylene itself. With increasing stability of the carbenes produced from substituted ethylenes this mechanism seems more probable, but such a direct dissociation has not yet been demonstrated even for electron-rich ethylenes (31).*

Photochemical isomerizations in the presence of a complexing metal carbonyl compound have also been reported (32).

Thermal *catalytic* isomerizations of olefins are often observed. Ground-state triplet molecules (O_2, S_2, Se_2) (4,33), nitrous oxides (34), and iodine (35) are effective catalysts. Both a mechanism which involves the triplet state and one involving a chemical mechanism which involves the triplet state and one involving a chemical intermediate have been proposed to account for the catalytic effect. The catalytic effect of acids for the isomerization of polarized ethylenes is mentioned in Sect. IV-B-4.

One should bear in mind that isomerization rates in substi-

*But the corresponding phenomenon is known in the dissociation of the N=N double bond in the dimeric nitroso compounds (see below).

tuted ethylenes are often investigated in solutions. Special solvent effects on barriers and entropies of activation must therefore be taken into account. On the other hand, theoretical calculations refer to the free molecule in the gas phase. One normally postulates that in very dilute solutions the properties of the molecules approximate those in the gas phase, if no specific solvent catalysis is observed.

B. Substituted Ethylenes

1. *General Remarks*

The rotational barrier about the double bond of ethylene itself is the highest one known. Substituents generally lower this barrier. As a first approximation, we may divide the effects on double-bond rotation as follows:

1. Steric effects
2. Electronic effects
   a. Stabilization of biradicals
   b. Polarization of the double bond

Of course this separation of "effects" is suspect from the theoretical point of view and is sometimes quite arbitrary (36), but it allows one to understand the effects of substituents on double-bond rotation and, in practice, it allows predictions of barriers to *Z,E* isomerization.

Under certain conditions a combination of these "effects" may lower the rotational barrier so markedly that fast thermal rotation occurs at room temperature, or even that a near-perpendicular conformation *P* or *M** is the ground state of the molecule and the planar ethylene represents the transition state of the enantiomerization or enantiotopomerization (rotation about the "C=C double bond").

2. *Steric Effects*

The steric hindrance of substituents in *Z* positions will lead to a destabilization of the planar ground state (GS) with respect to the perpendicular transition state (TS). This was proved in the case of the polarized ethylenes *1* (37) and *2* (38). With increasing bulkiness of the group R the energy barrier decreases (Table 1). Similar effects have been found for rotation about C=N bonds (see below).

*The notation of the chiral transition states as *P* or *M* follows the Cahn-Ingold-Prelog sequence rules (5).

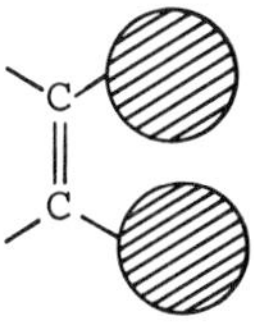

GS 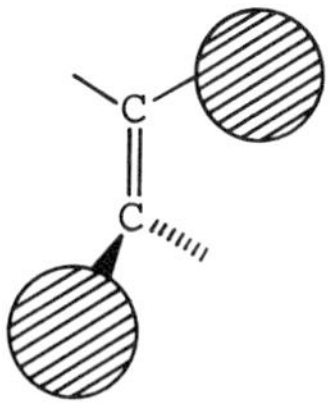 TS 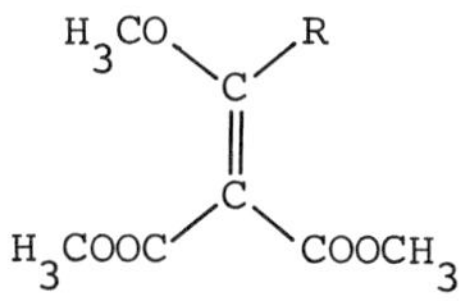 *1* 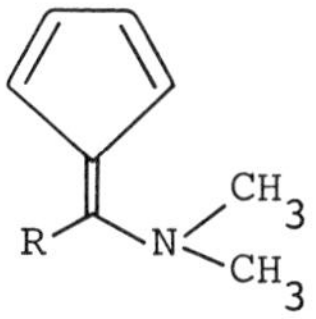 *2*

In *o*-substituted stilbenes (*3*) one observes the reverse steric effect. With increasing bulkiness of substituents in the ortho positions of the benzene ring the C=C double-bond rotation becomes more and more hindered (39). One of the dimethylamino groups has to pass the substituent R on rotation, and this becomes more difficult, the larger the groups are.

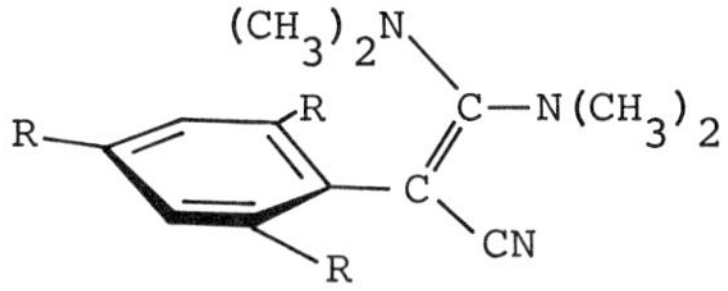

3

Interesting properties result from very large steric effects in bisfluorenylidenes *4*, and in similar compounds of the structural type *5*. Because of considerable overlap of the

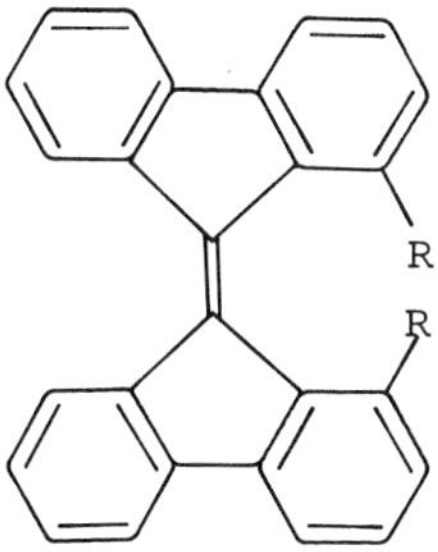

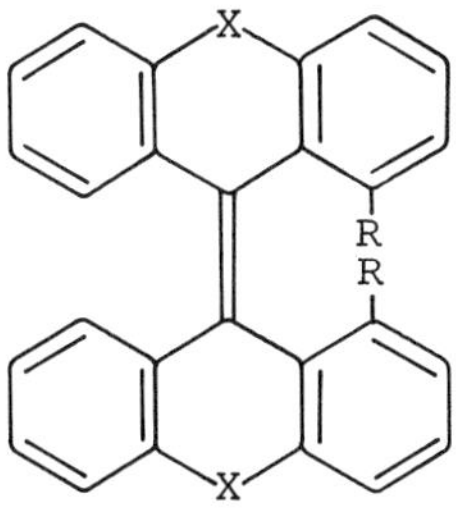

X = O, CH_2, CO

4 *5*

van der Waals radii, complete coplanarity of these molecules is not possible (40). A folded and a twisted structure of the double bond have been discussed (41). Each displays *Z*,*E* isomerism leading to four diastereoisomers, each of which is chiral. Therefore four enantiomeric pairs can exist in these compounds;

Table 1 Steric Effect on Double-Bond Rotation (37,38)

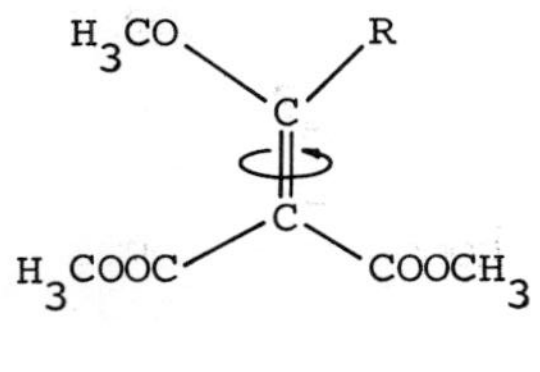

	1				2		
	R	Solvent	T_c °C	$\Delta G^\ddagger$ (37), kcal/mol	Solvent	T_c °C	$\Delta G^\ddagger$ (38), kcal/mol
a	H	Hexachlorbutadiene	> 206	~ 27.7	$(CH_3)_2SO$	148	22.1
b	CH_3		191	25.7	$(CH_3)_2SO$	38	16.4[a]
c	C_2H_5		162	24.7			
d	$i\text{-}C_3H_7$		152	23.3			
e	$t\text{-}C_4H_9$	Bromobenzene	61	18.3			

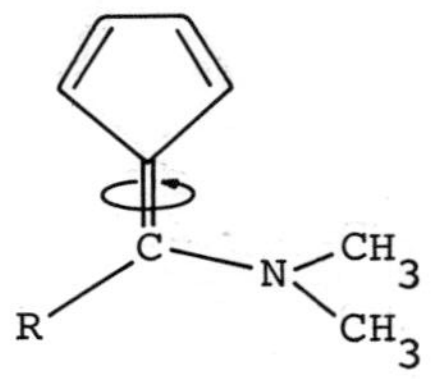

[a]The large effect of the C-methyl group ($\Delta\Delta G^\ddagger_{2a \rightarrow 2b}$ 5.7 kcal/mol) can be separated into a steric effect (about 2.8 kcal/mol) and an electronic effect on stabilization of the carbonium center (about 2.9 kcal/mol) (38).

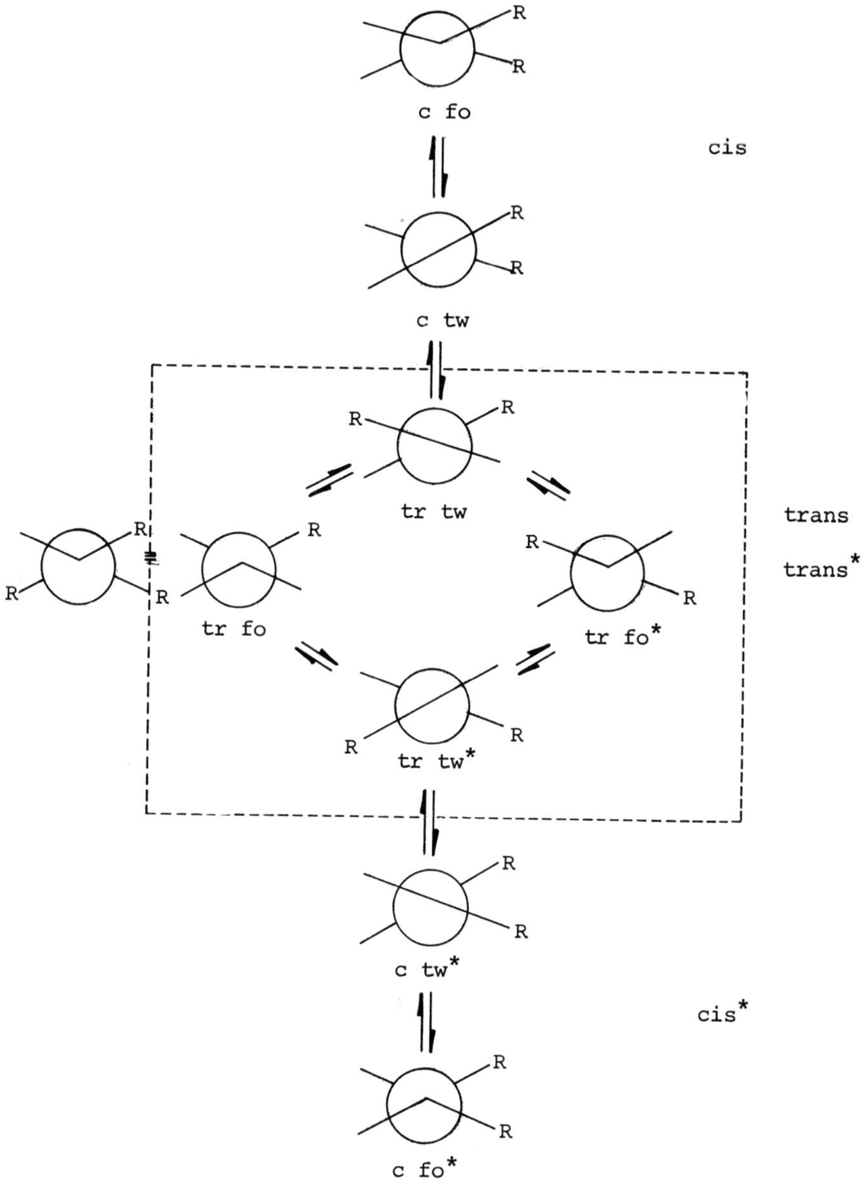
R
R
c fo
cis
R
R
c tw
R
R
tr tw
trans
trans*
R
R
R
R
R
R
tr fo
tr fo*
R
R
tr tw*
R
R
c tw*
cis*
R
R
c fo*

Scheme 1. Conformations of sterically crowded ethylenes (the conformations marked with an asterisk are enantiomeric to the unmarked ones).

they are shown in Scheme 1.

Two exchange processes have been found by dnmr studies: enantiomerization (by "labeling" the molecule with a prochiral R group) and diastereomerization. These processes may (41) or may not (42) have similar energies. It is not yet clear if the observed diastereomerization represents a *Z,E* isomerization (41) or the interconversion between twisted and folded ethylenes of the *E* conformation (42). The enantiomerization is interpreted as the exchange between the two chiral twisted molecules. It will be interesting to see the results of further studies on these compounds (5) with respect to their thermochromic and piezochromic properties (43).

3. *Electronic Effects*

a. Symmetric Ethylenes. In earlier publications, the low barriers for *Z,E* isomerization of butene-2, dimethyl maleate, and other compounds in connection with low frequency factors of the Arrhenius equation (eq. [2]) have been explained by the involvement of the triplet state (Fig. 2) (44). However, critically refined measurements have shown "normal" frequency factors (log A = 11-14) and the triplet mechanism can be ruled out for the thermal uncatalyzed isomerization (4a,24). It is noteworthy that the isomerization becomes the faster, the more stable the corresponding biradical of the olefin (compare compounds *6*-*9*).

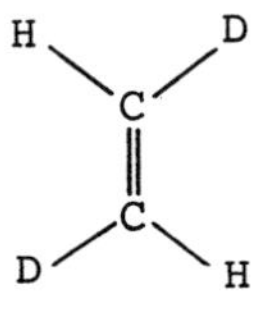

H D C C D H   H C_6H_5 C C H C_6H_5 O C_6H_5 O C_6H_5

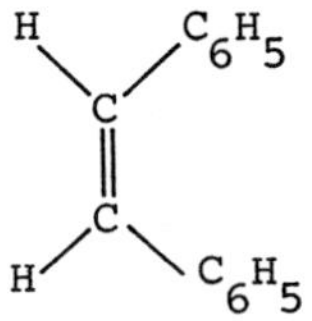

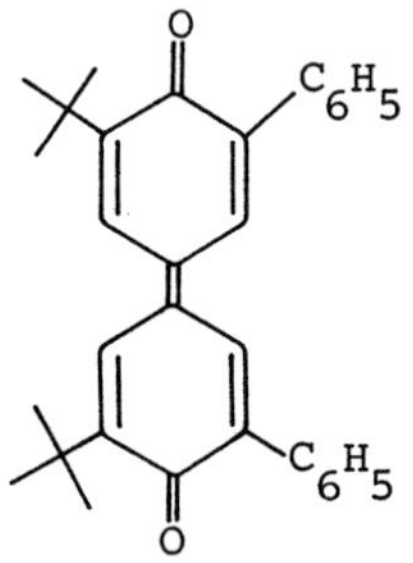

	6	*7*	*8a*
E_a, kcal/mol	65.0	42.8	21.1
log A	13	12.8	11.4
Ref.	23	44	45

One might imagine this biradical to be the triplet state of the olefin, but this cannot be true because even in the dipheno-

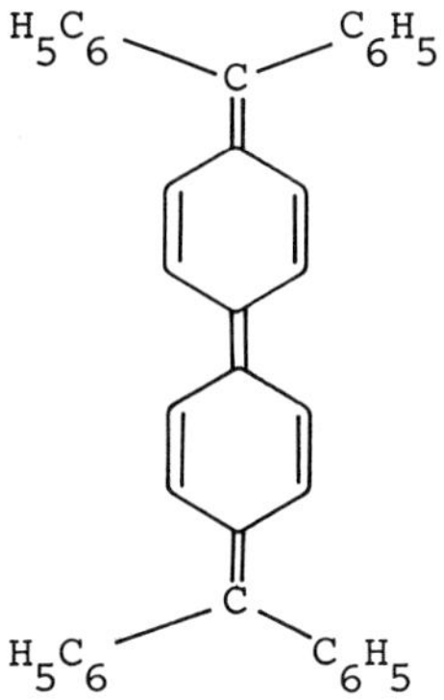

	9a	*9b*
E_a, kcal/mol	small*	strong negative
Ref.		47

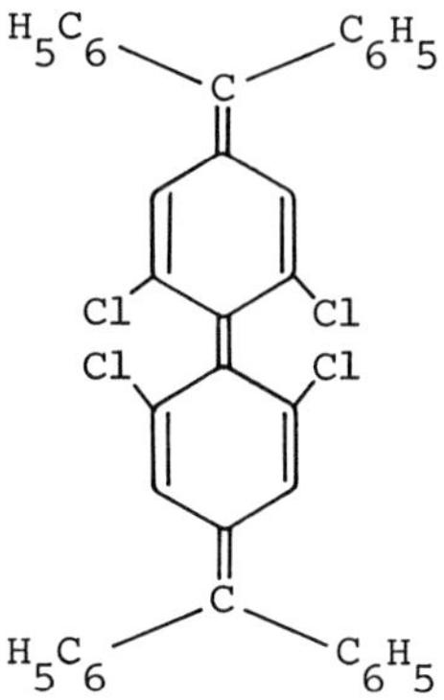

quinone *8a* a normal log *A* value of 11.4 was found experimentally (45). Therefore one must conclude that the low barrier results from the lowering in bond energy. If we combine this "radical" stabilization with steric effects, the perpendicular form is the most stable form, as has been shown for compound *9b* (47).

The MINDO/2 calculations of the barrier heights of cumulenes show a decreasing barrier with an increasing number of calculated double bonds (17) (Table 2). The experimental results (48) for certain substituted cumulenes agree well with these calculations.

Intramolecular motions of annulenes in which double-bond rotations may be involved are discussed in ref. 49.

b. Polarized Ethylenes. Substitution of one carbon atom of the double bond with electron-donating groups and of the other with electron-withdrawing groups diminishes the double-bond order by charge separation.

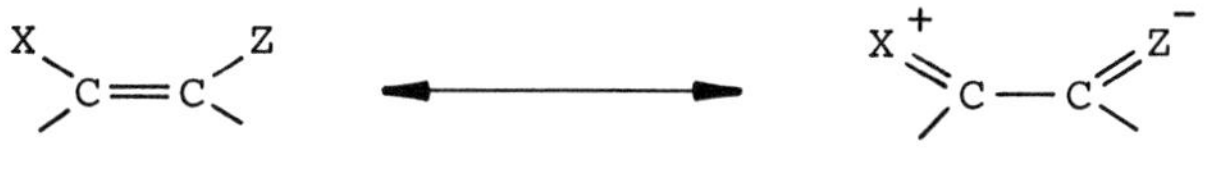

Several selected examples are shown in Table 3. Comparing the thresholds, one can order the substituents with increasing

*The low-lying triplet state ($\Delta G^o_{S\text{-}T}$ ≃5 kcal/mol (46)) will probably allow a fast rotation.

Table 2 Rotational Barriers in Cumulenes

Calculated MINDO/2 (17)			Observed			
			No.		Barrier, kcal/mol	Ref.
$H_2C(=C)_nH_2$	n = 1	54.1	*6*	DHC=CHD	65.0	23
	n = 2	35.2				
	n = 3	32.3	*10*	$(H_5C_6)(H_9C_4\text{-}t)C=C=C=C(t\text{-}C_4H_9)(C_6H_5)$	30	48
	n = 4	26.1				
	n = 5	22.7	*11*	$(H_5C_6)(H_9C_4\text{-}t)C=C=C=C=C=C(t\text{-}C_4H_9)(C_6H_5)$	20	48

Table 3 Selected Examples of Ethylenes with Low Rotational Barriers Determined by nmr Spectroscopy

Formula structure type	No.	Solvent[a]	T_c °C	$\Delta G^{\ddagger}$, kcal/mol	Other data	Ref.
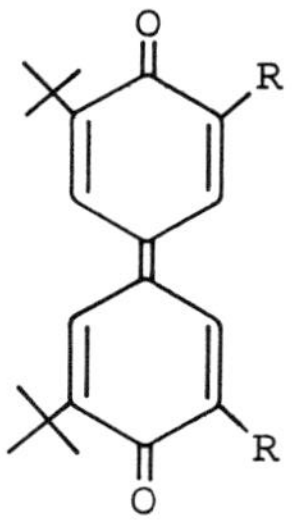	*8a*: R = C_6H_5	TCB	151	24.2	E_a = 21.1 kcal/mol	45
		$CDCl_3$	37	23.2[b]	log A = 11.4	
					ΔG^o = 0.1 kcal/mol	
	8b: R = OCH_3	TCE	113	22.0[c]	ΔG^o = 0.18 kcal/mol	51
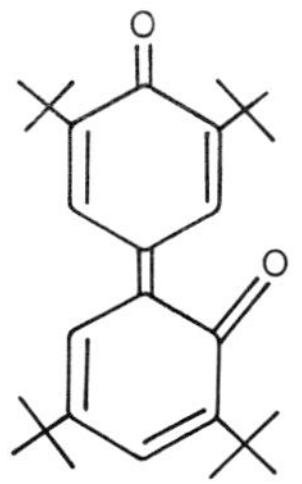	*12a*	Decalin	155	21.0		50

Table 3 Continued

Formula structure type	No.	Solvent[a]	T_c °C	$\Delta G^{\ddagger}$, kcal/mol	Other data	Ref.
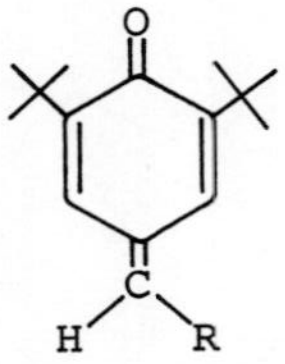	*12b*: R = 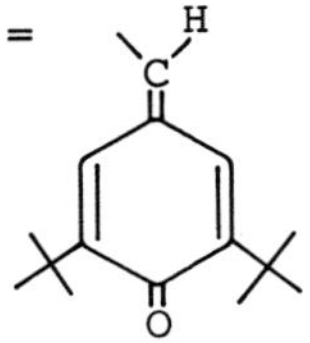	TCB	192	25.8		52
	12c: R =	Toluene	95	19.9[e]		52
		Decalin	135	20.0[f]		52
	12d: R = 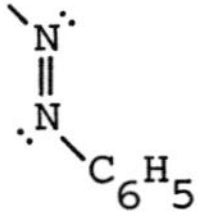	TCB	180	>25.6		52
	12e: R = OCH_3	TCB	180	>26		52

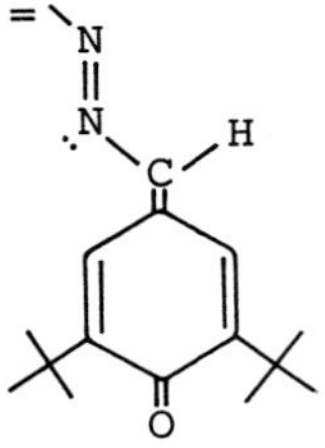

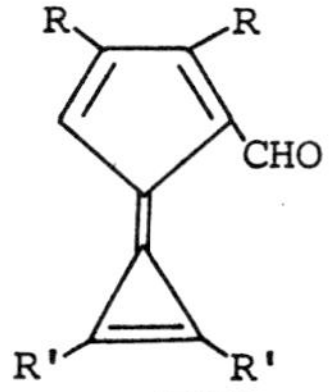

R, R, CHO, R', R'	*13*: R = $COOCH_3$; R' = *n*-C_3H_7	DCB	105	19.2		53
N-CH_3	*14*	Ac	100	19.6		54
N-CH_3	*15*	Ac	-50	11.4		54
H_3COOC, $COOCH_3$, H_3C, O, CH_3	*16*	$CDCl_3$	66	16.5		55

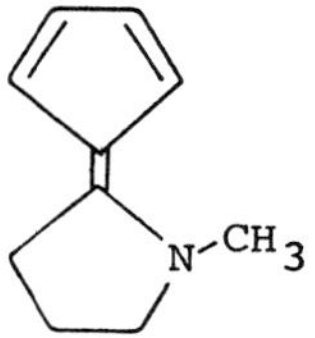

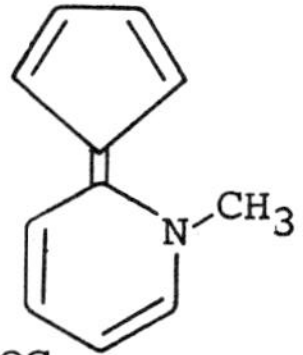

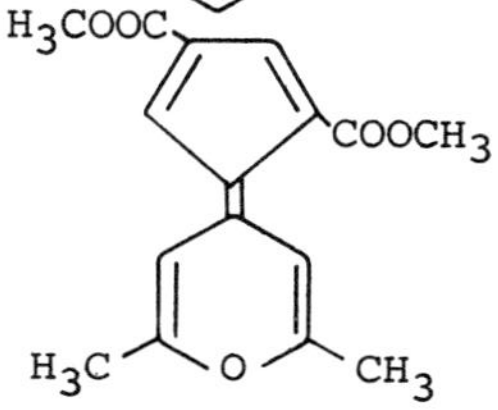

Table 3 Continued

Formula structure type	No.	Solvent[a]	T_c °C	$\Delta G^{\ddagger}$, kcal/mol	Other data	Ref.
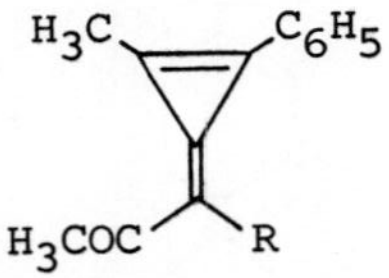	*17a*: R = $COCH_3$	NB	87	19.0		56
	17b: R = COC_6H_5	NB	130	22.6[d]		56
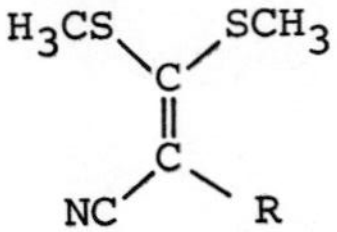	*18a*: R = $C_6H_4NO_2$*p*	DCB	>190	>25		57
	18b: R = $COOCH_3$	DCB	185	24.5		57
	18c: R = COC_6H_5	DCB	126	20.6		57
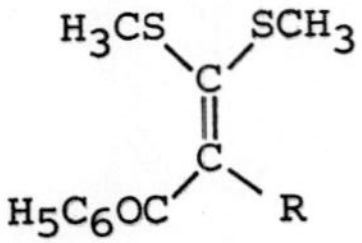	*19a*: R = $COOC_2H_5$	DCB	100	19.2		57
	19b: R = $COCH_3$	DCB	78	17.5		57
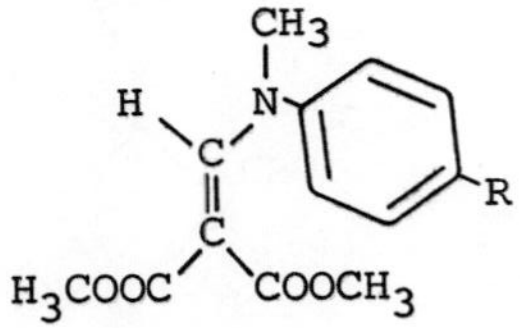	*20a*: R = OCH_3	$CHBr_3$	77	18.5		58
	20b: R = H	$CHBr_3$	90	19.4		58
	20c: R = NO_2	$CHBr_3$	137	22.1		58

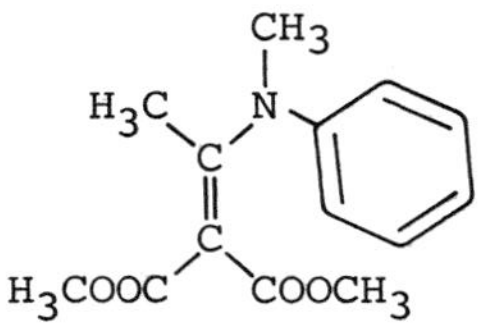 *20d*		Ac	-76	10.0	58
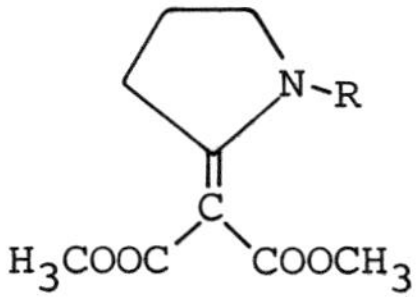 *20e*:	R = CH_3	Ac	-81	9.8	58
20f:	R = $CH_2C_6H_5$	Ac	-71	10.0	58
20g:	R = C_6H_5	Ac	5	13.7	58
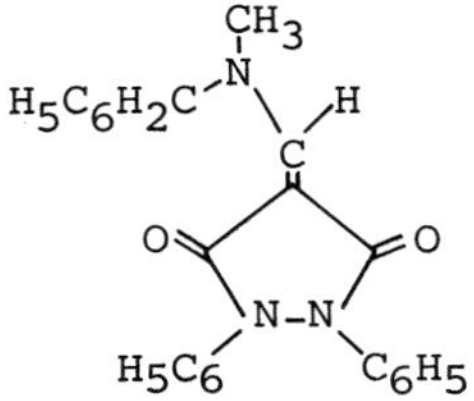 *21*		$CHCl_2$-$CHCl_2$	70	19.2	59

Table 3 Continued

Formula structure type	No.	Solvent[a]	T_c °C	$\Delta G^{\ddagger}$, kcal/mol	Other data	Ref.
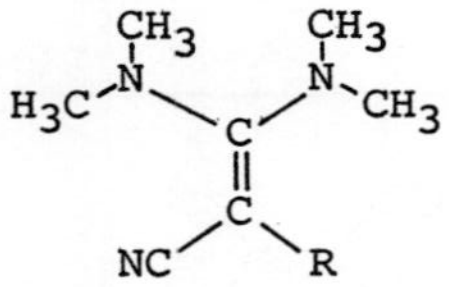 $[(CH_3)_2N]_2C{=}C(NC)R$	*22*: R = COC_6H_5	CHClF-CHClF	21	15.3		60
	23a: R  = $-C_6H_4-H$	TCB	138	21.0		39
	23b: R  = $-C_6H_4-F$	TCB	139	21.1		39
	23c: R 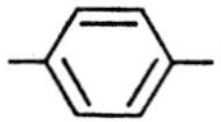 = $-C_6H_4-Cl$	TCB	121	20.1		39
	23d: R  = $-C_6H_4-Br$	TCB	122	20.2		39
	23e: R  = $-C_6H_4-COOCH_3$	TCB	83	18.2		39
	23f: R  = $-C_6H_4-COCH_3$	TCB	77	17.9		39
	23g: R  = $-C_6H_4-CN$	TCB	72	17.6		39
	23h: R  = $-C_6H_4-NO_2$	TCB	45	16.2		39

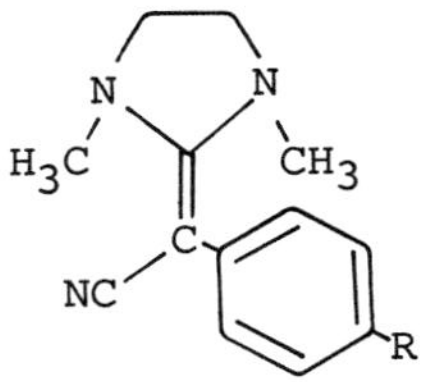

	24a: R = H	9.8	61
	24b: R = Br	9.8	62

[a]TCB = trichlorobenzene; TCE = tetrachloroethylene; DCB = *o*-dichlorobenzene; Ac = acetone-d_6; NB = nitrobenzene.
[b]By equilibration.
[c]*Z* (more stable) to *E*.
[d]More stable → less stable.
[e]*tert*.-Butyl signals used.
[f]Quinonoid protons used.

withdrawing efficiency as follows:

$$\mathrm{H, C_6H_5 < CN < COOR < COC_6H_5 < COCH_3}$$

The electron-releasing effect decreases in the order:

$$\mathrm{R_2N > RO > RS > Alkyl}$$

This relative order depends on the specific possibilities of conjugation. Thus, due to steric hindrance of the N-methyl groups in *23*, its conjugative ability is much less than that in the cyclic compounds *24* in which the nitrogens are in the optimum situation for conjugation with the exocyclic double bond. Sometimes such pictures are too simple for a complete representation of all facts. Thus, in contrast to the experimental observations, one would have anticipated a smaller barrier for *20g* than for *20d* (see Table 3). On the other hand, solvent effects and effects of concentration are frequently not studied carefully, and some individual measurements are uncertain. We refer in this article only to general trends.

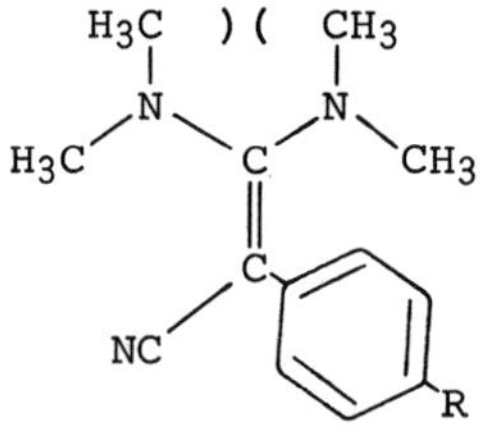

23 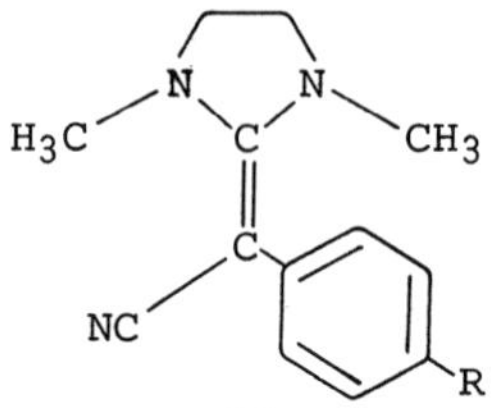 *24*

The linear Hammett correlation of $\Delta G^{\ddagger}$ for double-bond rotation in aryl-substituted ketene aminals *23* (39) clearly gives evidence for the importance of polarization effects.

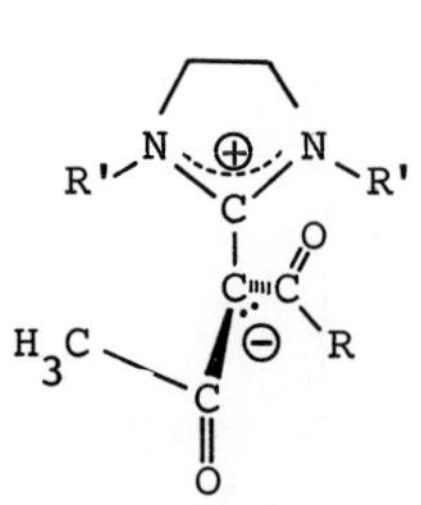

25 GS

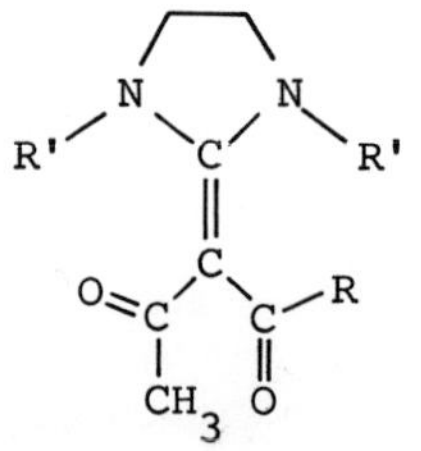

25 TS R' = CH_3, $CH_2C_6H_5$
R = CH_3, C_6H_5

The combination of the very strong electron-donating ability of the imidazolidine system and the strong electron-attracting acyl groups gives rise to olefins for which twisted ground states are to be expected. The competition between steric strain ($\Delta G_2^{\ddagger}$) and conjugation ($\Delta G_1^{\ddagger}$) may be compared with the case of the bisfluorenylidenes (see above). In contrast to the latter compounds the conjugation effects should be small in *25*. The proposed energy profile is shown in Figure 4.

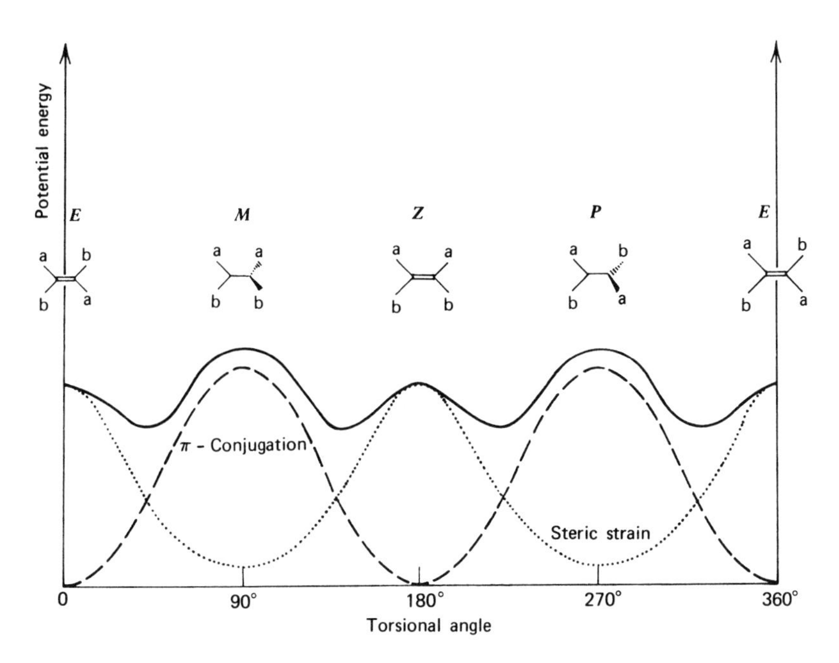

Fig. 4. Energy profile of the "double-bond" rotation in *25*.

The observed enantiomerization process (61) (magnetic nonequivalence in prochiral groups) represents the interconversion of the type $t \rightleftharpoons t^*$, whereas the Z,E isomerization in *25* is fast (Scheme 1). The exocyclic double bonds of nonalternate systems such as fulvenes often have a low bond order because of a partial dipolar ground state of the molecule. Even in the case where dipolar character in the ground state is not high, the dipolar transition state of the double-bond rotation is energetically favored (see *26*–*28*):

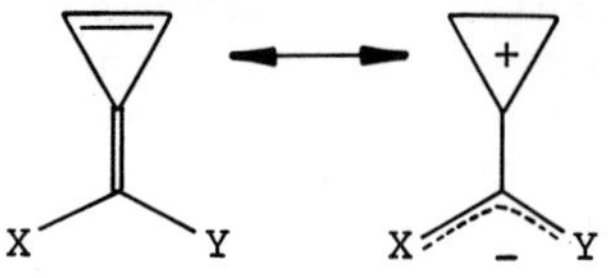

26

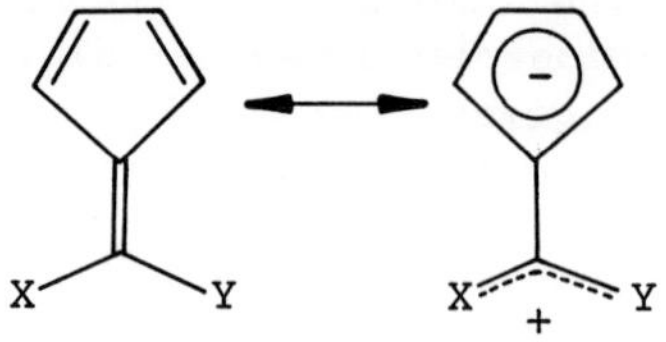

27

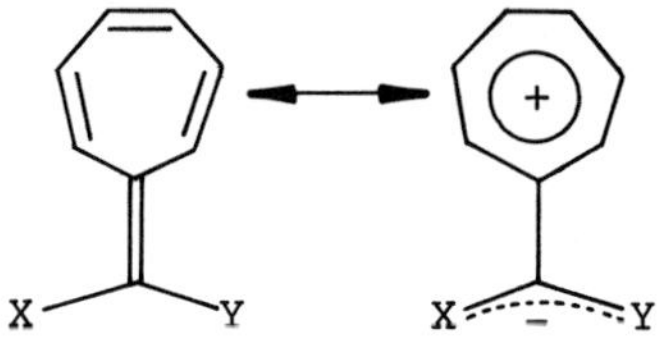

28

This was first shown in the calicene derivative *13* (53) in which the three-membered ring represents an electron-releasing group and the five-membered ring an electron-attracting substituent.

4. *Solvent Effects on Carbon-Carbon Double-Bond Rotations*

Normally Z,E isomerizations of substituted ethylenes are measured in solution. Solvation of the ethylenes in the ground state and in the transition state may be different; the effect of solvents therefore play an important role in the interpretation of the measured rates. Remarkable solvent effects have been observed for Z,E topomerization of polarized ethylenes (39,53,55). In accordance with expectations, a polar solvent increases the rate by stabilization of the dipolar transition state (Table 4). A linear Hammett correlation of barriers in *23d* and *23e* with E_T (64) values has been found recently (212). Protic solvents, on the other hand, may also have large effects on isomerization rates. Traces of acids or water in the solvents catalyze the isomerization of enamines (63) and keteneaminals (39) *29*:

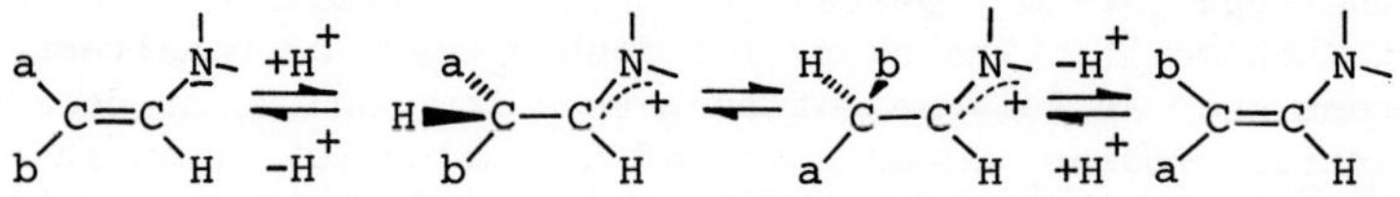

29

Table 4 Solvent Effects on Carbon-Carbon Double-Bond Rotation in 1,1-Bis(dimethylamino)-2-cyano-2-(*p*-bromophenyl)-ethylene (*23d*)

Solvent	Dielectric Constant	T, °C	$\Delta G^{\ddagger}$, kcal/mol
1,2,4-Trichlorobenzene	4.0	122	20.2
Pyridine	12.5	91	18.6
Nitrobenzene	36.1	89	18.5
Acetonitrile-d_3	38.8	75	17.9
Dimethyl sulfoxide-d_6	48.9	65	17.4

Therefore caution is necessary in the interpretation of measured rates in terms of a thermal uncatalyzed process.

A quite different effect has been found in the stilbazolium salts *30*, in which there is a positive charge in the molecule. Only a shift of this charge is required to produce double-bond rotation. Therefore, no influence of solvent polarity (represented as E_T values (64)) has been found (65). Protic solvents, on the other hand, decrease the thermal Z,E isomerization rate.

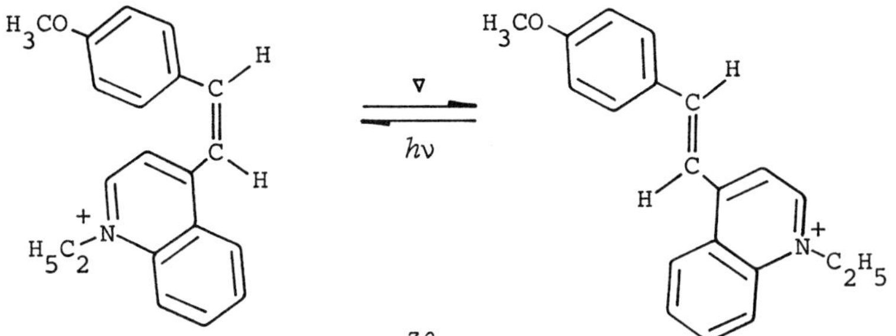

H_3CO H C C H N^+ H_5C_2 ⇌ (∇ / $h\nu$) H_3CO H C C H N^+ C_2H_5

30

C. Partial Carbon-Carbon Double Bonds

Normally one considers a bond as a double bond, if it is represented in the structural formula without charge separation. Following this convention, we described the exocyclic C-C bond in the cyclic keteneaminal *25* as a double bond, although we know that the twisted polar form in fact represents the energy minimum. On the other hand, formal single bonds often have considerable double-bond character due to polarization of the molecule. In such cases *Z,E* isomerization is often observed by spectroscopic methods (nmr spectroscopy, ultrasonic measurements (66). Examples of this type are discussed in refs. 13 and 67.

Other cases that cannot be classified unequivocally as single bonds or double bonds are represented by the allylic system (cation (*31*), annion (*32*) or radical (*33*)). Some barriers in

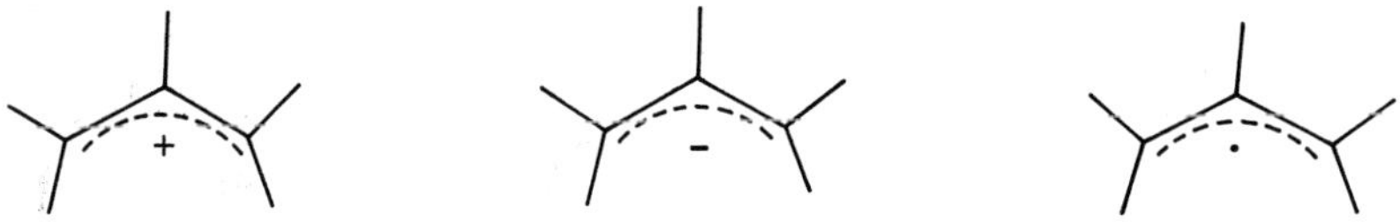

31 *32* *33*

allylic cations are collected in Table 5.

The barrier heights amount to about 12-20 kcal/mol for a number of methyl-substituted allyl cations. The steric hindrance of the methyl groups destabilizes the coplanar ground state, which can easily be seen by comparison of compounds *31a* and *31b* or *31e* and *31f*. Therefore one must expect a much higher barrier for the rotation in the simple allyl cation. This is in agreement with theoretical *ab initio* calculations which give a barrier of 32 kcal/mole for the C_3H_5 cation (70).

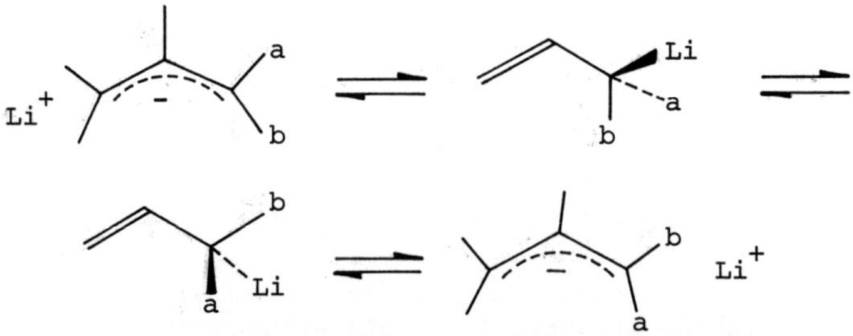

Allylic anions (71) and allylic radicals (72) also resist rotation about their partial double bond, as might be expected from simple Hückel theory, which indicates considerable double-bond order.

Generally the barriers in allylic species are high, although

Table 5 Rotational Barriers in Some Allylic Cations

No.	Compound	T_c °C	$\Delta G^{\ddagger}$, kcal/mol	Other Data	Ref.
31a	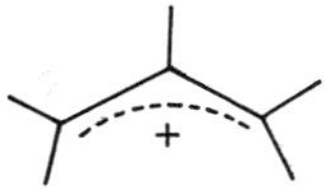	-11	13.8 ± 1		68
31b		55	15–17		69
31c	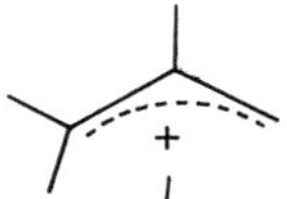	-49	11.7 ± 1[a]		68
31d	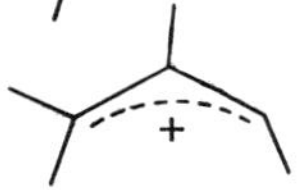	22	15.8 ± 1		68
31e				E_a = 17.5 kcal/mol log A = 11.8	70
31f				E_a = 24.0 kcal/mol log A = 14.0	70
31g	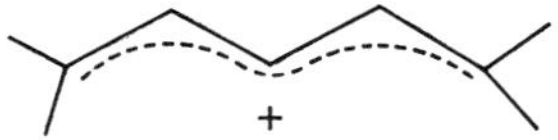	95	19		69

[a]The rotational barrier about the 1,2 bond of the 1,1,2 Arimethyl allyl cation was measured. The 2,3 barrier should be much higher.

in some cases low barriers for allylic anions (in allyllithium compounds) have been found (73). The question in these cases arises as to the extent to which these allyllithium compounds are really dissociated, and whether the observed process should be viewed as a rotation about the partial double bond or about the single bond of the covalent species.

D. Inversions at Carbon-Carbon Double Bonds

A double bond with three ligands (vinyl species) may have different molecular shapes: planar C_{2v}, planar C_s, or nonplanar C_s'. The ground-state conformation is easy to understand on the basis of simple MO considerations, which correlate the conformation and the number of valence electrons occupying the MO system (74) (Fig. 5). The potential energy of the $2b_2$-$7a'$ orbital is strongly dependent on the geometry; occupancy of this orbital will lead to the C_s shape.

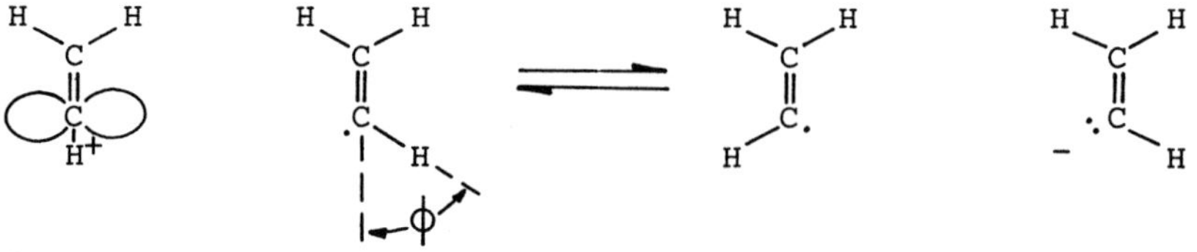

34 *35* *36*

The vinyl cation *34* (highest occupied orbital: $1b_1$) from this consideration should have the planar C_{2v} structure which is in accordance with chemical information (75,76) and *ab initio* calculations (77,78). In contrast, the vinyl radical *35* shows an energy minimum for the planar molecule with an angle of Φ = 49.2° (78). The barrier of inversion was determined to be 2 kcal/mol by esr measurements, (79,80) which should be compared to the value 4.3 kcal/mol calculated by the semiempirical MINDO method (81). Because of the small barrier and the small mass of the hydrogen atom, the latter need not pass the energy maximum, but tunneling (79) may be the rate-determining step. That the experimentally found rate is faster than that calculated can therefore be explained by tunneling. The same is not true for heavier atoms or groups such as alkyl groups, etc. The experimental barriers correspond to *inversion* across the saddle point in the energy surface (C_{2v} symmetry of the TS). The *rotation* about the C=C double bond is expected to have a much higher barrier in ethylene itself.

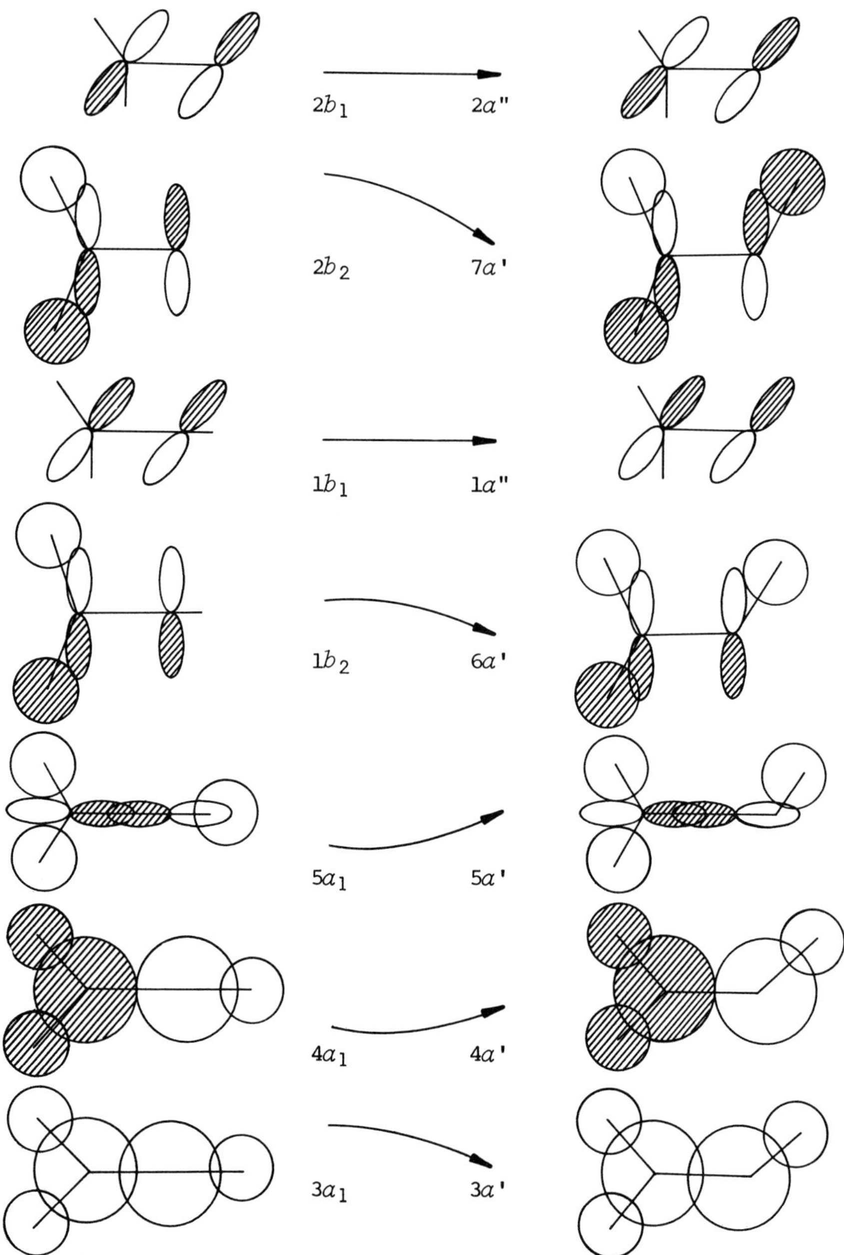

Fig. 5. Molecular orbital pictures and qualitative correlation diagram for planar H_2ABH molecules (74).

The influence of substituents Z on Z,E topomerization barriers is to be compared with the substituent effect on planar inversion of the nitrogen in imines (Chap. V); the barriers decrease in the order (82-84):

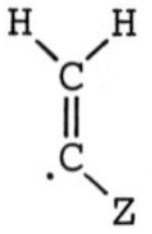

H H
C
‖
.C
Z

α-halovinyl radicals >α-alkylvinyl radicals >α-arylvinyl radicals

The α-arylvinyl radicals are very rapidly inverting species or have linear (C_{2v}) symmetry in the ground state (84).

The inversion barriers of vinyl anions *36* are much higher than those of vinyl radicals (85). The results of the MINDO calculation are shown in Table 6 (81). The stability of the vinyl species might be compared to the stability of the cyclopropyl analogues (see also the inversion on N (13,21)).

The configurational stability of vinyl anions has long been established experimentally (85). Normally, stereochemistry is retained if such species are intermediates in chemical reactions. Again, higher isomerization rates of α-arylvinyl carbanions than for α-alkylvinyl carbanions have been found.

Table 6 Inversion Barriers at Carbon (MINDO) (81)

Radicals		Anions	
Compound	V, kcal/mol	Compound	V, kcal/mol
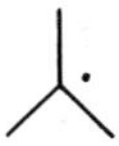 ·	-4.9	 – :	20.2
 C ‖ .C	4.3	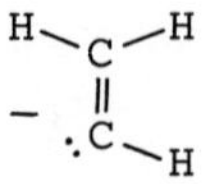 H–C–H ‖ – :C–H	31.1
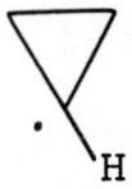 · H	4.8	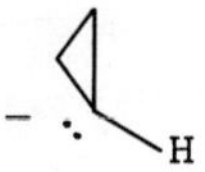 – : H	36.6

V. THE CARBON-NITROGEN DOUBLE BOND

A. *Z,E* Isomerism of the Immonium Double Bond

Z,E isomerism is to be expected for azaethylenes, i.e., structures with an immonium double bond. Many of the immonium

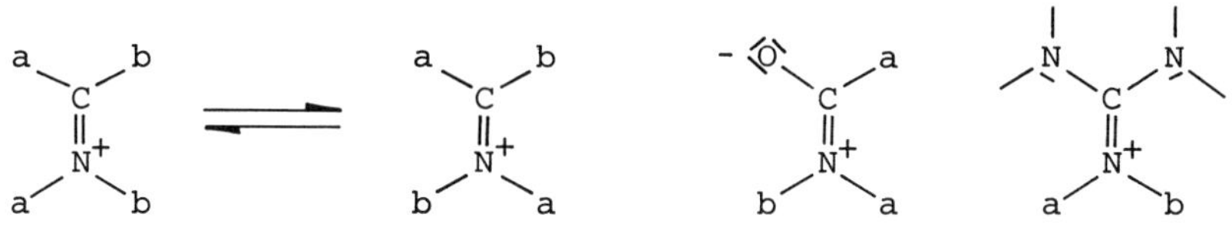

Z *E*

37 *38* *39*

C=N bonds so far studied represent "partial" double bonds. However, the differentiation between "partial" and "pure" double bonds is, as already mentioned, rather arbitrary. As a result of numerous nmr investigations in the last few years, partial C=N double bonds are among the most extensively studied types of bonds. The rotational barriers in amides *38*, with a double-bond order of about 40% (86), amount to 17-24 kcal/mol, depending on the nature of the substituents. Amides as well as derivatives have been extensively reviewed elsewhere (13,87), and will therefore not be discussed here.

The double-bond order in guanidinium salts *39* is expected to be less than about 33%, and rotational barriers of 10-20 kcal/mol have been determined for these compounds (88,89). Extrapolation to the "pure" $C=N^+$ double bond leads to a barrier of the order of 45 kcal/mol or more. This is in agreement with the findings of dnmr studies of *C,C*-dialkylimmonium salts (89,90) in which no rotation about the double bond was found. In spite of this high resistance to rotation, to our knowledge a successful separation of simple immonium salts has not yet been accomplished, although the opposite is true for the more easily rotating cyclopropenimmonium salts (90).

We may draw the following general conclusions about C=N rotation, conclusions which correspond closely to those relating to C=C double-bond rotation:

1. Increasing the size of the ligands lowers the barrier of C=N rotation. Examples of this have been shown for amides (13) and guanidinium salts (89).
2. Large groups in ortho positions to *N*-aryl or *C*-aryl

Table 7 Hammett Correlations for Rotation About Partial C=N Double Bonds

Formula structure type	No.	T_c, °C	$\Delta G^{\ddagger}$ kcal/mol	ρ	Refs.
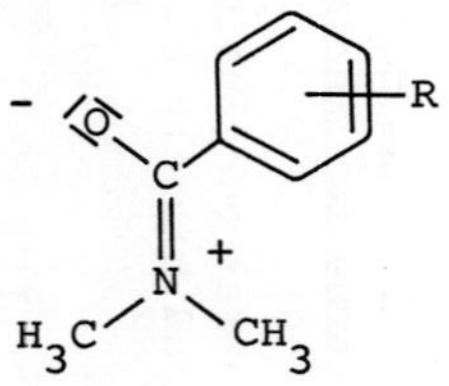 $^{-}$O–C(C$_6$H$_4$R)=N^{+}(CH$_3$)$_2$	*40*	25	15.5 R = H	negative	93
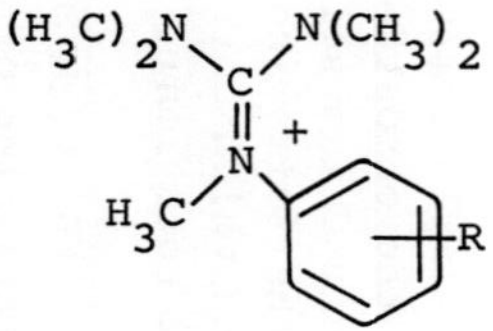 ((H$_3$C)$_2$N)$_2$C=N^{+}(CH$_3$)(C$_6$H$_4$R)	*41*	-46	11.1 R = H	+1.09	89
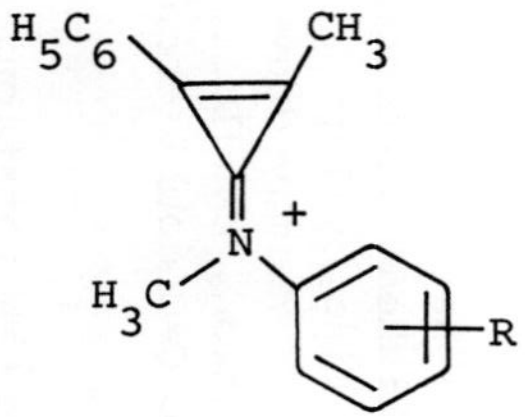 2-H$_5$C$_6$-3-CH$_3$-cyclopropenylidene=N^{+}(CH$_3$)(C$_6$H$_4$R)	*42a*			+1.14	90,94

Structure	Compound				Ref.
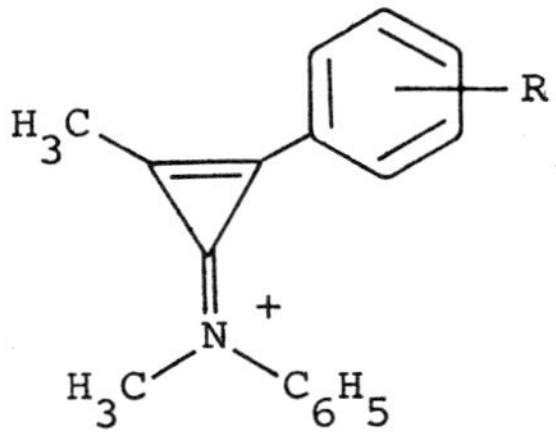 H_3C, aryl(R)-cyclopropenylidene=$N^+(CH_3)(C_6H_5)$	*42b*			-1.03	90,94
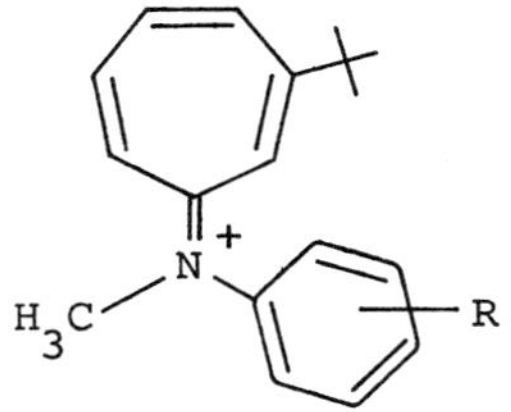 tert-butyl-cycloheptatrienylidene=$N^+(CH_3)$(aryl-R)	*43*	40	16.0 R = H	+1.80	95

substituted C=N bonds stabilize the C=N bond against rotation. This is true for amides (91,92) as well as guanidinium salts (88,89).

3. Electron-withdrawing substituents facilitate $C{=}N^+$ rotation when attached to nitrogen but inhibit it when attached to the immonium carbon, and vice versa for electron-donating groups.

Hammett correlations have been found for benzamides (93), *40*, guanidinium salts (89), *41*, cyclopropenimmonium salts (90,94), *42*, and tropenylidenimmonium salts (95), *43* (Table 7). Some more recent nmr investigations of partial C=N double-bond rotation are listed in ref. 96.

B. The Carbon-Nitrogen Double Bond of Imines

1. *Mechanistic Possibilities for Interconversion*

The syn-anti isomerization of imines which are isoelectronic with vinyl carbanions may occur by several different mechanistic pathways. First of all, we have to consider the rotation about the C=N double bond made possible by polarization, whereby a partial negative charge accumulation on the nitrogen (*44'*) or the carbon (*44"*) results. A biradical transition state is unlikely because of the different electronegativities of carbon and nitrogen, and is also incompatible with the linear Hammett correlations which were found for several imines. Because of the positive Hammett constant ρ for *N*-aryl substituted imines (97-102), only possibility *44'* need be taken into account.

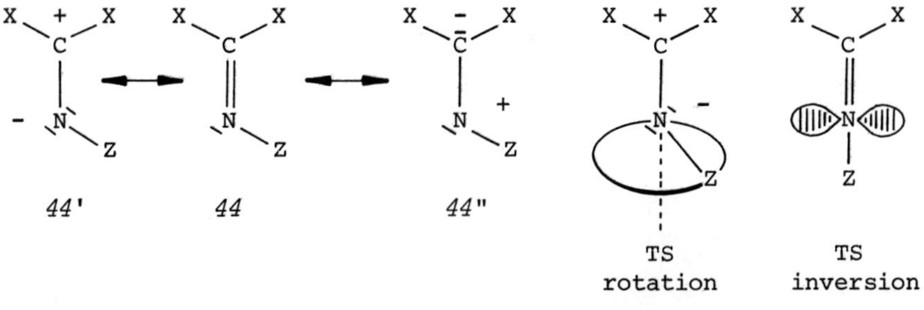

44' ⟷ *44* ⟷ *44"*

TS rotation

TS inversion

In addition to rotation, the planar inversion of nitrogen (lateral shift mechanism) has been discussed since 1930 (103). *Catalyzed* processes have been observed, and are especially important if the barrier for the *uncatalyzed* isomerization is high. This must be kept in mind when the extensively studied isomerization of oximes *45* (104) is considered. The possibility of rearrangement of an oxime to the tautomeric nitroso compound exists (104c,d); moreover, the formation of hydrogen bonds or protonation by the acidic hydrogen of the hydroxyl group may change the system,

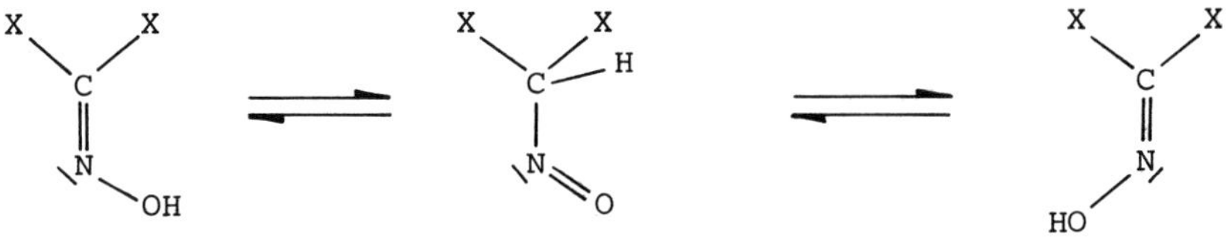

45

as was found to be the case in C=C double-bond isomerization of polarized ethylenes (see above).

Specific solvent effects will be discussed later on.

2. *Distinction between Rotation and Inversion*

The difference between inversion and rotation lies in the different pathways taken by the substituent on nitrogen during the isomerization (Fig. 6). For the case where $a = b$, the symmetry of the transition state (TS) is C_{2v} (linear arrangement CNZ) for inversion, but C_s (the CNZ plane is perpendicular to the XCN plane) for rotation. HMO calculations for the simple methyleneimine show a lower energy for C_{2v} than for C_s symmetry (74) (Fig. 7). This is in complete agreement with *ab initio* SCF-LCAO-MO-calculations which predict a barrier of 27.9 kcal/mol for the C_{2v} methyleneimine and 57.5 kcal/mol for the rotational transition state of C_s symmetry (105). Semiempirical CNDO/2 calculations give barriers of 31.1 kcal/mol (C_{2v}) and 61.1 kcal/mol (C_s) (106). Therefore there should be no doubt that inversion is favored over rotation in the simplest imine. No experimental data for this compound are available.

The effect of substituents on the rotational and inversional barriers, respectively, may be different; thus it remains to be proved which mechanism operates for substituted imines. For some imines, in fact, an intermediate* mechanism has been proposed (106,107). The dependence of the potential energy upon the angle of nonplanarity θ determines which mechanism prevails (Fig. 8). The following experimental findings are seen as indicative of an inversion mechanism or inversionlike mechanism.

*The meaning of *mixed* is that rotation and inversion occur simultaneously via transition states which differ only slightly in energy (Fig. 8d). The difference between *mixed* and *intermediate* was not clear from ref. 106.

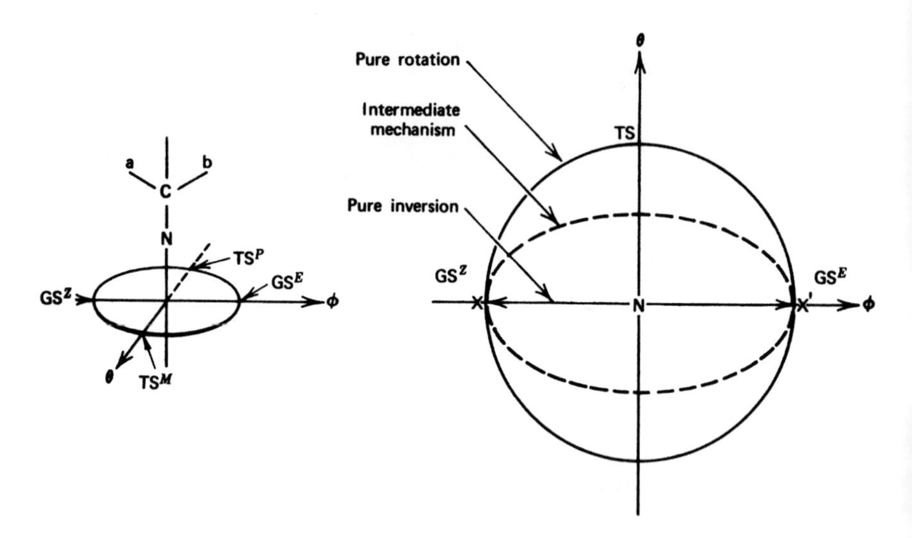

Fig. 6. Pathway of the substituent on nitrogen during *Z*,*E* isomerization (on right: viewed in the direction of the C=N double bond). The angle ϕ is between N-X and the extension of the C=N bond axis in the double-bond plane, whereas Θ is the same angle at right angles to that plane.

1. Rotation about a C=N double bond should have a barrier comparable to those to rotation about a C=C double bond. However, the experimental values are much smaller for imines than for the corresponding olefins (98):

E_a = 42.8 kcal/mol (44) E_a = 18.0 kcal/mol (108)

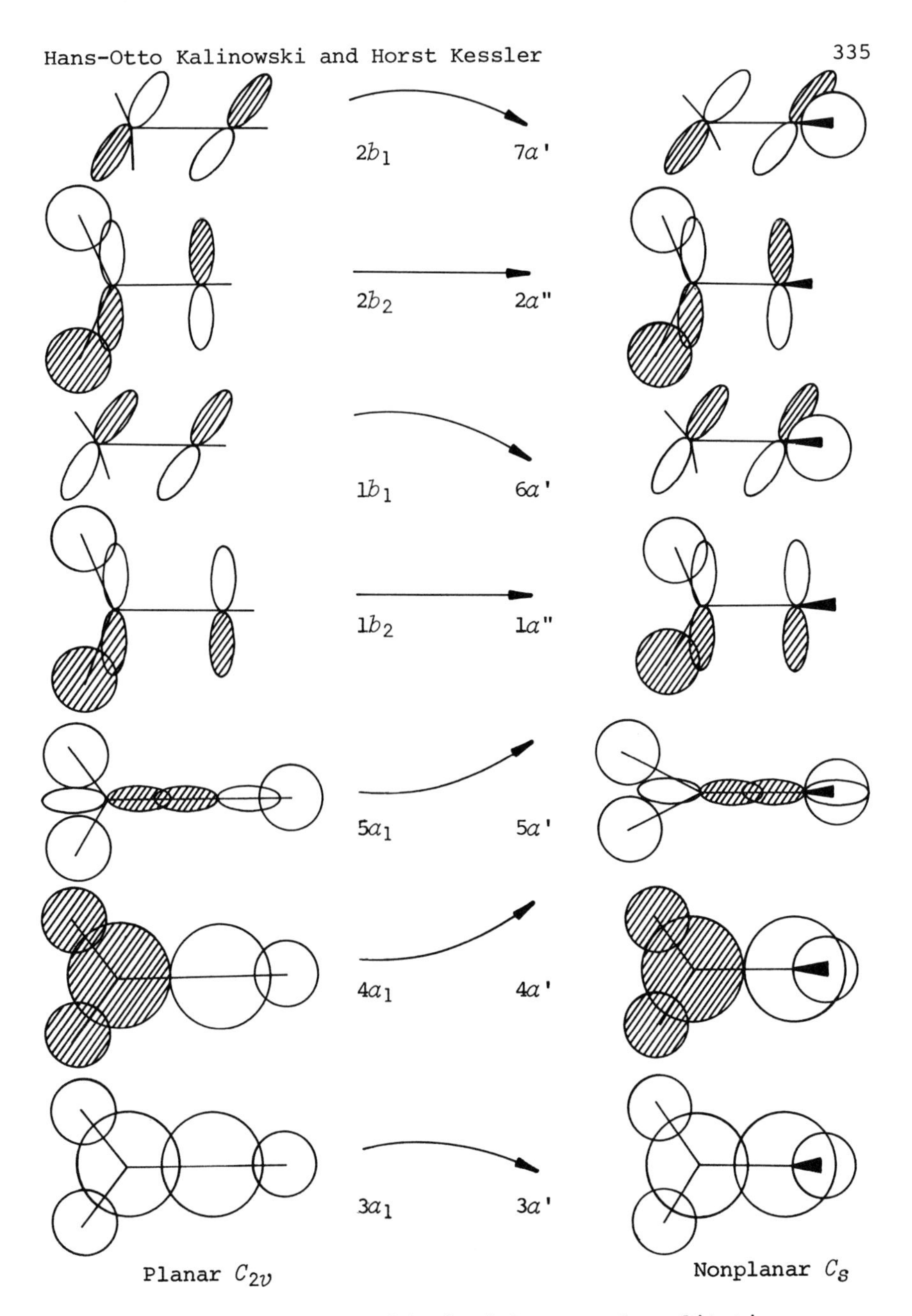

Fig. 7. Molecular orbital pictures and qualitative correlation diagram for H_2ABH molecules in planar C_{2v} and nonplanar C_s shapes (74), i.e. the TS's of the imine isomerization.

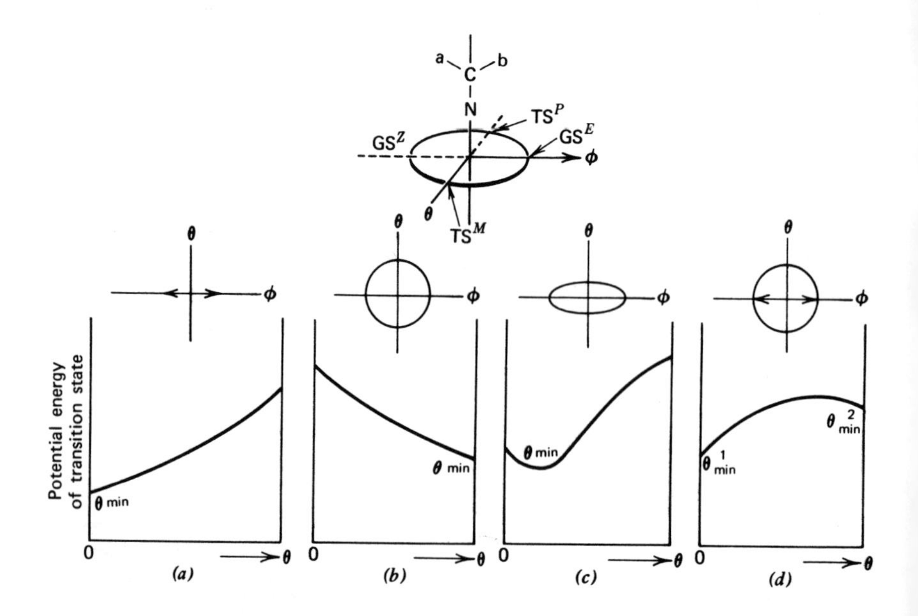

Fig. 8. Dependence of the potential energy of the transition state on θ as a function of the mechanism of syn-anti isomerization (see also Fig. 6). (*a*) Inversion; (*b*) rotation; (*c*) intermediate (inversionlike); (*d*) mixed: inversion as well as rotation occur. The "ideal" rotation corresponds to $\theta_{min} = \phi$. It is also possible that the CNZ valence angle is larger in the TS than in the GS ($\theta_{min} > \phi$). This is also called "rotation." If $\theta_{min} = 0$, pure inversion occurs; if $\theta_{min} < \phi$ it is called an intermediate mechanism, which is inversionlike for a small θ_{min} but rotationlike for θ_{min} almost as large as ϕ.

2. The influence of substituents at the imino nitrogen is greater than that of substituents at carbon:

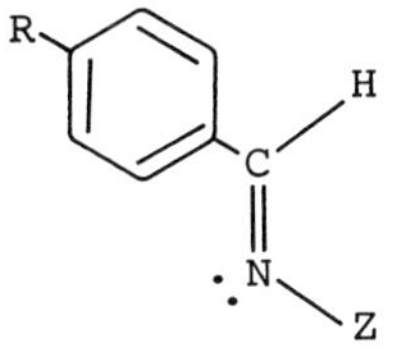

$\rho_C = -0.4$ (109) 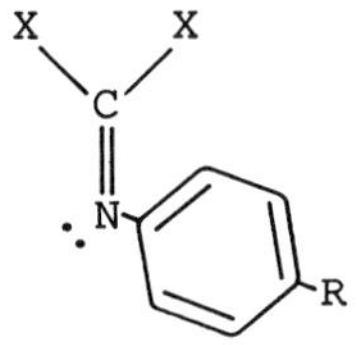 $\rho_N = +1.3$ to $+2.2$

For rotation one would expect ρ_N tc be nearly equal to ρ_C.

3. Large groups in ortho positions of *N*-aryl-substituted imines facilitate the syn-anti topomerization (99-101,111,112) (Fig. 9). The steric effect on rotation about double bonds in which no inversion is possible shows quite different behavior (see Sects. IV-B-2 and V-A).

4. No influence of solvent polarity on syn-anti topomerization rates for guanidines (88,100), imino esters, and thioimino esters (101) has been observed. The highly polar transition state of the rotation would be favored by polar solvents and thus the barrier should decrease, as was found for polarized ethylenes (Sect. IV-B-4). It should be stated that these arguments have been criticized recently, and that the facts cited have been taken as indication of an intermediate mechanism with an inversionlike TS (106,107).

5. The retarding influence of protic solvents on the syn-anti topomerization can be explained by protonation or by a hydrogen bond to the lone electron pair which is needed for inversion (88,100,101). The same effect is observed in pyramidal nitrogen inversion (21,113). Also complex formation with Lewis acids, such as $AlCl_3$, greatly affects the rate of syn-anti isomerization (114).

6. The syn-anti topomerization of guanidines requires about the same activation energy as does the rotation about the partial double bond in guanidinium salts, although in the latter the double-bond order is less than 33%. The bond order should correlate with energy barriers to C=N partial double-bond rotations (115) and thus, as a result of the higher bond order of the guanidine (0.80 by CNDO/2 calculation (106)), one would expect a much higher barrier of C=N rotation in the free base compared to the salt if a rotation mechanism were operative.

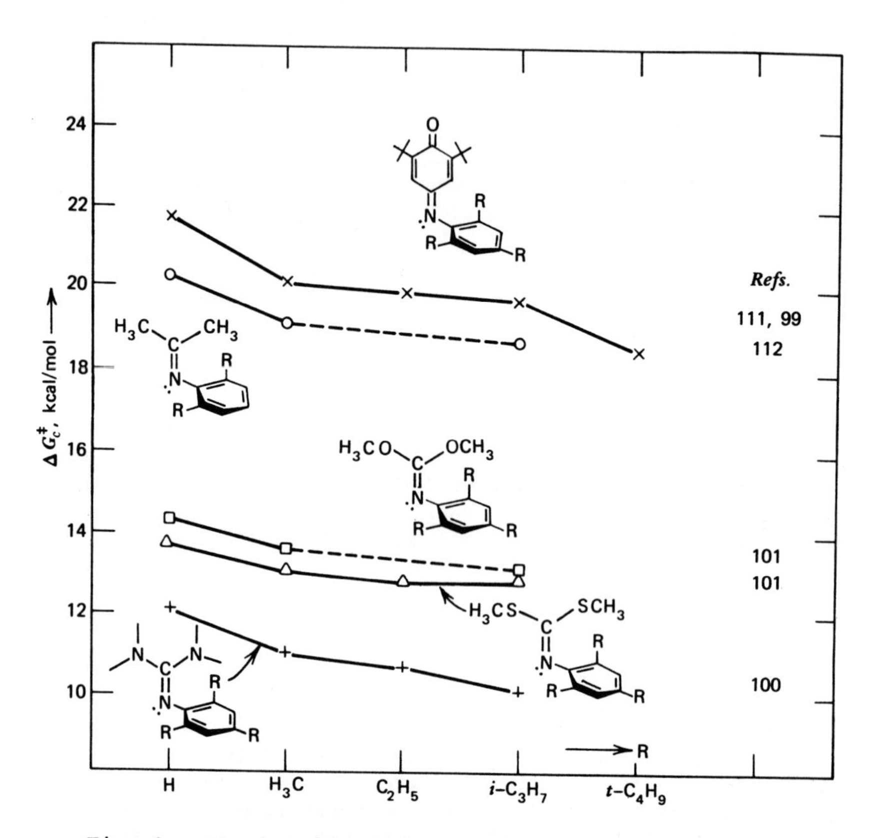

Fig. 9. Steric effects on topomerization barriers of *o*-substituted *N*-arylimines.

$(CH_3)_2N$ — C(=N — C_6H_5) — $N(CH_3)_2$

$\Delta G^{\ddagger}$ = 12.1 kcal/mol (100)

$[(CH_3)_2N$ ⋯ C ⋯ $N(CH_3)_2$; N(H_3C)(C_6H_5)$]^+$ I^-

$\Delta G^{\ddagger}$ = 11.1 kcal/mol (89)

7. The influence of substituents on imino nitrogen stereomutation is analogous to their influence on pyramidal nitrogen inversion (13). The order of decreasing stability for both

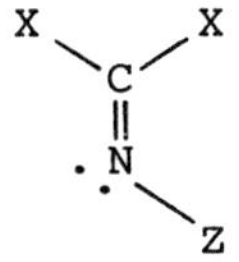

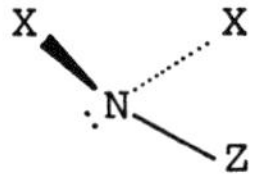

structural types is as follows:

Z = alkoxy > dialkylamino > halogen > alkyl > aryl > acyl

The quantitative correlation is discussed below.

8. The substitution of *N*-arylimines with prochiral groups allows the determination of the stereochemical nature of the intramolecular rearrangements shown in (Scheme 2) (116).

It was possible to show that the enantiotopomerization rate (exchange of prochirality of R and R') is the same as the syn-anti topomerization rate when X = X', as is to be expected for the nitrogen inversion process. When X ≠ X' it could be shown that *N*-aryl rotation *and* C=N double-bond rotation are more than 10^8 times slower than nitrogen inversion in compound *46* (116).

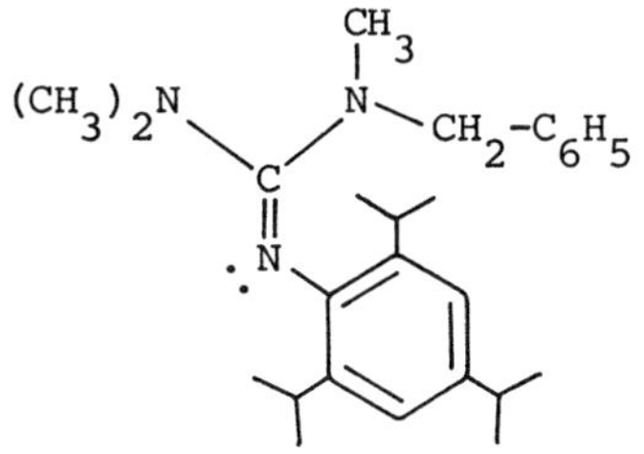

46

The following facts are used to support the torsion model of syn-anti isomerization:

1. The influence of substituents X on the imino carbon shows a dramatic decrease of the inversion barrier in the order (13,100,117):

quinoneimine > ketimine > *C*-arylimine > imino esters > guanidine

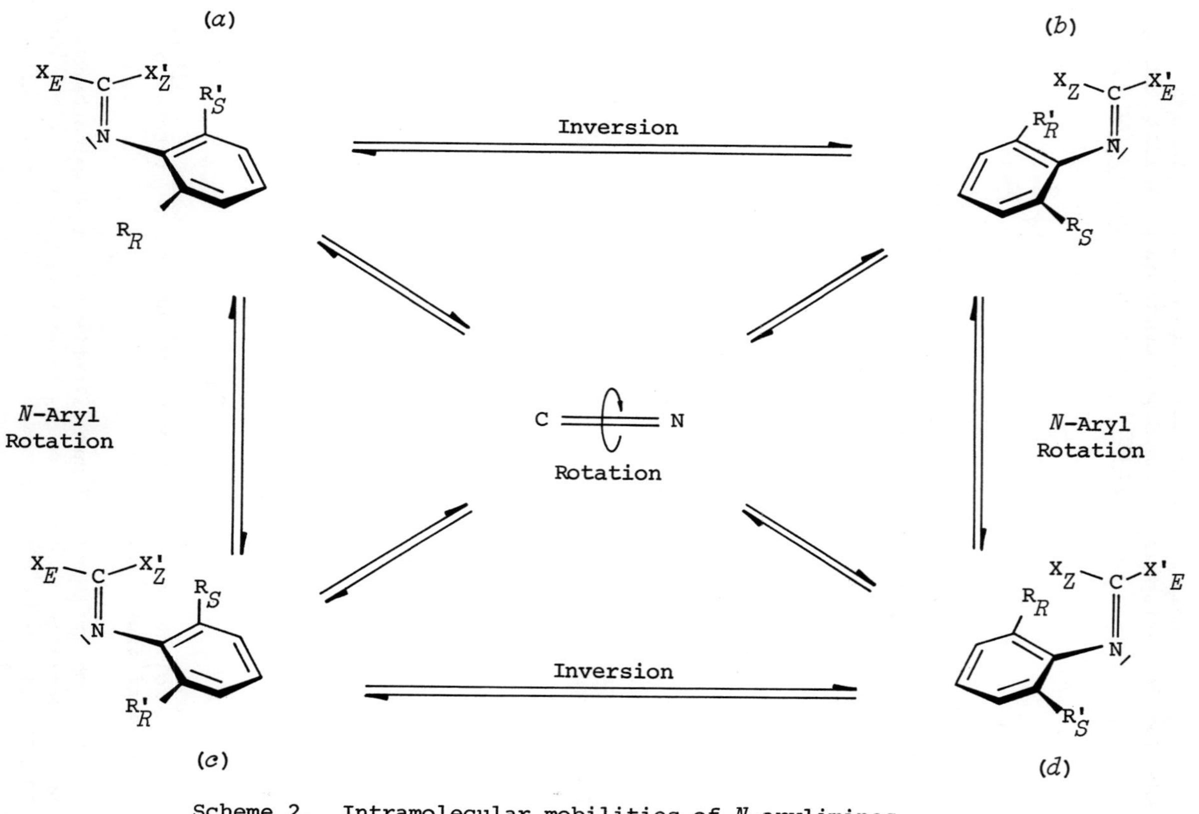

Scheme 2. Intramolecular mobilities of *N*-arylimines.

This is the expected order for rotation via a polarized transition state. However, it does not refute the inversion mechanism as long as the influence of X on the rate of nitrogen inversion is unknown.

2. The relatively low barriers of quinonediimine *N*-oxides (*47*) with respect to their corresponding quinonediimines (*48*) have been used to support the rotational model (118). In our opinion, the comparison of such different molecules as imines

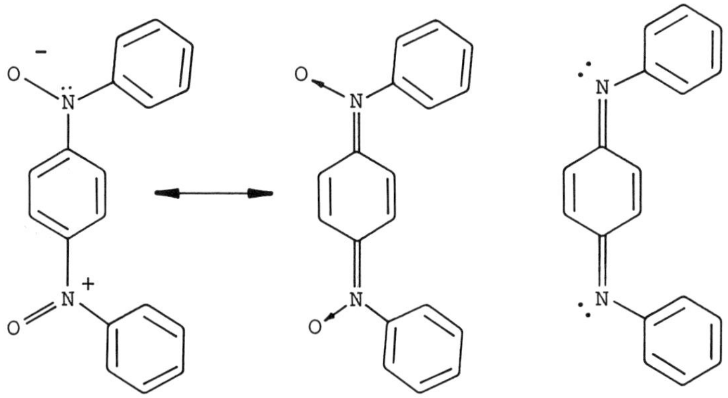

47

$\Delta G^{\ddagger} \simeq 12$ kcal/mol

48

$\Delta G^{\ddagger} \simeq 20$ kcal/mol

and nitrones does not allow the drawing of definite conclusions about the mechanism.

3. In the strongly polar *N*-aryl-hexafluoroacetonimines (*49*) a negative Hammett correlation has been found by dnmr studies (119). A plausible interpretation of this fact is only

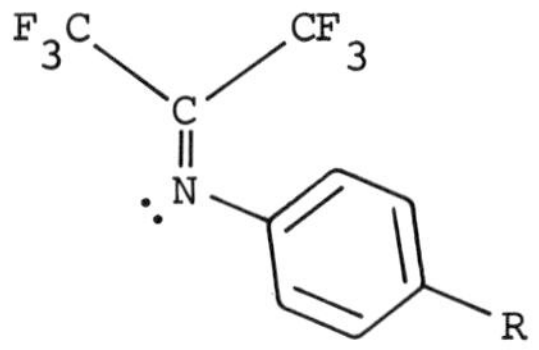

49

$\rho = -0.98$
($R = NO_2$ excluded)

possible assuming an inverse polarization (negative carbon) for syn-anti isomerization via rotation. The exception of the *p*-nitro compound can be explained by participation of an inversion. Therefore, *49* might represent the rare case in which both mechanisms are operating with similar rates.

The negative ρ value for the stereomutation of an imine is an exception. In all other *N*-arylimines positive ρ values of +1.3 to +2.2 have been found, which indicates that the same mechanism is operating for all these compounds.

Generally, in our opinion, the arguments for the inversion mechanism are much stronger than those for the torsional model, and the inversion, or at least an intermediate mechanism that is near to the inversion mode, is most probable.

The intermediate mechanism was proposed to explain the excessively low isomerization barriers of *N*-sulfonyl*imines* (99b,107,120) compared with *N*-sulfonyl*amines* (121). Preliminary CNDO/2 calculations indicate an intermediate inversionlike TS (106), but one should bear in mind that these calculations are not able to predict the barriers of syn-anti isomerization satisfactorily. Certainly the rotational contribution to the intermediate mechanism cannot be large in *N*-arylimines. The steric effect in *o*-substituted *N*-arylimines should drastically hinder the rotation and the barrier of such substituted imines should thus increase to the "pure" inversion barrier. The observed effect, however, is a decreasing barrier; i.e., the favorable steric effect on inversion is greater than the unfavorable effect on the diminishing rotational part.

The calculation of the isomerization of carbodiimide has shown similar barriers for rotation and inversion (105). A single TS (intermediate mechanism) therefore is possible for this type of compound.

3. *Substituent Effects on Carbon-Nitrogen Bond Isomerization*

a. Steric Effects. The substituents on carbon or nitrogen are responsible for steric strain in the ground state. This is completely analogous to the situation with other double bonds (C=C: Sect. IV-B-2; C=N: Sect. V-A). Examples are found in Table 8. Thus the barrier in the *N*-*t*-butyl-substituted diaziridinimine *73b* is about 4 kcal/mol lower than that in the *N*-methyl-substituted diaziridinimine *73a* (135). The steric effect in *o*-substituted *N*-arylimines is similar (Fig. 9). Because of the possibility of twisting the benzene ring plane and the C=N double-bond plane this effect is relatively small. The partial twisting of these planes is shown to exist by crystallographic (142) and uv measurements (143), even in unsubstituted *N*-arylimines.

Table 8 Activation Parameters for syn-anti Isomerization

Formula structure type	No.	Solvent[b]	T, °C	$\Delta G^{\ddagger}$, kcal/mol	Other Data	Parameters for empirical calculation[a] x	z	Ref.
Z = C_6H_5[c,d]	*50a*	TCB	140	22.0		37 ≡	0.59	99
Z = $SO_2C_6H_5$	*50b*	TCB	44	17.1			0.46	99
Z = OCH_3	*50c*	TCB	>180	>23			>0.62	99
Z = CN	*51a*	A		18.9			0.55	126
= CHRAryl	*51b*	Q	180	23			0.67	112
= C_6H_5[d]	*51c*	DPE	126	20.3		34 ≡	0.59	112

$(H_5C_6H_2C)(CH_2C_6H_5)C{=}N{-}S{-}C_6H_3(NO_2)_2$	*52*	CB	101	20.0		29-33[h]	123
$(H_3COC)(COCH_3)C{=}N{-}C_6H_5$	*53*	DPE	105	19.4		33 ≡ 0.59	112
$(H_3COOC)(COOCH_3)C{=}N{-}C_6H_5$	*54*	TCB	91	18.9		32 ≡ 0.59	100
$(H_3C)[(H_3C)_2C{=}C(CH_3)]C{=}N{-}SO_2C_6H_5$	*55*			16.3	E_a = 16.2 kcal/mol log A = 12.8 $\Delta S^\ddagger$ = −2 e.u.		107

Table 8 Continued

Formula structure type	No.	Solvent[b]	T, °C	$\Delta G^{\ddagger}$, kcal/mol	Other Data	Parameters for empirical calculation[a] x	z	Ref.
$(H_3C)(CH_3)C=N-N(CH_3)-C_6H_5$	*56*	HCB	134	21.1				129
		DPE	>190	>26				110
$(H_3C)(CH_3)C=N-N=C(CH_3)(H_3C)$	*57*	DPE	>180					112
		TCB	170	23.9				110

Cl, C=N, Z = CH_3	*58a*	CH	50[e]		E_a = 25±0.3 $\Delta S^‡$ = −3.6 e.u.	25	≡ 1	122
Z = Cl	*58b*	CH	60	>31			>1.24	98
Z = Br	*58c*	CH	60	>28			>1.12	
Z = OCH_3	*58d*	Decane	60	>39			>1.56	
O_2N, C=N, CH_3	*59*	CH	50[e]		E_a = 27.1±0.5	27	≡ 1	122
H_3CO, OCH_3, C=N, Z = C_6H_5[f]	*60*	CCl_4	62	18.1		31	≡ 0.59	98
H_3CO, C=N, Cl	*61*	CCl_4	62		E_a = 19.7±0.4 $\Delta S^‡$ = −2.1±1.5 e.u.			98

Table 8 Continued

Formula structure type	No.	Solvent[b]	T, °C	$\Delta G^{\ddagger}$, kcal/mol	Other Data	Parameters for empirical calculation[a] x	z	Ref.
$(4\text{-}H_3C\text{-}C_6H_4)_2C{=}N\text{-}Z$								
Z = S–C_6H_4–Cl	*62a*	$CDCl_3$	72	18.6			0.6–0.7[g]	123
Z = S–C_6H_3–$(NO_2)_2$	*62b*	$CDCl_3$	67	18.5			0.6–0.7[g]	123

Structure		Compound	Solvent						Ref.
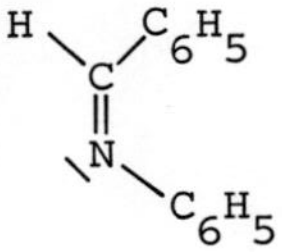 $H(C_6H_5)C{=}N{-}C_6H_5$		*63*	C_2H_5OH			E_a = 16.0			131 108
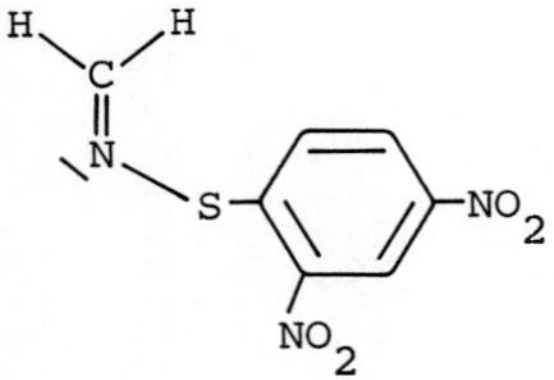 $H_2C{=}N{-}S{-}C_6H_3(NO_2)_2$ (2,4-dinitrophenyl)		*64*	CB	0	13.5		19–22[h]		123
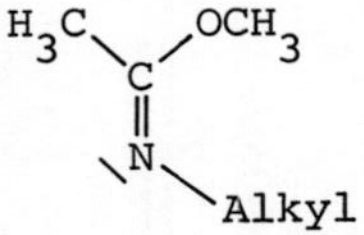 $H_3C(OCH_3)C{=}N{-}$Alkyl		*65*	CCl_4		>23				130
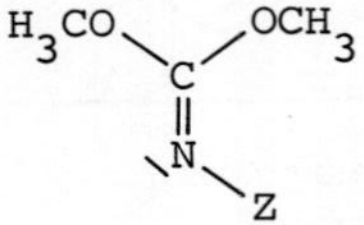 $H_3CO(OCH_3)C{=}N{-}Z$	Z = Cl	*66a*	DPE	>160	>23			>0.95	124
	Z = $CH_2C_6H_5$	*66b*	A	85	18.9		—	0.78	125
	Z = C_6H_5[d,i]	*66c*	A	−2	14.3		24 ≡	0.59	101 117 124 125
	Z = CN	*66d*	A	−16	14.1			0.58	126
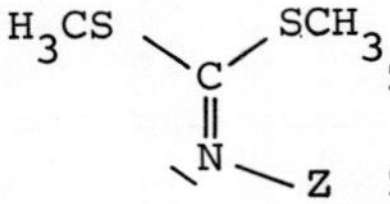 $H_3CS(SCH_3)C{=}N{-}Z$	Z = C_6H_5[d,k]	*67a*	A	−25	13.7				101 102
	Z = CN	*67b*	A	1	14.1				126

Table 8 Continued

Formula structure type	No.	Solvent[b]	T, °C	$\Delta G^{\ddagger}$, kcal/mol	Other Data	Parameters for empirical calculation[a] x	z	Ref.
$(CH_3)_2N$–C(=N–Z)–$N(CH_3)_2$; Z = OCH_3[n]	*68a*	DPE	>200	>25			>1.3	128
Z = $N(CH_3)_2$	*68b*	DPE	>200	>25			>1.3	128
Z = $CH_2C_6H_5$	*68c*	CCl_4	71	18.5			0.9	128
Z = C_6H_5[l]	*68d*	$CS_2/CDCl_3$	-35	12.1		20.5 ≡	0.59	100,128
Z = CN	*68e*	A	<-90	<10[m]				126 127
Z = $COCH_3$	*68f*	$CS_2/CDCl_3$	<-90	<8.5[m]				128 127
Z = CH_3	*68g*	DCB	78	18.8[p]			0.92	140
		CCl_4	89	19.1[p]			0.93	141
H_3C–N(D)–C(=N–CN)–N(D)–CH_3	*69*	A	-43	12.3				126

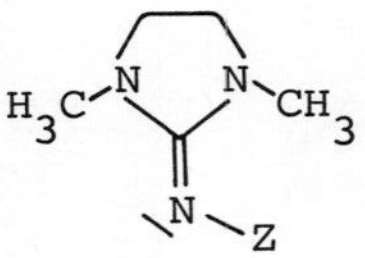 Z = OCH_3	*70a*	DPE	165	23.2	1.79	128
Z = $N(CH_3)_2$	*70b*	DPE	137	21.0	1.62	128
Z = $CH_2C_6H_5$	*70c*	CH_2Cl_2	-46	11.7	13 ≡ 0.9	128
Z = C_6H_5	*70d*	$CS_2/CDCl_3$	<-80	<10[q]	<0.8	128
Z = CH_3	*70e*	CH_2Cl_2	-32	12.2	0.94	141
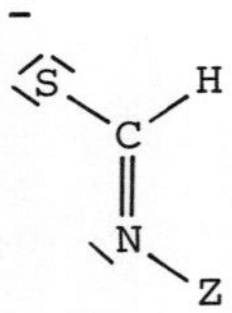 Z = CH_3	*71a*	CD_3OD		<20.0		133
Z = $CH_2CH(CH_3)_2$	*71b*	CD_3OD		21.4		133
Z = $CH_2C_6H_5$	*71c*	CD_3OD		20.9		133
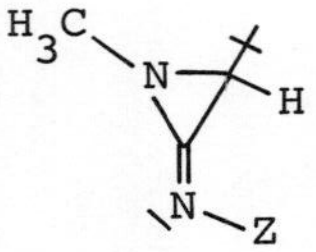 Z = CH_3	*72a*	CCl_4		23.3		134
Z = t-C_4H_9	*72b*	CCl_4		21.4		
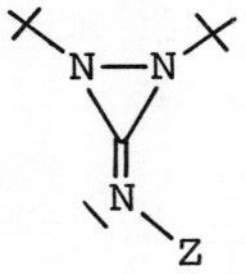 Z = CH_3	*73a*	Neat	109	21.4		135
Z = t-C_4H_9	*73b*	Neat	51	17.5		
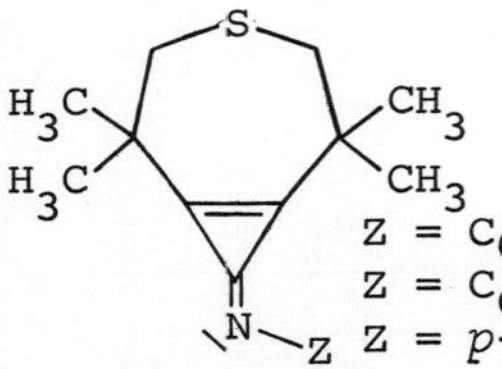 Z = C_6H_{11}	*74a*	DPE	145	23.8		136
Z = C_6H_5	*74b*	Tol	79	18.7		
Z = p-$O_2NC_6H_4$	*74c*	Tol	8	14.7		

Table 8 Continued

Formula structure type		No.	Solvent[b]	T, °C	$\Delta G^{\ddagger}$, kcal/mol	Other Data	Parameters for empirical calculation[a] x	z	Ref.
$(F_3C)_2C{=}N{-}Z$	$Z = C_6H_5$[o]		Pyridine		15.45±0.019	E_a = 14.58±0.18			119

[a]see text p. 354; $\Delta G^{\ddagger}_{CH_3} = x$ calculated from $z_{C_6H_5} \equiv 0.59$.

[b]see footnote a in Table 3; Q = quinoline; CB = chlorobenzene; CH = cyclohexane; Tol = toluene-d_8.

[c]Hammett correlation of twelve derivatives gives ρ = 1.5±0.3 (99b).

[d]Steric effect of substituents in the aromatic ring was proved (see Fig. 9).

[e]Equilibration temperature.

[f]Hammett correlation of four derivatives gives $\rho \approx +1.5$ at 62°C (108).

[g]Assuming an x constant of 26-31.

[h]Assuming a z value of 0.6-0.7; see compound *62b*.

[i]Hammett reaction constant ρ = 1.63 at -10°C.

[k]Hammett reaction constant ρ = 1.56 at -25°C.

[l]Hammett reaction constant ρ = 2.95 at -50°C.

[m]At low temperature the rotation about the carbon-nitrogen "single" bonds could be observed (127).

[n]The low barrier of this compound has been found to be caused by impurities in the sample and solvents. The highly purified substance shows no coalescence up to 180°C (128,140).

[o]Hammett reaction constant $\rho = -0.98$ except for the *p*-nitro compound. No $z_{C_6H_5}$ value is calculated because the rotation mechanism probably operates.

[p]The disagreement with the expected barrier (20.5 kcal/mol) might be explained by the catalytic effect of the halogenated solvents (110).

[q]Taken from compound *68c*.

*b. Electronic Effects.** The electronic effects on planar and pyramidal inversion of nitrogen can be divided in a σ-effect and a π-effect (21). Increasing electronegativity of the substituent on nitrogen causes increasing *s*-character of the nitrogen lone pair in the ground state, resulting in increasing stability toward inversion, as has been found both by theoretical calculations (144) and experimentally (e.g., see compounds *70a* (128), *70b* (128), and *70c* (141).

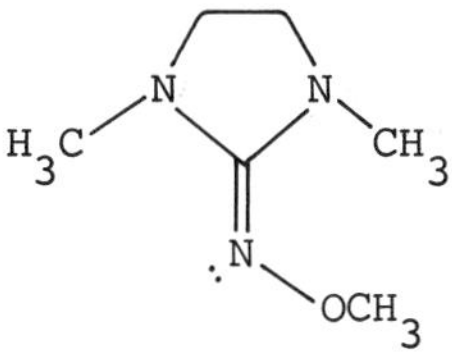

	70a	*70b*	*70c*
Z on imine N	OCH_3	$N(CH_3)_2$	CH_3
$\Delta G^{\ddagger}$	= 23.2	21.0	12.2 kcal/mol

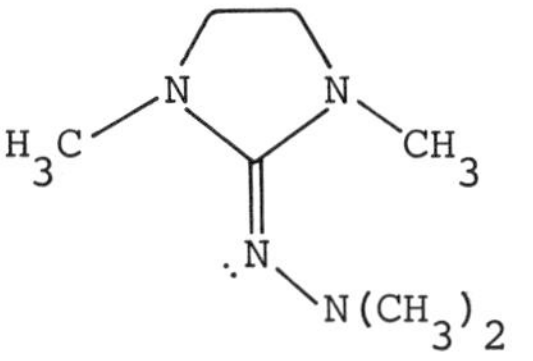

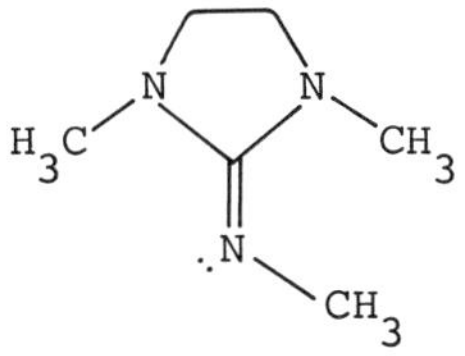

The *N*-fluoro compounds always have the highest stability. In the *N*-chloro compounds, the effect of electronegativity is larger than the effect of p_{π}-d_{π} conjugation because *N*-chloro compounds have higher stabilities than *N*-alkyl ones.

The effect of π-orbitals in the substituent Z on *N*-inversion may be explained by the better conjugation of the occupied *p*-orbital in the transition state with empty *p*- or *d*-orbitals† on Z, as compared to the conjugation of the occupied sp^2 orbital in the ground state. The effect of the π- and σ-orbitals on Z are complementary. A π-donor or σ-acceptor will increase the barrier, and vice versa. The barrier decreases in the following order of Z:

$$\text{alkyl} > \text{aryl} \simeq \text{cyano} > \text{acyl}$$

The C=N-Z bond angle also increases in this order. The bond angle can be correlated quantitatively with inversion barriers (145). The conjugation effect is clearly demonstrated by the Hammett correlation in *N*-arylimines (13,98-102), which corresponds completely with the effect on pyramidal inversion of *N*-arylaziridines (146). This qualitative consideration is suffi-

*A detailed discussion regarding the theoretical background and the origin of the inversional barrier is found in ref. 21.

†The effect of *d*-orbitals on *N*-inversion is dubious and is hardly to separate from the effect of electronegativity.

cient to understand the effects on barriers by the substituent Z. Most of these effects could be verified by *ab initio* and semiempirical MO-calculations (21), although the quantitative agreement with measured barriers is often poor.

The basis for the comparison of experimental barriers with calculated ones is the use of $\Delta H^{\ddagger}$ values. Very often, however, only free energies of activation $\Delta G^{\ddagger}$ are known. As a first approximation, these values have been used, since in most cases where a complete kinetic analysis has been made, low entropies of activation (range -4 to 0 e.u.) have been found (107,122) (but see ref. 102). Because of the similarity of planar and pyramidal inversion - both involve a change of the hybridization of the lone electron pair from sp^x to p - the substituent effects are expected to be similar for compounds displaying these two types of inversion. An empirical method has been developed to calculate barriers for pyramidal nitrogen inversion by substituent constants using eq. [3] (147).

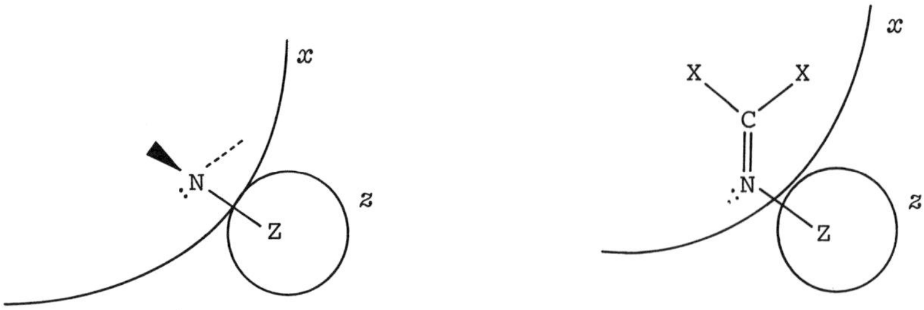

$$\Delta G^{\ddagger}_{Z} \quad x \cdot z \qquad [3]$$

The substituent constant z depends only on the nature of Z, whereas the x constant represents the structure of the remainder of the molecule. The barrier of the *N*-methyl compound is as reference ($z_{CH_3} = 1$). The corrected* barriers can be plotted against the $\Delta G^{\ddagger}$ values of the methyl compounds (x value) to give, at least as a first approximation, straight lines as expected from eq. [3].

For planar inversion it is better to use *N*-phenyl compounds as reference, because many inversion barriers of anils are known

*For all compounds the entropy of activation is assumed to be zero. $\Delta G^{\ddagger}$ values are used neglecting their temperature dependence. If Arrhenius activation energy was determined the barrier is corrected assuming a log A value of 13 calculated by a linear E_a - log A correlation (see ref. 13).

but only a few measurements for *N*-methylimines are available. To allow a direct comparison with z values of pyramidal inversion we used the $z_{C_6H_5}$ value of 0.59, which was determined in amines. The x values obtained by this procedure ($x = \Delta G^{\ddagger}_{C_6H_5}/z_{C_6H_5}$) are listed in Table 8. The z values for substituents z other than C_6H_5 and CH_3 correspond almost exactly those for pyramidal inversion.

Generally it is possible to estimate a number of inversional rate constants by this procedure. Exceptions have been found for *N*-tosyl compounds: the z value of this group is 0.55-0.58 (from pyramidal inversion (147), but for compound *50b* it is only 0.46, and for compound *55* about 0.48. This deviation has been used as an indication of the mixed mechanism (107).

Another very interesting difference is represented by the barriers of acetone hydrazones (110,129) and acetone azine (110,112). A relatively low barrier (21 kcal/mol) has been found for *56* when hexachlorobutadiene is used as solvent (110, 129). From the empirical estimate (eq. [3]) the barrier in acetone hydrazone should be about 50 kcal/mol.*

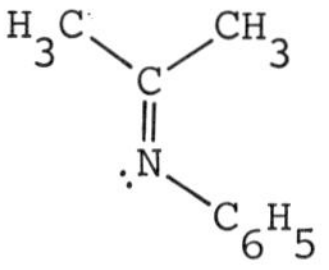

$(CH_3)_2C{=}N{-}C_6H_5$ $\qquad$ $(CH_3)_2C{=}N{-}CH_3$ $\qquad$ $(CH_3)_2C{=}N{-}NMe_2$ $\qquad$ $(CH_3)_2C{=}N{-}N(Me)C_6H_5$

$\Delta G^{\ddagger}_{C_6H_5} = 20.3$ kcal/mol

51c $\qquad$ *51d* 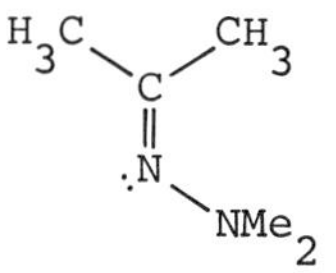 $\qquad$ *51e* 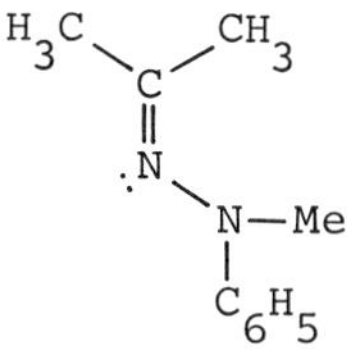 $\qquad$ *56*

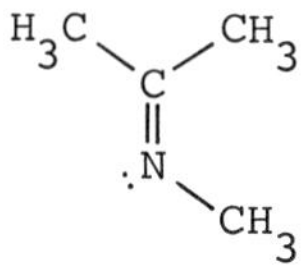

$$\Delta G^{\ddagger}_{NMe_2} = \Delta G^{\ddagger}_{CH_3} \cdot z_{NMe_2}$$

$z_{NMe_2} = 1.45$ (ref. 147)

$z_{C_6H_5} = 0.59$ (ref. 147)

With $\Delta G^{\ddagger}_{CH_3} = \Delta G^{\ddagger}_{C_6H_5}/z_{C_6H_5}$ it follows that

$$\Delta G^{\ddagger}_{NMe_2} = \frac{\Delta G^{\ddagger}_{C_6H_5}}{z_{C_6H_5}} \cdot z_{NMe_2} = 50.0 \text{ kcal/mol} \simeq \Delta G^{\ddagger}_{NMeC_6H_5}$$

*The barrier of the corresponding *N*-phenylimine is not known. Therefore acetone anil is used as reference.

More detailed investigation has demonstrated that no syn-anti topomerization in *56* is observed in diphenyl ether, ethylenetetracarboxylic acid tetramethyl ester, or diethyl maleate up to 200°C ($\Delta G^{\ddagger}$ > 26 kcal/mol) but a rather fast isomerization rate in chlorine containing solvents such as tetrachloroethane, tetrachloroethylene, hexachlorocyclopentadiene, and tetrachlorocyclobutenone (110). Effects of acid as well as dimerization could be excluded as being responsible for the easy topomerization. The nature of the solvent effect is not yet clear* but there is much evidence for the catalytic nature of the isomerization of hydrazones and azines. Many earlier measurements have not excluded the possibility of catalytic effects. This is especially important when barriers of inversion are relatively high and competition reactions could be faster, e.g., in oximes and hydrazones in which active hydrogens are present (104). The empirical calculation helps to recognize such differences and discrepancies.

Quite a different mechanism has been considered for *Z,E* isomerization in certain *O*-alkyl oximes *76* (149).

$$(H_5C_6)(H)C{=}N{-}OC(C_6H_5)_3 \;\underset{+\cdot C\phi_3}{\overset{-\cdot C\phi_3}{\rightleftharpoons}}\; \left[(H_5C_6)(H)C{=}N{-}\dot{O}{:} \longleftrightarrow (H_5C_6)(H)C{=}\overset{+}{N}{-}O^{-}\right]$$

76

$$(H_5C_6)(H)C{=}N{-}OC(C_6H_5)_3 \;\underset{+\, C\phi_3}{\overset{-\, C\phi_3}{\rightleftharpoons}}\; (H_5C_6)(H)C{=}N{-}\dot{O}{:} \longleftrightarrow (H_5C_6)(H)C{=}\overset{+}{N}{-}O^{-}$$

The dissociation of the C-O bond to the iminoxy radical which can rapidly undergo inversion, followed by recombination, leads to the observed isomerization. The easy inversion of iminoxy radicals, which has been observed by esr spectroscopy (150), may be explained by the lower electron density of the nitrogen orbital which is *p*-hybridized in the transition state (cf. the inversion barriers of vinyl carbanions and of vinyl radicals).

The C=N double-bond isomerization of azacumulenes has been studied by semiempirical CNDO calculations (151,152) as well as

*An explanation has been given in terms of complex formation between the C=N double bond as donor and the chlorine of the solvent as acceptor with formation of a sigma bond to the imino carbon (110). For similar halo complexes see ref. 148.

by nmr measurements of suitable derivatives (137,138) (Table 9). The stability of α,ω-diazacumulenes has been calculated to alternate between a relatively high barrier for those with an even number of carbon atoms (interconversion of geometric isomers or topomers; diastereomers or diastereotopomers) and low barriers for those with an odd number of carbons (interconversion of enantiomers). Recent nmr measurements on carbodiimides which indicate a low barrier to enantiomerization have confirmed these predictions and explain the unsuccessful attempts to resolve carbodiimides (153).

It is interesting that in carbodiimides the rotational and the inversional transition state both have about the same energy.* The rotational transition state (trans is favored over cis) has a larger CNH valence angle (135°) than in the ground state (115°), indicating an intermediate between rotation and inversion (180°).

The description of the barrier in carbodiimide involves the transfer of the electron population of the inverting nitrogen into the π-orbital of the C=N double bond of the noninverting nitrogen (21,105). In agreement with this explanation, the isomerization barrier decreases in the order keteneimine, carbodiimide, and we may anticipate an even lower barrier in isocyanates, because the increasing electronegativity of X in going from X = C to C = O leads to increasing electron flow from the inverting nitrogen into the corresponding π-orbital (no measurements for isocyanates are available; a possibility of studying this process might be provided by the observation of the magnetic nonequivalence in prochiral ortho substituents of *N*-arylisocyanates; see Sect. VI-B.)

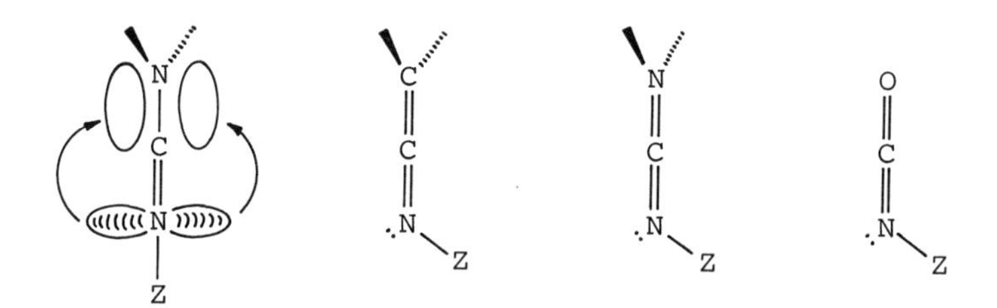

inversion barrier decreases

*Only one of the nitrogens inverts. The simultaneous inversion of both nitrogens (TS: $D_{\infty h}$-symmetry) requires much more energy.

Table 9 *Z,E* Isomerization of the C=N Double Bond in Azacumulenes

Compound	Kind of Isomers	Calculated barriers, kcal/mol		Compound	No.	Experimental barriers, kcal/mol	Ref.
$HN(=C)_n=NH$		INDO (151)	*ab initio* (21)				
HN=NH	Diastereomers	46.2	50.7 (105) 72.5 (154)				
HN=C=NH	Enantiomers	8.0 (155)	8.4	$i\text{-}C_3H_7N=C=N\text{-}i\text{-}C_3H_7$	*77*	6.7	138
HN=C=C=NH	Diastereomers	23.9					
HN=C=C=C=NH	Enantiomers	6.9		$R(i\text{-}C_3H_7)C=C=NC_6H_5$, R = C_6H_5	*78a*	9.1	137
				R = CH_3	*78b*	12.2	137

VI. THE NITROGEN-NITROGEN DOUBLE BOND

A. *Z*,*E* Isomerization of Azo Compounds

The thermal (156-164) and photochemical (162-164) isomerization of azobenzene and derivatives has been the subject of much experimental investigation (1c). Normally azobenzene exists in the thermodynamically more stable *E* ground state, which can be easily photoisomerized to the *Z* conformer. The thermal reversion to the *E* conformer has been studied, and it has been shown that this process requires much less activation energy than the *Z*→*E* isomerization of stilbene (Table 10).

The following arguments have been used to favor inversion at nitrogen ("in-plane isomerization") over N=N double-bond rotation:

1. The barrier in azobenzene is much lower than that in stilbene (25,160,165), yet rotation about the N=N double bond should be at least as difficult as about the C=C double bond (25,151).

2. Azobenzene and azomesitylene (163) show no difference in barrier due to steric hindrance. Rotation should, however, be hindered by the *o*-methyl groups (see Sects. IV-B-2 and V-B-3-a).

3. Semiempirical MO calculation of diimides have led to a barrier of 46 kcal/mol for inversion of one nitrogen, but 72 kcal/mol for N=N rotation (151). The synchronous inversion of both nitrogens requires 116 kcal/mol and is therefore not expected to be involved. These values are valid for the singlet isomerization itinerary which is supported by the "normal" Arrhenius preexponential factor of $A = 10^{12}$-10^{15} for isomerization of azo compounds. Rotation is also excluded by *ab initio* (105) calculations and by symmetry considerations using the extended Hückel treatment (172).

All substituents in the para position of the aryl rings of azobenzene increase the isomerization rate (160). This is in contrast to earlier results which were linearly correlated by Hammett constants (159,173). The general increase of the rates upon substitution is to be expected for the inversion and is, in our opinion, further evidence for the inversion pathway. An electron-withdrawing substituent A in the aryl ring facilitates the inversion of the nitrogen *1*, but an electron-donating substituent D should facilitate the inversion of nitrogen *2* if we assume a similar influence of substituents as found in the inversion in imines (Sect. V-B). The nonlinear Hammett plot therefore results from the fact that the inverting nitrogen is

Table 10 $Z \rightarrow E$ Isomerizations of the Nitrogen-Nitrogen Double Bond

Compound		No.	Solvent	E_a, kcal/mol	Other Data	Refs.
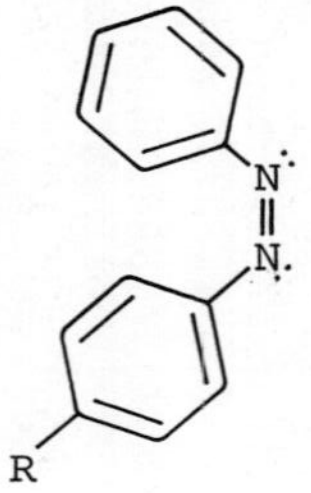	R = H	*79a*	Ac	23		156
			Benzene	23.7	log A = 11.65	159
	R = $N(CH_3)_2$	*79b*	Benzene	21.0	log A = 11.1	158
	R = NO_2	*79c*	Benzene	20.9	log A = 11.0	158
				23.2		159
	R = CH_3	*79d*	Benzene	23.5		167
	R = OCH_3	*79e*	Benzene	21.0	log A = 10.25	158
	R = Cl	*79f*	Benzene	22.4	log A = 11.0	158,159
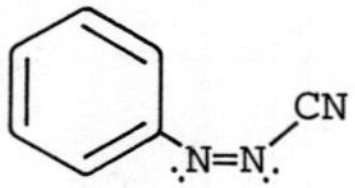		*80*	Benzene	23.0	log A = 15	167
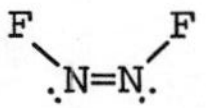		*81*	Argon	35.2		166
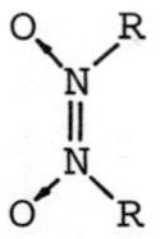	R = CH_3	*82a*	Ethanol	19.1	log A = 11.6	168
	R = $CH_2CH(CH_3)_2$	*82b*	Ethanol	22.1	log A = 13.9	168

Table 10 Continued

Compound		No.	Solvent	E_a, kcal/mol	Other Data	Refs.
R–N(:)=N(→O)–R	R = C_6H_5	*83*		22.8	$\Delta S^{\ddagger}$ = +5.1 e.u.	176
R_2N^+=N–O$^-$	R = CH_3	*84*	Pure	23.3[a]		170
			Gas phase	21.1[a]		170

[a]This value represents $\Delta G^{\ddagger}$.

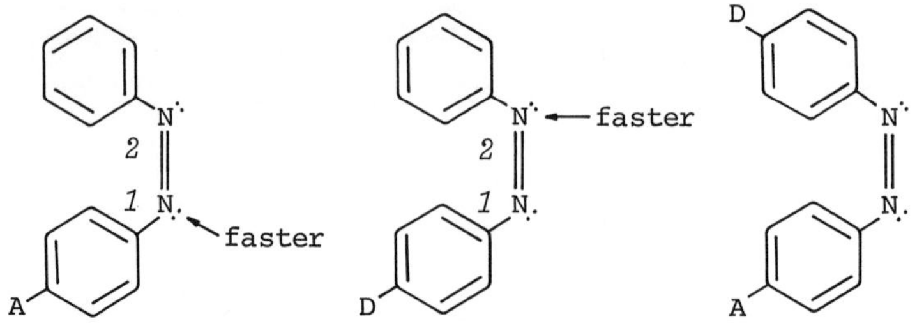

not the same for azobenzenes substituted, with donors and acceptors, and only the fastest process is observed in these kinetic experiments. The rotational path should give a linear Hammett plot.

The mechanism for azobenzenes of the type *85*, which should be polarized by the push-pull effect, is not yet clear (161). The presence of a significant solvent effect suggests that isomerization of these azobenzenes may involve a rotational mechanism in which dipolar structures are important (161). On the other hand, the similarly substituted stilbenes do not readily undergo thermal isomerization (169).

That the barriers in diazocyanides (167) are similar to those in azobenzenes may now be explained by the similarity in inversion substituent constants z for the phenyl group and the cyano group (Table 8). The influence on the inversion should therefore be similar.

Z,*E*-isomerizational barriers of aliphatic azo compounds have not yet been reported. The empirical estimation of the inversion barrier in azomethane yields a value of about 39 kcal/mol ($\Delta G^{\ddagger}_{C_6H_5}$ = 23 kcal/mol; $z_{C_6H_5}$ = 0.59). This is in agreement with the observed high stability of the *Z* and *E* isomers of 1,2-diazacyclooctene on distillation (174).

B. Other Double Bonds on Nitrogen

Some other N=N double bonds (or, as the case may be, partial double bonds) are found in *O*,*N*-dialkylnitramines *86*, (175) azoxybenzene *87* (176), dimeric nitrosoalkanes *88* (177), and nitrosamines *89* (170,171). There are no detailed investigations concerning the mechanism of isomerization in these compounds. In *86*, *87*, and *89*, both rotation about the double bond and inversion of the nitrogen bearing the lone electron pair are possible paths of isomerization (13). The measured barriers in these

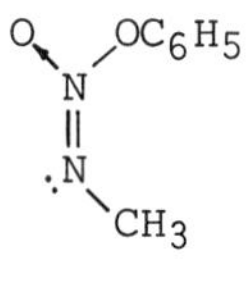

86 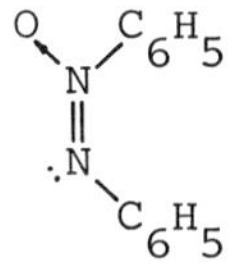 87 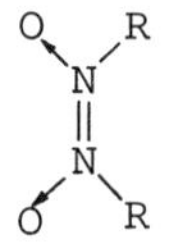 88 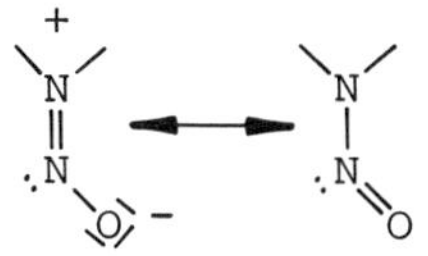 89

"Stable" $E_a = 22.8$ kcal/mol (see Table 10)
$\Delta S^{\ddagger} = + 5.1$ e.u.

cases represent minimum barriers for each process.

In this connection it is interesting to compare the alkyl nitrites *90*. The delocalization of electrons over three centers leads to a planar molecular shape; in other words, the *Z* and *E* isomers represent minima of the potential energy surface (178, 179). The isomerization can occur either by rotation about the

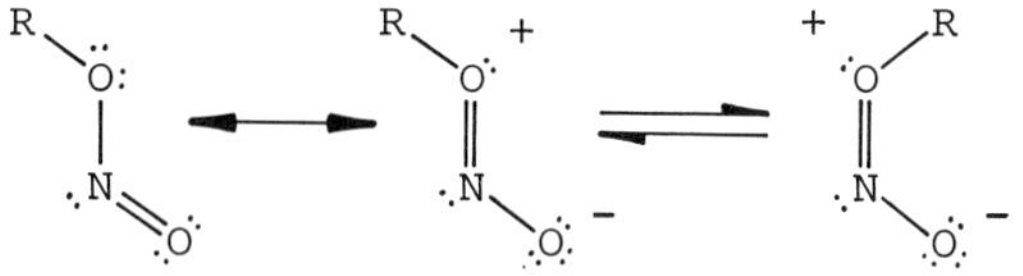

90

partial N=O double bond, or by inversion on nitrogen or on oxygen. Of the latter two possibilities, the oxygen inversion should be favored over nitrogen inversion (Sect. VII), but no data concerning these points are available for alkyl nitrites.

Dimeric nitroso compounds *88* (for their structures, see ref. 180), on the other hand, may dissociate with the possibility of interconversion upon recombination (181), although direct rotation about the N=N bond is also possible (177). There is no

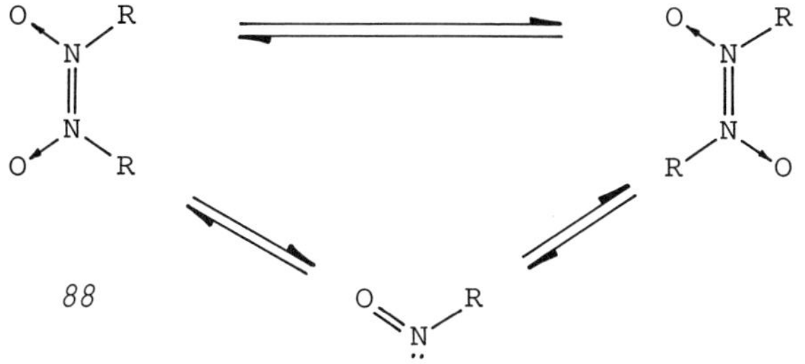

88

evidence available regarding the preferred pathway, but dissociation is involved, at least in part, during *Z,E* isomerization.

Planar inversion is possible in monomeric nitroso compounds

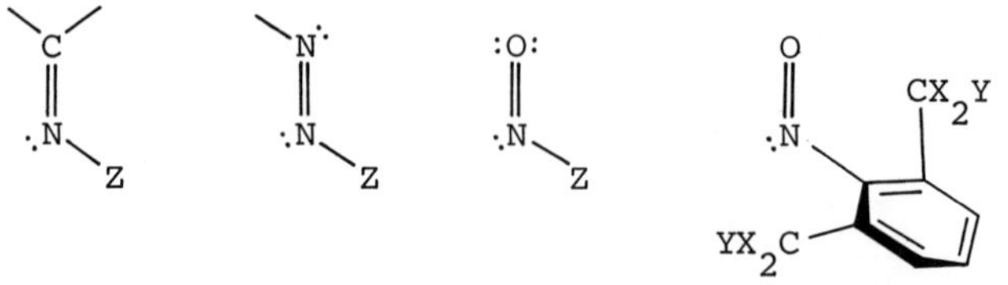

91

which are isoelectronic with imines. In nitrosamines and alkyl nitrites we have already discussed the possibility of planar nitrogen inversion as a competition reaction to thermal rotation (see above). There is also another means of experimentally studying the planar inversion on nitrogen in nitroso compounds: certain substituted nitrosobenzenes *91* that bear prochiral groups ($-CX_2Y$ groups) in the ortho positions will show magnetic nonequivalence of X if rotation about the *N*-aryl bond as well as nitrogen inversion is slow. The enantiotopomerization process represents the minimum energy barrier for nitrogen inversion.

VII. *Z,E* ISOMERIZATION OF OTHER DOUBLE BONDS

Protonated carbonyl compounds are isoelectronic with imines and vinyl anions. Therefore the question arises once again as

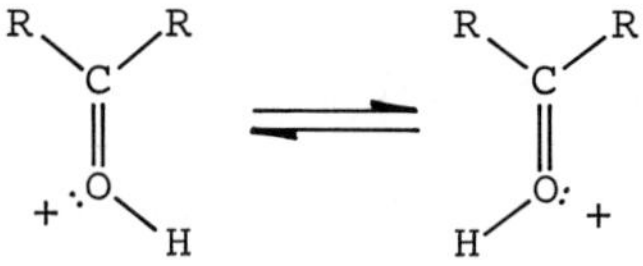

to whether the mechanism of *Z,E* isomerization is rotation or inversion. The kind of mechanism depends on the double-bond order of the C=O bond (Fig. 10).

It is evident that rotation is facilitated by lower double-bond order of the C=O bond. Inversion should not, to a first approximation, be affected by the bond order* (although the sp^3 hypridized oxygen of ethers has, in fact, a higher barrier to

*The effect of the substituents at carbon on the inversion rate is neglected.

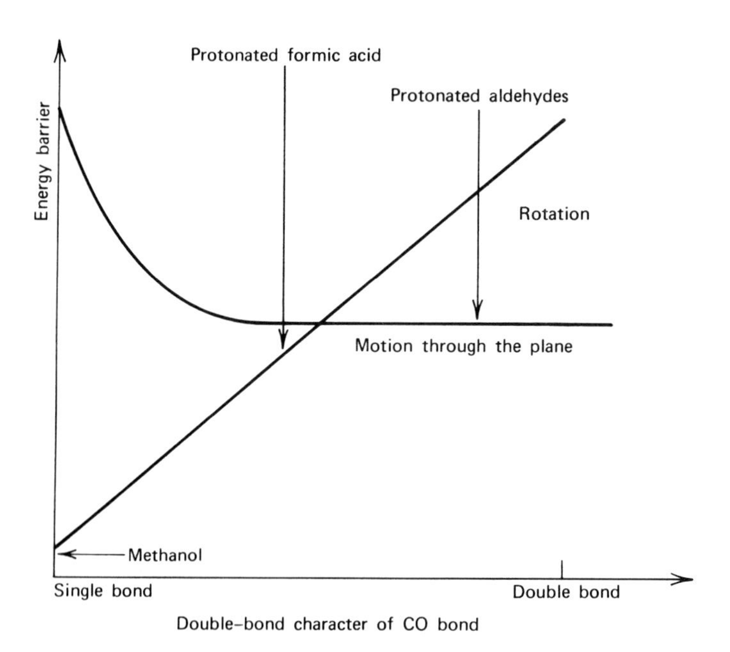

Fig. 10. Schematic comparison of energy barriers for rotation and in-plane movement as a function of double-bond order (182).

inversion than the sp^2 oxygen*). In accordance with this concept (182), the inversion barrier of protonated aldehydes is lower than the rotational one, whereas the reverse should be true of protonated formic acid, because the $\overset{+}{C}$-OH double-bond order is lower in the latter.

The barriers of protonated formaldehyde were calculated to be 17.2 kcal/mol (inversion) and 31.4 kcal/mol (rotation). The corresponding values for protonated acetaldehyde are 18.0

*The inversional barrier of oxygen in ethers was calculated to be higher than 34 kcal/mol (183). Experimentally the barrier must be higher than that in a diphenyl ether (184) (17.8 kcal/mol) and in a cyclophane (183) (20 kcal/mol). In both compounds oxygen inversion is one of the possible pathways to explain the observed topomerization process.

kcal/mol (inversion) and 27.9 kcal/mol (rotation), but for protonated formic acid *95a* 17.1 kcal/mol (inversion) and 14.3 kcal/mol (rotation). The optimum COH valence angle for the rotational transition state in the latter compound is larger (133°) than for the ground state (120°). This situation corresponds to the intermediate mechanism discussed for the imines (see Fig. 8c) with the difference that it is "rotationlike" in the oxygen case but "inversionlike" in the nitrogen case. Similar results, which are in general agreement, have been obtained by semiempirical SCF-LCAO-MO calculations of the CNDO/2 variety (185), but here the barriers of rotation and inversion in protonated aldehydes are found to be quite similar. Experimentally, several nmr measurements of *Z,E* isomerizational barriers have been carried out on protonated carbonyl species (186). They are listed, together with those in other compounds with partial CO double bonds, in Table 11.

In the protonated species, deprotonation-reprotonation may be responsible for the observed isomerization (187). This is also true for *C*-protonated anisoles. In some cases hydrogen transfer has been excluded by the line shape of the OH nmr peak. Such an alternative dissociation mechanism is not likely in *99-103* but from the experimental data a decision about the preferred mechanism can hardly be made. The low barrier of the *t*-butylated compound *105c* gives evidence that planar inversion of oxygen takes place with free enthalpy of activation less than 15 kcal/mol in this compound. The rotation about the partial C=O double bond is very strongly hindered by the bulkiness of the *t*-butyl groups, as may be seen by comparison with the analogous thiouronium salt *109d* (Table 12; barrier 18.5 kcal/mol). In going from sulfur to oxygen, steric hindrance should increase, and thus the rotational barrier in *105c* should be higher than 18 kcal/mol. Further evidence for the possibility of inversion at oxygen is obtained by studying the magnetic nonequivalence of prochiral substituents in *o*-substituted uronium salts *106* (203). The syn-anti topomerization and isomerization are fast in *106* and *107* ($\Delta G^{\ddagger} > 10$ kcal/mol) but magnetic nonequivalence with a high barrier is found only in *107* ($\Delta G^{\ddagger} = 24.6$ kcal/mol). In analogy with the explanation of the same effects in imines (see Sect. V-B-2), this can be explained by oxygen inversion, which changes the prochirality of the isopropyl group simultaneously with syn-anti topomerization in *106* On the other hand, prochirality is retained in *107* during inversion.

Finally, we shall discuss briefly the *Z,E* isomerism of C=S double bonds. The effect of ortho substituents of *S*-aryl thiouronium salts *108* (Table 12) gives clearest evidence for the

Table 11 *Z,E* Isomerization at Carbon-Oxygen Double Bonds (nmr Parameters)

Compound			No.	Solvent	$\Delta G^{\ddagger}$, kcal/mol	E_a, kcal/mol	Other Data	Refs.
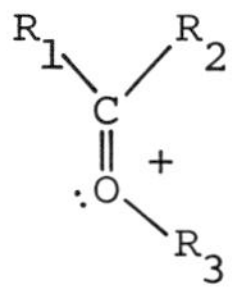 $R_1R_2C=\ddot{O}^{+}-R_3$								
R_1	R_2	R_3						
C_2H_5	C_2H_5	BF_3	*92*	CH_2ClF	> 8			188
CH_3	C_2H_5	H	*93*	HF/SbF_5	>17			189
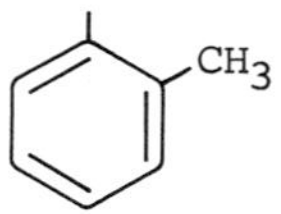 CH_3 (ring)	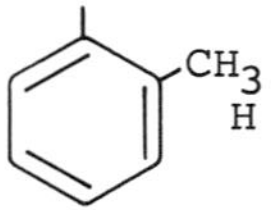 CH_3 (ring)	H	*94*	HSO_3F/SbF_5	13.6		$\Delta H^{\ddagger}$ = 9.3 kcal/mol $\Delta S^{\ddagger}$ = -14.5 e.u.	190
H	OH	H	*95a*	HF/SbF_5		15.3	log A = 15.7	191
CH_3	OH	H	*95b*	HF/SbF_5		11.2	log A = 12.0	191
CH_3	OH	H	*95b*	HF/BF_3		13.5	log A = 14.4	191
CH_3	OH	H	*95b*	$HSO_3F/SbF_5/SO_2$		12.6	log A = 13.2	191,192
H	CH_2Cl	H	*95c*	$HSO_3F/SbF_5/SO_2$			T_c = -30°C	193
 O (ring)		H	*96*	$HSO_3F/SbF_5/SO_2$			T_c = -50°C	194

OCH_3	OCH_3 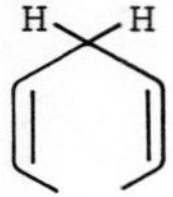	H	*97*			8±2	T_c = -43°C	195
		CH_3	*98a*	HF/BF_3			T_c = -15°C	196
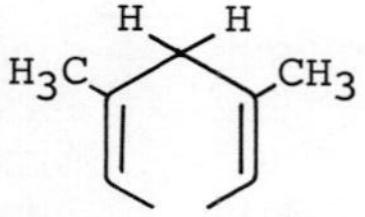 H_3C	CH_3	CH_3	*98b*	HF			T_c = -40°C	196
OCH_3	OCH_3	CH_3	*99a*	CF_3COOH	15	11±4		197
OCH_3	OCH_3	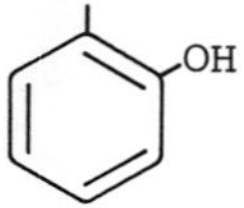 OH	*99b*	HSO_3F/SO_2			T_c = -42°C	198
OCH_3	H	CH_3	*100a*	CH_3NO_2/CH_2Cl_2			Δν = 0.32 ppm (-30°C)	199
OCH_3	CH_3	CH_3	*100b*	CH_3NO_2/CH_2Cl_2			Δν = 0.22 ppm (-30°C)	199
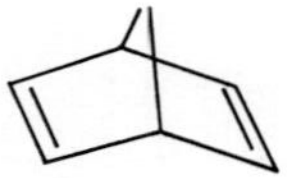		CH_3	*101*	HSO_3F	11.7			200
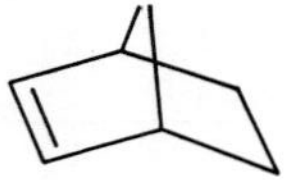		CH_3	*102*	HSO_3F	14.7			201
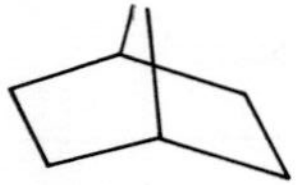		CH_3	*103*	HSO_3F	18.4			201

Table 11 Continued

Compound			No.	Solvent	$\Delta G^{\ddagger}$, kcal/mol	E_a, kcal/mol	Other Data	Refs.
$N(CH_3)_2$	$N(CH_3)_2$	CH_3	*104*	CH_2Cl_2	<10[a]			202
H_3C–N	N–CH_3	CH_3	*105a*	CH_2Cl_2	<10[b]			202
		C_6H_5	*105b*	CH_2Cl_2	<10[b]			202
		CH_3	*105c*	$CDCl_3$	14.8			202

[a]Rotation about the C=N bond is observed in the nmr spectrum below -40°C.

[b]A splitting of 20 Hz for the *N*-methyl groups is assumed. An accidentally equal chemical shift for both of them is excluded by the comparison with guanidines, guanidinium salts (88, 89), ketane aminals (39), etc., of similar structure.

Table 12 nmr Parameters for the *Z*,*E* Isomerization in *S*-Arylisothiuronium Perchlorates (209)

Compound		No.	Solvent	T_c, °C	$\Delta G^{\ddagger}$, kcal/mol
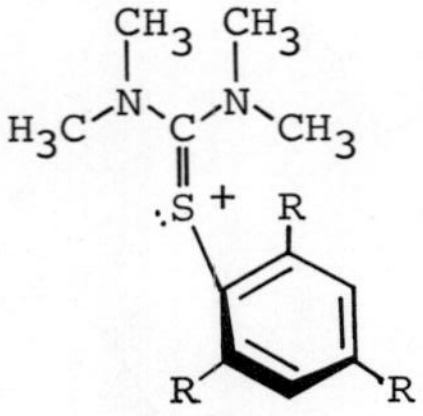 H_3C, CH_3, CH_3, CH_3 (N, C, N, :S+), R, R, R; ClO_4^-	R = H	*108a*	CH_2Cl_2	-83	9.1
	R = CH_3	*108b*	CH_2Cl_2	-64	10.0
	R = i-C_3H_7	*108c*	CH_2Cl_2	-29	12.0
	R = *t*-C_4H_9	*108d*	$(CD_3)_2SO$	+89	17.9
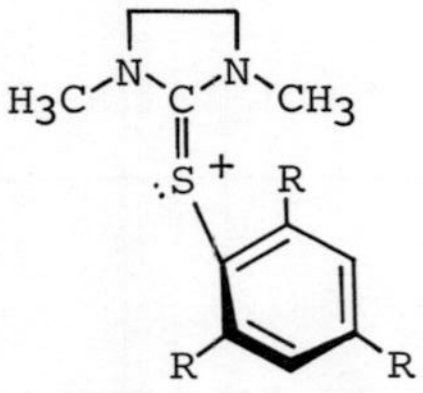 H_3C, CH_3 (N, C, N, :S+), R, R, R; ClO_4^-	R = H	*109a*	CH_2Cl_2	<-115	<7.6
	R = CH_3	*109b*	CH_2Cl_2	-107	8.0
	R = i-C_3H_7	*109c*	CH_2Cl_2	-80	9.3
	R = *t*-C_4H_9	*109d*	Trichlorobenzene	+106	18.5

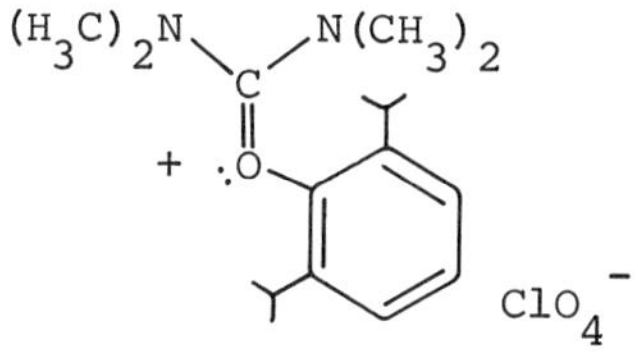

$(H_3C)_2N$ $N(CH_3)_2$ ClO_4^-

106

$(H_3C)_2N$ $N-CH_2C_6H_5$ inversion

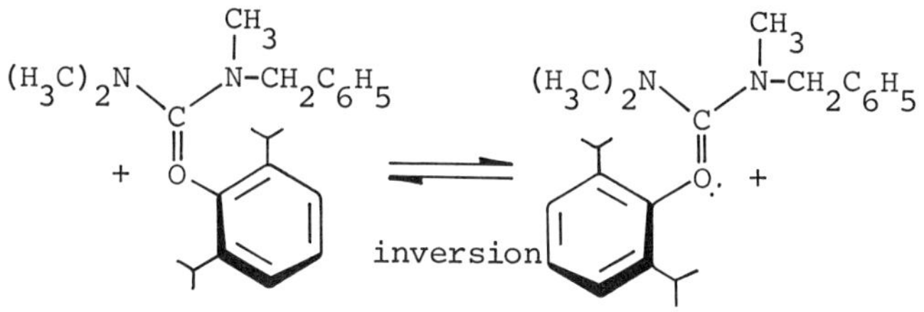

107

rotation pathway for syn-anti topomerization of these compounds (209). The steric hindrance to topomerization increases with increasing size of the substituent. Especially the *t*-butylated compounds show relatively high barriers which are near the limit beyond which a separation at room temperature is possible. Much higher stabilities are observed for sulfines (204). It was possible to isolate derivates of phenylchlorosulfine *110a* (205), diarylsulfines *110b* (206), as well as naphthyl dithiocarboxylmethyl ester *S*-oxide *110c* (207). Faster isomerization is observed in thiourea-*S*-monoxide *110d* (208).

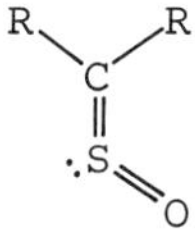

a: R = Cl, R' = C_6H_5
b: R = R' = aryl
c: R = *S*-CH_3, R' = naphthyl
d: R = R' = NHR"

110

VIII. CONCLUSIONS

A large number of barriers for fast *Z,E* isomerizations has been measured by modern physical methods. The dependence of barriers as well as double-bond order on substituents can be rationalized by empirical, semiempirical, and *ab initio* calculations. The isomerization of imines has been proved to occur by an "inversionlike" mechanism. The planar inversion barrier of X in $R_2C = XR'$ should generally decrease in the order X = $C^{\ominus}$, N, $O^{\oplus}$ as has been found for pyramidal inversion (13,21,210). The rotational barrier should also decrease in the same order but in the simple derivatives (R = H) rotational barriers are higher than those for inversion (211). Comparing the inversion barriers of elements of the first row of the periodic system with those of the second, a large increase is found experimentally in going from oxygen compounds such as *105* to the corresponding sulfur compounds (e.g., *109*). This is also true for pyramidal inversion of oxonium and sulfonium (13) compounds as well as that of amines and phosphines (13,211). A further characteristic dependence of the geometry and inversion barrier on the electron population of the inverting orbital is observed (vinyl anions > vinyl radicals > vinyl cations) and this is a consequence of the origin of the inversional barrier. The main part of the barrier results from the promotion of the unbonded orbital from sp^2 to p hybridization. Of course the attendant promotion energy depends on the occupancy of the corresponding orbital.

ACKNOWLEDGMENT

We thank the "Deutsche Forschungsgemeinschaft" and the "Fonds der Chemischen Industrie" for financial support.

REFERENCES

1. (a) K. Mackenzie in *The Chemistry of Alkenes*, S. Patai, Ed. Interscience, New York, 1964, p. 387; (b) L. Crombie, *Quart. Rev.*, *6*, 106 (1952); (c) G. M. Wyman, *Chem. Rev.*, *55*, 625 (1955).
2. J. E. Blackwood, C. L. Gladys, K. L. Loening, A. E. Petrarca, and J. E. Rush, *J. Amer. Chem. Soc.*, *90*, 509 (1968).
3. (a) M. J. S. Dewar, E. A. C. Lucken, and M. A. Whitehead, *J. Chem. Soc.*, *1960*, 2423; (b) W. B. Jennings and R. Spratt, *Chem. Commun.*, *1970*, 1418.
4. (a) A. J. Merer and R. S. Mulliken, *Chem. Rev.*, *69*, 639 (1969); (b) G. S. Hammond and N. J. Turro, *Science*, *142*, 1541 (1963); (c) S. Patai and Z. Rappoport in *The Chemistry of Alkenes*, S. Patai, Ed., Interscience, New York, 1964, p. 565.
5. R. S. Cahn, C. Ingold, and V. Prelog, *Angew. Chem.*, *78*, 413 (1966); *Angew. Chem. Intern. Ed. Engl.*, *5*, 385 (1966).

6. K. R. Hanson, *J. Amer. Chem. Soc.*, *88*, 2731 (1966).
7. Compare, e.g., the path via vinyl cations: C. A. Grob, *Chimia*, *25*, 87 (1971).
8. "IUPAC-Nomenclature," *J. Org. Chem.*, *35*, 2849 (1970).
9. G. Binsch, E. L. Eliel, and H. Kessler, *Angew. Chem.*, *83*, 618 (1971); *Angew. Chem. Intern. Ed. Engl.*, *10*, 570 (1971).
10. A. A. Frost and R. G. Pearson, *Kinetics and Mechanism*, Wiley, New York, 1961.
11. *Technique of Organic Chemistry*, S. L. Friess, E. S. Lewis, and A. Weissberger, Eds., Vol. VIII/1, Interscience, New York, 1961.
12. G. Binsch, in *Topics in Stereochemistry*, E. L. Eliel and N. L. Allinger, Eds., Vol. 3, Interscience, New York, 1968, p. 937.
13. H. Kessler, *Angew. Chem.*, *82*, 237 (1970); *Angew. Chem. Intern. Ed. Engl.*, *9*, 219 (1970).
14. G. Porter in *Technique of Organic Chemistry*, S. L. Friess, E. S. Lewis, and A. Weissberger, Eds., Vol. VIII/2, Interscience, New York, 1963, p. 1055.
15. E. Wyn-Jones and R. A. Pethrick, in *Topics in Stereochemistry*, E. L. Eliel and N. L. Allinger, Eds., Vol. 5, Wiley-Interscience, New York, 1970, p. 205.
16. J. M. Lehn in *Conformational Analysis*, G. Chiurdoglu, Ed., Academic Press, New York, 1971.
17. M. J. S. Dewar and E. Haselbach, *J. Amer. Chem. Soc.* *92*, 590 (1970).
18. O. Sinanoglu and K. B. Wiberg, *Sigma Molecular Orbital Theory*, Yale University Press, New Haven, 1970.
19. M. Menzinger and R. L. Wolfgang, *Angew. Chem.* *81*, 446 (1969); *Angew. Chem. Intern. Ed. Engl.*, *8*, 438 (1969).
20. H. M. Frey and R. Walsh, *Chem. Rev.*, *69*, 103 (1969).
21. J. M. Lehn, *Fortschr. Chem. Forsch.*, *15*, 311 (1970).
22. Ref. 21, p. 322. See also J. Reisse in *Conformational Analysis*, G. Chiurdoglu; Ed., Academic Press, New York, 1971.
23. J. E. Douglas, B. S. Rabinovitch, and F. S. Looney, *J. Chem. Phys.*, *23*, 315 (1955).
24. (a) M. C. Lin and K. J. Laidler, *Can. J. Chem.*, *46*, 973 (1968). (b) R. B. Cundall and T. F. Palmer, *Trans. Faraday Soc.*, *57*, 1936 (1961).
25. I. G. Murgulescu and Z. Simon, *Z. Physik Chem. (Leipzig)*, *221*, 29 (1962); *Rev. Romaine Chim.*, *11*, 21 (1966).
26. (a) R. J. Buenker, *J. Chem. Phys.*, *48*, 1368 (1968); (b) U. Kaldor and I. Shavitt, *J. Chem. Phys.*, *48*, 191 (1968).
27. R. Hoffmann, *Tetrahedron*, *22*, 521 (1966).
28. E. B. Moore, *Theoret. Chim. Acta*, *7*, 144 (1967).

29. P. Ganis and J. D. Dunitz, *Helv. Chim. Acta*, *50*, 2379 (1967).
30. M. C. Sauer and L. M. Dorfman, *J. Chem. Phys.*, *35*, 497 (1961). H. Okabe and J. R. McNesby, *J. Chem. Phys.*, *36*, 601 (1962). J. R. McNesby and H. Okabe, *Advan. Photochem.*, *3*, 157 (1964).
31. N. Wiberg, *Angew. Chem.*, *80*, 809 (1968); *Angew. Chem. Intern. Ed. Engl.*, *7*, 766 (1968). R. W. Hoffmann, *Angew. Chem.*, *80*, 823 (1968); *Angew. Chem. Intern. Ed. Engl.*, *7*, 754 (1968).
32. M. Wrighton, G. S. Hammond, and H. B. Gray, *J. Amer. Chem. Soc.*, *92*, 6068 (1970), and references cited therein.
33. D. F. Evans, *J. Chem. Soc.*, *1960*, 1735.
34. B. S. Rabinovitch and F. S. Looney, *J. Chem. Phys.*, *23*, 2439 (1955).
35. R. G. Dickinson et al., *J. Amer. Chem. Soc.*, *59*, 472 (1937); *ibid.*, *61*, 3259 (1939); *71*, 1238 (1949). H. Steinmetz and R. M. Noyes, *J. Amer. Chem. Soc.*, *74*, 4141 (1952).
36. A. Streitwieser, p. 161 in ref. 18.
37. Y. Shvo, *Tetrahedron Lett.*, *1968*, 5923.
38. A. P. Downing, W. D. Ollis, and I. O. Sutherland, *Chem. Commun.*, *1967*, 143; *ibid.*, *1968*, 1053; *J. Chem. Soc.*, *B*, *1969*, 111.
39. H. Kessler, *Chem. Ber.*, *103*, 973 (1970).
40. S. C. Nyburg, *Acta Cryst.*, *7*, 779 (1954). V. Franzen and H. Joschek, *Ann. Chem.*, *648*, 63 (1961).
41. I. R. Gault, W. D. Ollis, and I. O. Sutherland, *Chem. Commun.*, *1970*, 269.
42. W. D. Ollis, personal communication.
43. E. Fischer, *Fortschr. Chem. Forsch.*, *7*, 605 (1967); G. Kortüm and W. Zoller, *Chem. Ber.*, *103*, 2062 (1970).
44. G. B. Kistiakowsky and M. Z. Nelles, *Z. Physik Chem. (Leipzig), Bodenstein-Festband*, 369 (1931). G. B. Kistiakowsky and W. R. Smith, *J. Amer. Chem. Soc.*, *56*, 638 (1934); *58*, 766 (1936).
45. A. Rieker and H. Kessler, *Chem. Ber.*, *102*, 2147 (1969).
46. H. Stieger and H. D. Brauer, *Chem. Ber.*, *103*, 3799 (1970).
47. Eu. Müller and H. Neuhoff, *Ber. Deut. Chem. Ges.*, *72*, 2063 (1939). Eu. Müller and E. Tietz, *Ber. Deut. Chem. Ges.*, *74*, 807 (1941).
48. R. Kuhn, B. Schulz, and J. C. Jochims, *Angew. Chem.*, *78*, 449 (1966); *Angew. Chem. Intern. Ed. Engl.*, *5*, 420 (1966).
49. R. C. Haddon, V. R. Haddon, and L. M. Jackman, *Fortschr. Chem. Forsch.*, *16*, 103 (1971); J. F. M. Oth, *Pure Appl. Chem.*, *25*, 573 (1971).
50. H. Kessler and A. Rieker, *Tetrahedron Lett.*, *1966*, 5257.

51. P. Boldt, W. Michaelis, H. Lackner, and B. Krebs, *Chem. Ber.*, *104*, 220 (1971).
52. A. Rieker and H. Kessler, *Tetrahedron*, *24*, 5133 (1968).
53. A. S. Kende, P. T. Izzo, and W. Fulmor, *Tetrahedron Lett.*, *1966*, 3697.
54. J. H. Crabtree and D. J. Bertelli, *J. Amer. Chem. Soc.*, *89*, 5384 (1967).
55. G. Seitz, *Tetrahedron Lett.*, *1968*, 2305.
56. T. Eicher and N. Pelz, *Chem. Ber.*, *103*, 2647 (1970).
57. G. Isaksson, J. Sandström, and I. Wennerbeck, *Tetrahedron Lett.*, *1967*, 2233. J. Sandström and I. Wennerbeck, *Acta Chem. Scand.*, *24*, 1191 (1970).
58. Y. Shvo, E. C. Taylor, and J. Bartulin, *Tetrahedron Lett.*, *1967*, 3259. Y. Shvo and H. Shanan-Atidi, *J. Amer. Chem. Soc.*, *91*, 6683 (1969).
59. A. Mannschreck and U. Kölle, *Angew. Chem.*, *81*, 540 (1969); *Angew. Chem. Intern. Ed. Engl.*, *8*, 528 (1969).
60. H. Kessler, *Angew. Chem.*, *80*, 971 (1968); *Angew. Chem. Intern. Ed. Engl.*, *7*, 898 (1968). J. Sandström and I. Wennerbeck, *Chem. Commun.*, *1969*, 306.
61. J. Sandström and I. Wennerbeck, *Chem. Commun.*, *1971*, 1088.
62. H. O. Kalinowski and H. Kessler, to be published.
63. K. Herbig, R. Huisgen, and J. Huber, *Chem. Ber.*, *99*, 2546 (1966).
64. C. Reichardt, *Lösungsmittel-Effekte in der organischen Chemie*, Chemische Taschenbücher, Vol. *4*, Verlag Chemie, Weinheim, 1969. K. Dimroth, C. Reichardt, T. Siepmann, and F. Bohlmann, *Ann. Chem.*, *661*, 1 (1963); C. Reichardt, *Angew. Chem.*, *77*, 30 (1965); *Angew. Chem. Intern. Ed. Engl.*, *4*, 29 (1965).
65. H. Güsten and D. Schulte-Frohlinde, *Tetrahedron Lett.*, *1970*, 3567.
66. R. A. Pethrick and E. Wyn-Jones, *J. Chem. Soc.*, *A*, *1969*, 713.
67. J. Dabrowski and L. Kozetski, *J. Chem. Soc.*, *B*, *1971*, 345. B. Roques, C. Jaureguiberry, M. C. Fournié-Zaluski, and S. Combrisson, *Tetrahedron Lett.*, *1971*, 2693.
68. J. M. Bollinger, J. M. Brinick, and G. A. Olah, *J. Amer. Chem. Soc.*, *92*, 4025 (1970).
69. N. C. Deno, R. C. Haddon, and E. N. Nowak, *J. Amer. Chem. Soc.*, *92*, 6692 (1970).
70. P. v. R. Schleyer, T. M. Su, M. Saunders, and J. C. Rosenfeld, *J. Amer. Chem. Soc.*, *91*, 5174 (1969); P. v. R. Schleyer, personal communication.
71. D. H. Hunter and D. J. Cram, *J. Amer. Chem. Soc.*, *86*, 5478 (1964).
72. J. K. Kochi and P. J. Krusic, *J. Amer. Chem. Soc.*, *90*, 7157

(1968). C. Walling and W. Thaler, *J. Amer. Chem. Soc.*, *83*, 3878 (1961).

73. P. West, J. I. Purmort, and S. V. McKinley, *J. Amer. Chem. Soc.*, *90*, 797 (1968). H. H. Freedman, V. R. Sandel, and B. P. Thill, *J. Amer. Chem. Soc.*, *90*, 1762 (1968).
74. B. M. Gimarc, *J. Amer. Chem. Soc.*, *93*, 815 (1971).
75. W. D. Pfeifer, C. A. Bahn, P. v. R. Schleyer, S. Bocher, C. E. Harding, K. Hummel, M. Hanack, and P. J. Stang, *J. Amer. Chem. Soc.*, *93*, 1513 (1971).
76. D. R. Kelsey and R. G. Bergman, *J. Amer. Chem. Soc.*, *92*, 228 (1970).
77. R. Sustmann, J. E. Williams, M. J. S. Dewar, L. C. Allen, and P. v. R. Schleyer, *J. Amer. Chem. Soc.*, *91*, 5350 (1969).
78. W. A. Lathan, W. J. Hehre, and J. A. Pople, *J. Amer. Chem. Soc.*, *93*, 808 (1971).
79. R. W. Fessenden and R. H. Schuler, *J. Chem. Phys.*, *39*, 2147 (1963); *J. Phys. Chem.*, *71*, 74 (1967). P. H. Kasai and E. B. Whipple, *J. Amer. Chem. Soc.*, *89*, 1033 (1967).
80. E. L. Cochran, F. J. Adrian, and V. A. Bowers, *J. Chem. Phys.*, *40*, 213 (1964). F. J. Adrian and M. Karplus, *J. Chem. Phys.*, *41*, 56 (1964).
81. M. J. S. Dewar and M. Shanshal, *J. Amer. Chem. Soc.*, *91*, 3654 (1969).
82. L. A. Singer and N. P. Kong, *Tetrahedron Lett.*, *1967*, 643; *J. Amer. Chem. Soc.*, *89*, 5251 (1967).
83. J. A. Kampmeier and R. M. Fantazier, *J. Amer. Chem. Soc.*, *88*, 1959 (1966). L. A. Singer and N. P. Kong, *Tetrahedron Lett.*, *1966*, 2089; *J. Amer. Chem. Soc.*, *88*, 5213 (1966). G. D. Sargent and M. W. Brown, *J. Amer. Chem. Soc.*, *89*, 2788 (1967).
84. R. M. Kopchik and J. A. Kampmeier, *J. Amer. Chem. Soc.*, *90*, 6733 (1968). L. A. Singer and J. Chen, *Tetrahedron Lett.*, *1969*, 4849.
85. D. J. Cram, *Fundamentals of Carbanion Chemistry*, Academic Press, New York, London, 1965, pp. 130-135 and references cited therein.
86. L. Pauling, *The Nature of the Chemical Bond*, 3rd ed., Cornell University Press, Ithaca, N. Y., 1960.
87. W. E. Stewart and T. H. Siddall, III, *Chem. Rev.*, *70*, 517 (1970).
88. (a) H. Kessler and D. Leibfritz, *Tetrahedron Lett.*, *1969*, 427; (b) *Tetrahedron*, *25*, 5127 (1969).
89. H. Kessler and D. Leibfritz, *Chem. Ber.*, *104*, 2158 (1971).
90. A. Krebs and J. Breckwoldt, *Tetrahedron Lett.*, *1969*, 3797.
91. H. Kessler and A. Rieker, *Z. Naturforsch.*, *22b*, 456 (1967); *Ann. Chem.*, *708*, 57 (1967).

92. H. A. Staab and D. Lauer, *Chem. Ber.*, *101*, 864 (1968).
93. L. M. Jackman, T. E. Kavanagh, and R. C. Haddon, *Org. Magn. Resonance*, *1*, 109 (1969).
94. A. Krebs, personal communication.
95. A. Krebs, *Tetrahedron Lett.*, *1971*, 1901.
96. C. H. Bushweller, J. W. O'Neil, M. H. Halford, and F. H. Bisset, *Chem. Commun.*, *1970*, 1251; S. van der Werf and J. B. F. N. Engberts, *Rec. Trav. Chim.*, *89*, 423 (1970); W. Walter and K. P. Ruesz, *Ann. Chem.*, *743*, 167 (1971); W. Walter and E. Schaumann, *Ann. Chem.*, *743*, 154 (1971); H. Hermann, R. Huisgen, and H. Mäder, *J. Amer. Chem. Soc.*, *93*, 1779 (1971); A. E. Lemire and J. C. Thompson, *Can. J. Chem.*, *48*, 824 (1970).
97. G. Wettermark, J. Weinstein, J. Sousa, and L. Dogliatti, *J. Phys. Chem.*, *69*, 1584 (1965).
98. D. Y. Curtin, E. J. Grubbs, and C. G. McCarty, *J. Amer. Chem. Soc.*, *88*, 2775 (1966).
99. (a) A. Rieker and H. Kessler, *Tetrahedron*, *23*, 3723 (1967); (b) H. Kessler, *Angew. Chem.*, *79*, 997 (1967); *Angew. Chem. Intern. Ed. Engl.*, *6*, 977 (1967); for other quinone anils see refs. 139, 132.
100. (a) H. Kessler, *Tetrahedron Lett.*, *1968*, 2041; (b) H. Kessler and D. Leibfritz, *Tetrahedron*, *26*, 1805 (1970).
101. D. Leibfritz and H. Kessler, *Chem. Commun.*, *1970*, 655; H. Kessler, P. F. Bley, and D. Leibfritz, *Tetrahedron*, *27*, 1687 (1971).
102. A. Liden and J. Sandstrom, *Tetrahedron*, *27*, 2893 (1971).
103. E. Hückel, *Z. Physik*, *60*, 455 (1930).
104. (a) J. Meisenheimer and W. Theilacker in *Stereochemie*, K. Freudenberg, Ed., Franz Deuticke, Leipzig, Wien, 1933, pp. 964ff; (b) R. J. W. LeFevre and J. Northcott, *J. Chem. Soc.*, *1949*, 2235; (c) H. Uffmann, *Z. Naturforsch.*, *22b*, 491 (1967); *Tetrahedron Lett.*, *1966*, 4631; (d) R. K. Norris and S. Sternhell, *Aust. J. Chem.*, *19*, 841 (1966); *Tetrahedron Lett.*, *1967*, 97.
105. J. M. Lehn and B. Munsch, *Theoret. Chim. Acta (Berlin)*, *12*, 91 (1968).
106. M. Raban, *Chem. Commun.*, *1970*, 1415.
107. M. Raban and E. Carlson, *J. Amer. Chem. Soc.*, *93*, 685 (1971).
108. D. G. Anderson and G. Wettermark, *J. Amer. Chem. Soc.*, *87*, 1433 (1965).
109. G. Wettermark, J. Weinstein, J. Souza, and L. Dogliatti, *J. Phys. Chem.*, *69*, 1584 (1965).
110. H. Kessler and A. Pfeffer, unpublished results.
111. A. Rieker and H. Kessler, *Z. Naturforsch.*, *21b*, 939 (1966).

112. D. Wurmb-Gerlich, F. Vögtle, A. Mannschreck, and H. A. Staab, *Ann. Chem.*, *708*, 36 (1967).
113. J. B. Lambert, in *Topics in Stereochemistry*, E. L. Eliel and N. L. Allinger, Eds., Vol. 6, Wiley, New York, 1971, p. 19.
114. E. A. Jeffery, A. Meisters, and T. Mole, *Tetrahedron*, *25*, 741 (1969).
115. J. Sandström, *J. Phys. Chem.*, *71*, 2318 (1967); G. Isaksson and J. Sandström, *Acta Chem. Scand.*, *24*, 2565 (1970).
116. H. Kessler and D. Leibfritz, *Tetrahedron Lett.*, *1970*, 1423; H. Kessler and D. Leibfritz, *Chem. Ber.*, *104*, 2143 (1971).
117. N. P. Marullo and E. H. Wagener, *Tetrahedron Lett. 1969*, 2555.
118. R. W. Layer and C. J. Carman, *Tetrahedron Lett.*, *1968*, 1285.
119. G. E. Hall, W. J. Middleton, and J. D. Roberts, *J. Amer. Chem. Soc.*, *93*, 4778 (1971).
120. E. Carlson, F. B. Jones, and M. Raban, *Chem. Commun.*, *1969*, 1235.
121. J. B. Lambert, B. S. Packard, and W. L. Oliver, *J. Org. Chem.*, *36*, 1309 (1971).
122. D. Y. Curtin and J. W. Hausser, *J. Amer. Chem. Soc.*, *83*, 3474 (1961).
123. C. Brown, B. T. Grayson, and R. F. Hudson, *Tetrahedron Lett.*, *1970*, 4925.
124. F. Vögtle, A. Mannschreck, and H. A. Staab, *Ann. Chem.*, *708*, 51 (1967).
125. N. P. Marullo and E. H. Wagener, *J. Amer. Chem. Soc.*, *88*, 5034 (1966).
126. C. G. McCarty and D. M. Wieland, *Tetrahedron Lett.*, *1969*, 1787.
127. D. Leibfritz, Dissertation, Tübingen, Germany, 1970.
128. H. Kessler and D. Leibfritz, *Ann. Chem.*, *737*, 53 (1970).
129. Y. Shvo and A. Nahlieli, *Tetrahedron Lett.*, *1970*, 4273.
130. R. M. Moriarty, C. Yeh, K. C. Ramey, and P. W. Whitehurst, *J. Amer. Chem. Soc.*, *92*, 6360 (1970).
131. E. Fischer and Y. Frei, *J. Chem. Phys.*, *27*, 808 (1957).
132. A. Rieker and H. Kessler, *Tetrahedron Lett.*, *1967*, 153.
133. W. Walter and E. Schaumann, *Chem. Ber.*, *104*, 4 (1971).
134. H. Quast and E. Schmitt, *Angew. Chem.*, *82*, 395 (1970); *Angew. Chem. Intern. Ed. Engl.*, *9*, 381 (1970).
135. H. Quast and E. Schmitt, *Chem. Ber.*, *103*, 1234 (1970).
136. A. Krebs and H. Kimling, *Angew. Chem.*, *83*, 401 (1971); *Angew. Chem. Intern. Ed. Engl.*, *10*, 409 (1971).
137. J. C. Jochims and F. A. L. Anet, *J. Amer. Chem. Soc.*, *92*, 5524 (1970).
138. F. A. L. Anet, J. C. Jochims, and C. H. Bradley, *J. Amer.*

Chem. Soc., *92*, 2557 (1970).
139. B. D. Gesner, *Tetrahedron Lett.*, *1965*, 3559.
140. V. J. Bauer, W. Fulmor, G. O. Morton, and S. R. Safir, *J. Amer. Chem. Soc.*, *90*, 6846 (1968).
141. H. Kessler and H. O. Kalinowski, unpublished results.
142. H. B. Bürgi and J. D. Dunitz, *Chem. Commun.*, *1969*, 472.
143. E. Haselbach and E. Heilbronner, *Helv. Chim. Acta*, *51*, 16 (1968); V. I. Minkin, Y. A. Zhdanov, E. A. Medyantzeva, and Y. A. Ostroumov, *Tetrahedron*, *23*, 3651 (1967).
144. F. Kerek, G. Ostrogovich and Z. Simon, *J. Chem. Soc.*, *B*, *1971*, 541, and references cited therein.
145. J. F. Kincaid and F. C. Henriques, *J. Amer. Chem. Soc.*, *62*, 1474 (1940); C. C. Costain and G. B. B. M. Sutherland, *J. Phys. Chem.*, *56*, 321 (1952); R. E. Weston, *J. Amer. Chem. Soc.*, *76*, 2645 (1954); G. W. Koeppl, D. S. Sagatys, G. S. Krishnamurthy, and S. I. Miller, *J. Amer. Chem. Soc.*, *89*, 3396 (1967).
146. J. D. Andose, J. M. Lehn, K. Mislow, and J. Wagner, *J. Amer. Chem. Soc.*, *92*, 4050 (1970).
147. H. Kessler and D. Leibfritz, *Tetrahedron Lett.*, *1970*, 4289, 4293, 4297.
148. D. P. Stevenson and G. M. Coppinger, *J. Amer. Chem. Soc.*, *84*, 149 (1962); J. P. Lorand, *Tetrahedron Lett.*, *1971*, 2511.
149. E. J. Grubbs and J. A. Villarreal, *Tetrahedron Lett.*, *1969*, 1841.
150. J. R. Thomas, *J. Amer. Chem. Soc.*, *86*, 1446 (1964); R. O. C. Norman, *Chem. Brit.*, *6*, 66 (1970) and references cited therein.
151. M. S. Gordon and H. Fischer, *J. Amer. Chem. Soc.*, *90*, 2471 (1968).
152. Z. Simon, F. Kerek, and G. Ostrogovich, *Rev. Roumaine Chim.*, *13*, 381 (1968).
153. L. J. Rolls and R. Adams, *J. Amer. Chem. Soc.*, *54*, 2495 (1932).
154. L. J. Schaad and H. B. Kinser, *J. Phys. Chem.*, *73*, 1901 (1969).
155. D. R. Williams and R. Damrauer, *Chem. Commun.*, *1969*, 1380.
156. G. S. Hartley, *Nature*, *140*, 281 (1937); *J. Chem. Soc.*, *1938*, 633.
157. D. Schulte-Frohlinde, *Ann. Chem.*, *612*, 131 (1958).
158. D. Schulte-Frohlinde, *Ann. Chem.*, *612*, 138 (1958).
159. R. J. W. LeFevre and J. Northcott, *J. Chem. Soc.*, *1953*, 867; W. R. Brode, J. H. Gould, and G. M. Wyman, *J. Amer. Chem. Soc.*, *74*, 4641 (1952).
160. E. R. Talaty and J. C. Fargo, *Chem. Commun.*, *1967*, 65.
161. P. D. Wildes, J. G. Pacifici, G. Irick, and D. G. Whitten,

J. Amer. Chem. Soc., *93*, 2004 (1971).

162. E. Fischer, *J. Amer. Chem. Soc.*, *82*, 3249 (1960); G. Gabor, Y. F. Frei, and E. Fischer, *J. Phys. Chem.*, *72*, 3266 (1968).
163. D. Gegiou, K. A. Muszkat, and E. Fischer, *J. Amer. Chem. Soc.*, *90*, 3907 (1968).
164. G. Gabor and E. Fischer, *J. Phys. Chem.*, *75*, 581 (1971); G. Zimmermann, L. Chow, and U. Parik, *J. Amer. Chem. Soc.*, *80*, 3528 (1958).
165. Z. Simon, *Rev. Roumaine Chim.*, *11*, 35 (1966).
166. J. Binenboym, A. Burcat, A. Lifshitz, and J. Shamir, *J. Amer. Chem. Soc.*, *88*, 5039 (1966).
167. R. J. W. LeFèvre and J. Northcott, *J. Chem. Soc.*, *1949*, 944; *ibid.*, *1949*, 333; R. J. W. LeFèvre and H. Vine, *J. Chem. Soc.*, *1938*, 431.
168. A. U. Chaudhry and B. C. Gowenlock, *J. Chem. Soc.*, *B*, *1968*, 1083.
169. D. Schulte-Frohlinde, H. Blume, and H. Güsten, *J. Phys. Chem.*, *66*, 2486 (1962).
170. C. E. Looney, W. D. Phillips, and E. L. Reilly, *J. Amer. Chem. Soc.*, *79*, 6136 (1957); R. K. Harris and R. A. Spragg, *Chem. Commun.*, *1967*, 362; for other nitrosamines see ref. 171.
171. C. L. Bumgardner, K. S. McCallum, and J. P. Freeman, *J. Amer. Chem. Soc.*, *83*, 4417 (1961); S. Andreades, *J. Org. Chem.*, *27*, 4163 (1962); H. W. Brown and D. P. Hollis, *J. Mol. Spectry.*, *13*, 305 (1964); D. J. Blears, *J. Chem. Soc.*, *1964*, 6256; A. Mannschreck, H. Münsch, and A. Mattheus, *Angew. Chem.*, *78*, 751 (1966); *Angew. Chem. Intern. Ed. Engl.*, *5*, 728, (1966); A. Mannschreck and H. Münsch, *Angew. Chem.*, *79*, 1004 (1967); *Angew. Chem. Intern. Ed. Engl.*, *6*, 6984 (1967); T. Axenrod, P. S. Pregosin, and G. W. A. Milne, *Tetrahedron Lett.*, *1968*, 5293; T. Axenrod, M. J. Wieder, and G. W. A. Milne, *Tetrahedron Lett.*, *1968*, 401.
172. B. M. Gimarc, *J. Amer. Chem. Soc.*, *92*, 266 (1970).
173. H. H. Jaffé, *Chem. Rev.*, *53*, 191 (1953).
174. G. Quinkert and G. Vitt, personal communication.
175. A. H. Lamberton and G. Newton, *J. Chem. Soc.*, *1961*, 1797; *J. Chem. Soc.*, *C*, *1969*, 397.
176. P. Luner, *Can. J. Res.*, *30*, 674 (1952); C. A. Winkler, *Can. J. Res.*, *28*, 140 (1950).
177. A. U. Chaudhry and B. G. Gowenlock, *J. Chem. Soc.*, *B*, *1968*, 1083.
178. L. H. Piette, J. D. Ray, and R. A. Ogg, *J. Chem. Phys.* *26*, 1341 (1957); W. D. Phillips, C. E. Looney, and C. P. Spaeth, *J. Mol. Spectry.*, *1*, 35 (1957).

179. P. Gray and M. W. T. Pratt, *J. Chem. Soc.*, *1958*, 3403; R. N. Hazeldine and J. Jander, *J. Chem. Soc.*, *1954*, 691; R. N. Hazeldine and B. J. H. Mattinson, *J. Chem. Soc.*, *1955*, 4172; P. Tarte, *J. Chem. Phys.*, *20*, 1570 (1952); *Bull. Soc. Chim. Belges.*, *60*, 227, 240 (1951); *ibid.*, *62*, 401 (1953).
180. B. G. Gowenlock and W. Lüttke, *Quart. Rev.*, *12*, 321 (1958); G. Germain, P. Piret, and M. van Meerssche, *Acta Cryst.*, *16*, 109 (1963); and references cited therein.
181. R. Hoffmann, R. Gleiter, and F. B. Mallory, *J. Amer. Chem. Soc.*, *92*, 1460 (1970).
182. P. Ros, *J. Chem. Phys.*, *49*, 4902 (1968).
183. A. J. Gordon and J. P. Gallagher, *Tetrahedron Lett.*, *1970*, 2541.
184. H. Kessler, A. Rieker, and W. Rundel, *Chem. Commun.*, *1968*, 475.
185. K. F. Purcell and J. M. Collins, *J. Amer. Chem. Soc.*, *92*, 465 (1970).
186. G. A. Olah and A. M. White, *Chem. Rev.*, *70*, 561 (1970).
187. H. Hogeveen, A. F. Bickel, C. W. Hilbers, E. L. Mackor, and C. Maclean, *Rec. Trav. Chim.*, *86*, 687 (1967); H. Hogeveen, *Rec. Trav. Chim.*, *86*, 696 (1967); D. M. Brower, *Rec. Trav. Chim.*, *86*, 879 (1967).
188. U. Henriksson and S. Forsén, *Chem. Commun.*, *1970*, 1229.
189. D. M. Brouwer, *Rec. Trav. Chim.*, *86*, 879 (1967).
190. R. van der Linde, J. W. Dornseiffen, J. U. Veenland, and T. J. de Boer, *Tetrahedron Lett.*, *1968*, 525; *Spectrochim. Acta*, *24*, *A*, 2115 (1968).
191. H. Hogeveen, *Rec. Trav. Chim.*, *87*, 1313 (1968).
192. M. Brookhart, G. C. Levy, and S. Winstein, *J. Amer. Chem. Soc.*, *89*, 1735 (1967).
193. L. Thil, J. J. Riehl, P. Rimmelin, and J. M. Sommer, *Chem. Commun.*, *1970*, 591; R. Jost, P. Rimmelin, and J. M. Sommer, *Tetrahedron Lett.*, 1971, 3005.
194. G. A. Olah, D. P. Kelly, and N. Sucui, *J. Amer. Chem. Soc.*, *92*, 3133 (1970).
195. G. A. Olah and A. M. White, *J. Amer. Chem. Soc.*, *90*, 1884 (1968).
196. D. M. Brouwer, E. L. Mackor, and C. Maclean, *Rec. Trav. Chim.*, *85*, 114 (1966); see also G. A. Olah, *J. Amer. Chem. Soc.*, *89*, 5259 (1967).
197. B. G. Ramsay and R. W. Taft, *J. Amer. Chem. Soc.*, *88*, 3058 (1966).
198. H. Volz and G. Zimmermann, *Tetrahedron Lett.*, *1970*, 3597.
199. R. F. Borch, *J. Amer. Chem. Soc.*, *90*, 5303 (1968).
200. M. Brookhart, R. K. Lustgarten, and S. Winstein, *J. Amer.*

Chem. Soc., *89*, 6354 (1967).
201. R. K. Lustgarten, M. Brookhart, and S. Winstein, *Tetrahedron Lett.*, *1971*, 141.
202. H. O. Kalinowski and H. Kessler, unpublished results.
203. H. Kessler and H. O. Kalinowski, *Ann. Chem.*, *743* (1971); the nmr-spectral behavior is not yet published.
204. B. Zwanenburg and J. Strating, *Quart. Rept. Sulfur Chem.*, *5*, 579 (1970).
205. J. F. King and T. Dust, *J. Amer. Chem. Soc.*, *85*, 2676 (1963); *Can. J. Chem.*, *44*, 819 (1966).
206. S. Ghersetti, L. Lunazzi, G. Maccagnani, and A. Mangini, *Chem. Commun.*, *1969*, 834.
207. B. Zwanenburg, L. Thijs, and J. Strating, *Tetrahedron Lett.*, *1967*, 3453.
208. W. Walter and G. Randau, *Ann. Chem.*, *722*, 52 (1969).
209. H. Kessler and H. O. Kalinowski, *Angew. Chem.*, *82*, 666 (1970); *Angew. Chem. Intern. Ed. Engl.*, *9*, 641 (1970).
210. A. Rauk, L. C. Allen, and K. Mislow, *Angew. Chem.* *82*, 453 (1970); *Angew. Chem. Intern. Ed. Engl.*, *9*, 428 (1970).
211. J. M. Lehn, B. Munsch, and P. Millie, *Theoret. Chem. Acta (Berlin)*, *16*, 351 (1970).
212. H. Kessler and A. Walter, unpublished results.
213. L. Salem and C. Rowland, *Angew. Chem.* *84*, 86 (1972); *Angew. Chem. Intern. Ed. Engl.* *11*, 92 (1972).

RECENT LITERATURE

Nonplanar double bonds:

W. L. Mock, *Tetrahedron Lett.*, *1972*, 475.
L. Radom, J. A. Pople, and W. L. Mock, *Tetrahedron Lett.*, *1972*, 479.

Rotations in polarized ethylenes:

H. D. Hartzler, *J. Amer. Chem. Soc.*, *93*, 4961 (1971);
H. Shanan-Atidi and Y. Shvo, *Tetrahedron Lett.*, *1971*, 603;
A. Mannschreck and B. Kolb, *Chem. Ber.*, *105*, 696 (1972).

Mindo/2-SCF-MO-Calculations of rotation barriers of Carbonium ions:

M. Shanshal, *J. Chem. Soc. Perkin II*, *1972*, 335.

Inversion of vinyl radicals:

G. M. Whitesides, C. P. Casey, and J. K. Krieger, *J. Amer. Chem. Soc.*, *93*, 1379 (1971);
M. Shanshal, *Z. Naturforsch.*, *26a*, 1336 (1971).

Theoretical calculations on C=N-double bonds:

CNDO: C. H. Warren, G. Wettermark, and K. Weiss, *J. Amer. Chem. Soc.*, *93*, 4658 (1971);
MINDO: M. Shanshal, *Tetrahedron*, *28*, 61 (1972).

Rotational barriers in nitrones:

L. W. Boyle, M. J. Peagram, and G. H. Whitham, *J. Chem. Soc. B*, *1971*, 1728.

Substituent effects on C=N-isomerization:

W. Walter and E. Schaumann, *Chem. Ber.*, *104*, 4 (1971);
W. Walter and C. O. Meese, *Ann. Chem.*, *753*, 169 (1971);
D. R. Boyd, C. G. Watson, W. B. Jennings, and D. M. Jerina, *Chem. Commun.*, *1972*, 183;
F. L. Scott, F. A. Groeger, and A. F. Hegarty, *J. Chem. Soc. B*, *1971*, 1411;
F. L. Scott, T. M. Lambe, and R. N. Butler, *Tetrahedron Lett.*, *1971*, 2909;
A. Mannschreck and B. Kolb, *Chem. Ber.*, *105*, 696 (1972);
R. J. Cook and K. Mislow, *J. Amer. Chem. Soc.*, *93*, 6703 (1971).

Z,E-isomerization of azo-compounds:

CNDO/2-calculation: S. Ljundggren and G. Wettermark, *Acta Chem. Scand.*, *25*, 1599 (1971);
Activation energies of diazocyanides: J. Suszko and T. Ignasiak, *Bull. Acad. Pol. Sci. Ser. Sci. Chim.*, *1970*, 663.

Isomerization in N-alkyl-N-nitrosoanilines:

J. T. D' Agostino and H. H. Jaffe, *J. Org. Chem.*, *36*, 992 (1971).

Isomerization barriers in alkoxy-carbene complexes:

C. G. Kreiter and E. O. Fischer, *XXIII rd International Congress of Pure and Applied Chemistry*, *6*, 151 (1971).

SUBJECT INDEX

* (ep.) means "epoxidation of"

* (ep.) means "epoxidation of"

* (ep.) means "epoxidation of"

* (ep.) means "epoxidation of"

* (ep.) means "epoxidation of"

* (ep.) means "epoxidation of"

* (ep.) means "epoxidation of"

* (ep.) means "epoxidation of"

* (ep.) means "epoxidation of"

CUMULATIVE INDEX, VOLUMES 1–7